DEVELOPMENTS IN CRYSTALLINE POLYMERS—2

CONTENTS OF VOLUME 1

DEVELOPMENTS IN CRYSTALLINE POLYMERS—2

Edited by

D. C. BASSETT

J. J. Thomson Physical Laboratory, University of Reading, UK

ELSEVIER APPLIED SCIENCE
LONDON and NEW YORK

ELSEVIER APPLIED SCIENCE PUBLISHERS LTD
Crown House, Linton Road, Barking, Essex IG11 8JU, England

Sole Distributor in the USA and Canada
ELSEVIER SCIENCE PUBLISHING CO., INC.
52 Vanderbilt Avenue, New York, NY 10017, USA

WITH 22 TABLES AND 134 ILLUSTRATIONS

Softcover reprint of the hardcover 1st edition 1988

British Library Cataloguing in Publication Data

Developments in crystalline polymers.
—2–
1. Polymers and polymerization—
Periodicals 2. Crystals—Periodicals
547.7 QD380

ISBN-13: 978-94-010-7096-6 e-ISBN-13: 978-94-009-1341-7
DOI: 10.1007/978-94-009-1341-7

Library of Congress CIP data applied for

PREFACE

Modern society makes increasing demands for novelty in materials and their properties which are ever more exacting. Crystalline polymers are in the forefront of this demand and improvements are constantly occurring across the entire range from existing materials of high tonnage to novel materials with application in information technology. The developments recorded in this volume reflect this situation.

Chapter 1 is a comprehensive review of the polymer PHB, poly(hydroxybutyrate), which is new to industrial manufacturing but is a naturally occurring substance. It has potentially valuable properties but has excited interest especially because it is biodegradable. It may, therefore, provide one means of reducing environmental pollution.

Improvements in existing materials, beyond those which are obtainable by optimization of known variables, are most likely to come from understanding of structure–property relationships. Polymer science has now reached the stage where it is able to make effective synthesis of information from complementary techniques, leading to rapidly deepening understanding. Chapters 2, 3 and 4 are all concerned with technical developments which are contributing substantially to this synthesis. The possibilities of electron microscopy, specifically the characterization of lamellar microstructure, have been transformed by permanganic etching. Now real organization (which can be very different from what had previously been inferred) can be used as a basis for explaining polymeric properties.

In Chapter 3, Mitchell and Windle give a critical account of the assessment of orientation in liquid crystalline polymers, a rapidly developing new field in which they have played a leading part.

Solid-state NMR is making a notable impact on polymer science as Professor Kitamaru's chapter illustrates. It seems probable that the integration of this spectroscopy with other techniques of examination will provide a major stimulus to polymer science in the near future.

Gelation is important in several aspects of polymer science. It is a complex matter, but it is now much more understood because of the application of a combination of modern techniques. It also illustrates very well the breadth over which principles of polymer science apply, from natural through to synthetic materials.

The final chapter by Dr Coles concerns a topic at the very forefront of active development. In the few years since their initial synthesis, side-chain liquid crystal polymers have excited interest firstly because of their response to electric and magnetic fields and thence to potential applications in areas such as information storage, which is discussed here. What has been opened up, however, is a powerful means of tailoring properties to applications by synthesis of suitable molecules. This has long been an aim of materials science and this field will surely be the source of future developments.

D. C. Bassett

CONTENTS

LIST OF CONTRIBUTORS

D. C. Bassett

University of Reading, Physics Department, J. J. Thomson Physical Laboratory, Whiteknights, PO Box 220, Reading RG6 2AF, UK

Harry J. Coles

Department of Physics, Schuster Laboratory, The University, Oxford Road, Manchester M13 9PL, UK

P. A. Holmes

New Science Group, ICI plc, PO Box No. 11, The Heath, Runcorn, Cheshire WA7 4QE, UK

Ryozo Kitamaru

Institute for Chemical Research, Kyoto University, Uji, Kyoto 611, Japan

M. J. Miles

AFRC Institute of Food Research, Norwich Laboratory, Colney Lane, Norwich NR4 7UA, UK

G. R. Mitchell

University of Reading, Physics Department, J. J. Thomson Physical Laboratory, Whiteknights, PO Box 220, Reading RG6 2AF, UK

A. H. Windle

Department of Metallurgy and Materials Science, University of Cambridge, Pembroke Street, Cambridge CB2 3QZ, UK

Chapter 1

BIOLOGICALLY PRODUCED (*R*)-3-HYDROXY-ALKANOATE POLYMERS AND COPOLYMERS

P. A. HOLMES

New Science Group, ICI plc, Runcorn, Cheshire, UK

SUMMARY

Microorganisms are capable of producing a wide range of polymers and copolymers based on 3-hydroxypropionic acid substituted with various alkyl groups in the 3-position. The most common homopolymer is poly(3-hydroxybutyrate), PHB, which has a 3-methyl substituent, but monomers having C_2–C_5 alkyl side groups are found in natural copolyesters. The monomers are all optically active in the R absolute configuration. PHB can be produced from carbon substrates as diverse as glucose, ethanol, acetate, methane and even gaseous mixtures of carbon dioxide and hydrogen. The polymer occurs as discrete granules within the cell cytoplasmic space and can represent up to 80% of the dry cell weight. After extraction and purification it behaves as a normal crystalline thermoplastic with a melting point around 180°C and can be processed by conventional extrusion and moulding equipment. PHB and its copolymers with 3-hydroxypentanoic acid are now available commercially and are being evaluated in numerous potential applications. The copolymers are all genuinely biodegradable in that their rate of chemical hydrolysis is extremely slow but microorganisms produce both specific and non-specific esterase enzymes capable of degrading the materials rapidly to non-toxic monomers. Most of the potential uses exploit this property in medical, veterinary, horticultural and general disposable products.

INTRODUCTION

Poly(3-hydroxybutyrate), PHB, is a natural storage product of bacteria and algae. The polymer occurs as discrete granules within the cell cytoplasmic space and these can be seen in phase-contrast optical microscopy of PHB containing cultures and, more clearly, as optically empty areas in electron micrographs of ultrathin sections of bacteria (Fig. 1). Unlike other biological polymers such as proteins and polysaccharides, however, PHB was found to be thermoplastic with a high degree of crystallinity and a well-defined melting endotherm at around 180°C. The prospect of a novel plastic derived from renewable raw materials rather than oil or coal is clearly very exciting from a long-term viewpoint and PHB has been evaluated extensively by a number of companies including ICI.[1]

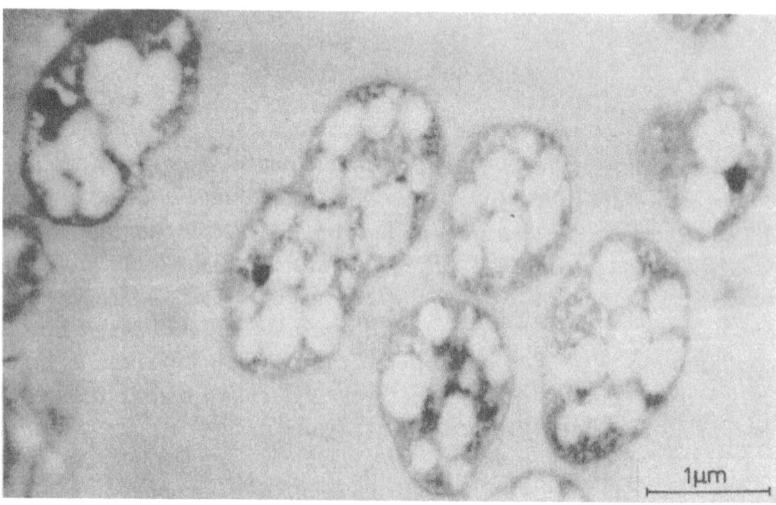

FIG. 1. Electron micrograph of ultrathin sections of *Alcaligenes eutrophus* showing PHB granules as optically empty areas.

POLYMER SYNTHESIS

The polymer was first isolated and characterized by Lemoigne in 1925 at the Pasteur Institute in Paris.[2] The chemical formula was found to be:

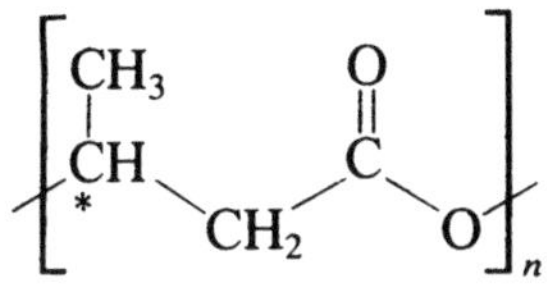

* Chiral centre in *R* absolute configuration.

with the chiral centre always in the *R* absolute configuration (or D(−) in traditional nomenclature). It occurs in a wide variety of microorganisms[3] including Gram-negative and Gram-positive bacteria as well as blue–green algae, or cyanobacteria as they are now called. Fungi do not synthesize PHB but have the ability to degrade it and use it as a carbon source for growth.[4]

The fundamental reason for PHB accumulation by these microorganisms has been the subject of much debate in the literature[5–8] with the polymer being involved in cell sporulation,[9] encystment[10] and gene expression.[11] However, the simplest view is that PHB acts as a carbon and energy storage medium for bacteria in the same way that glycogen and fats are stored by plants and animals respectively.[12] Certainly the level of PHB in cells can be dramatically increased from a few percent to over 80% of cell dry weight if, under otherwise ideal growth conditions, an essential nutrient such as nitrogen, oxygen, phosphorus or sulphur is limited.[13] Under these circumstances the cell, deprived of the means to grow and multiply, puts all of its energy into PHB production so that, when the nutrient limitation is lifted, it has an immediately accessible endogenous carbon source which may be used to ensure that growth restarts as quickly as possible.

Biosynthetic Pathway

The biochemical pathways of PHB biosynthesis are reasonably well understood and start with the conversion to acetate of an appropriate carbon substrate, such as low alcohols and acids, sugars or even gaseous mixtures of hydrogen and carbon dioxide.[14,15] The bacteria put a ‘handle’ on the acetate molecules in the form of an enzyme cofactor linked by a thioester bond. Two molecules of this acetyl coenzyme-A can then be condensed to give acetoacetyl coenzyme-A which is subsequently reduced to 3-hydroxybutyryl coenzyme-A. The final polymerase enzyme joins these units together to form PHB, and recycles the cofactor as shown in Fig. 2.

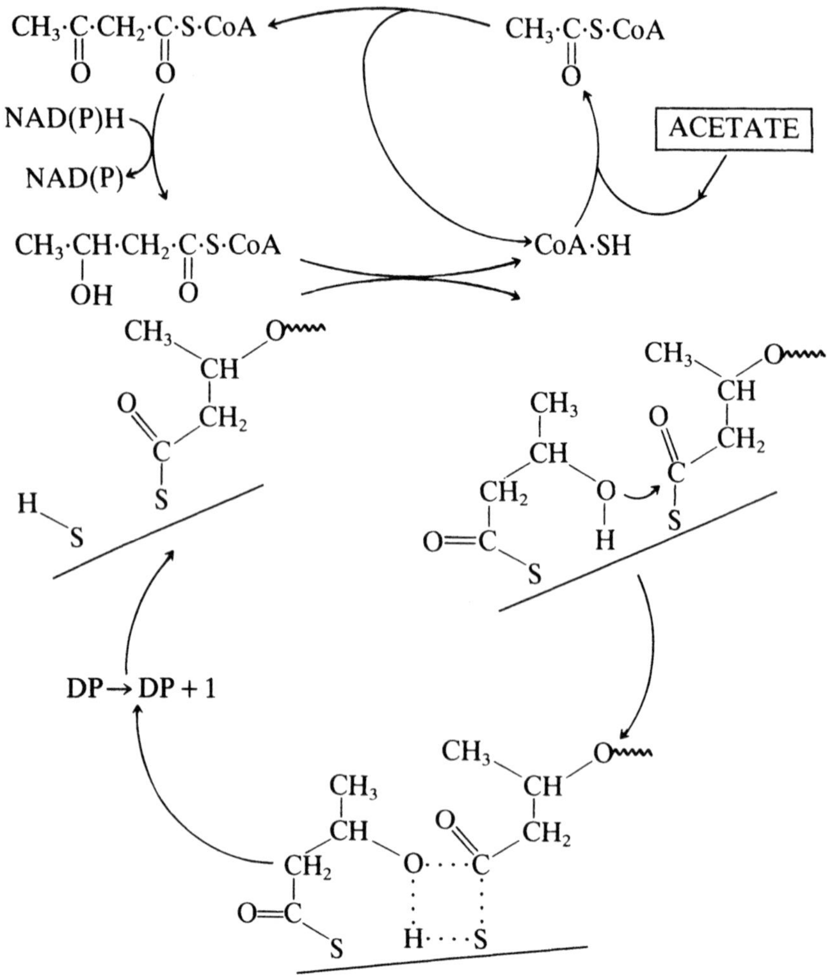

FIG. 2. Biosynthetic pathway of PHB production.

The precise action of the polymerase is not known. Indeed any mechanism by which a water-soluble monomer is polymerized inside a living cell and transported from the aqueous cytoplasm into a hydrophobic, highly crystalline granule is difficult to imagine. However, one theory is that the polymerizing enzyme locates the growing polymer chain and a new hydroxybutyrate unit at two thiol groups on its surface and condenses them via a four-membered ring transition state.

The polymer molecule then flip-flops back and forth between these thiol groups with the new monomer being coordinated at the free site. This mechanism is illustrated in Fig. 2.

Copolymer Biosynthesis

Enzymes are usually thought of as catalysing only very specific reactions such as the isomerization of glucose to fructose or the degradation of D- but not L-amino acids in racemic mixtures.[16] Even in PHB metabolism, an enzyme has been isolated and purified that will only degrade hydroxybutyrate dimer to monomer and will not touch trimers or higher oligomers.[17] It is therefore a little surprising to find that the PHB polymerase enzyme has a 'wobbly' specificity and can incorporate other 3-hydroxy alkyl esters into its growing polymer chain. In fact a whole range of poly(3-hydroxyalkanoate) copolymers have recently been identified in activated sewage sludges,[18,19] estuarine sediments[20] and in various freshwater and marine blue–green algae.[21]

It is very difficult to study the biochemical synthesis of such polymers produced in the environment. Fortunately, however, ICI found that well-defined copolymers could be produced by feeding bacterial monocultures with precisely specified carbon substrates.[22,23] For example, if *Alcaligenes eutrophus* is fed with a mixture of acetic and propionic acids under conditions of nitrogen or phosphorus depletion, then a copolymer of (*R*)-3-hydroxybutyrate and (*R*)-3-hydroxypentanoate is produced. The biochemical pathway of acetate to hydroxybutyrate has already been given in Fig. 2 and it requires only a small modification to see how one molecule of propionate can condense with one molecule of acetate to give the precursor of 3-hydroxypentanoate as shown in Fig. 3. Substitution of propionate by other members of the alkyl ester homologous series, of course, give rise to the other monomer units seen in the environmentally produced samples of poly(hydroxyalkanoates).

However, there are certain exceptions since it is known that some organisms metabolize some carboxylic acids, notably *n*-butyric, crotonic and 3-hydroxybutyric acids, directly to PHB and do not produce copolymers by the pathway shown in Fig. 3. Moreover, one very recent report suggests that bacteria can produce homopolymers of the higher hydroxyalkanoates from *n*-alkanes,[24] again avoiding the pathway given in Fig. 3. The most common biopolyesters, though, are the

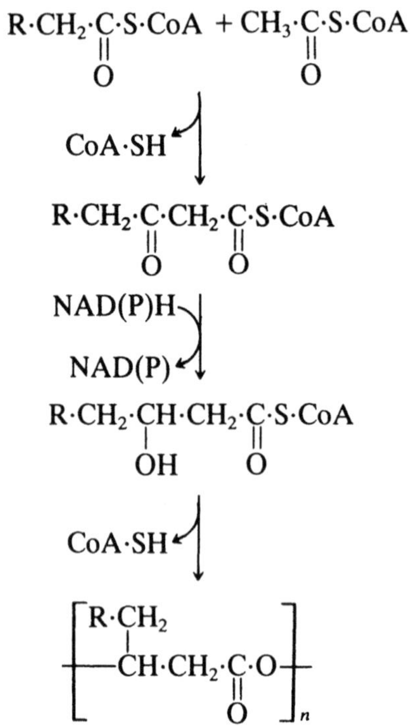

FIG. 3. Copolymer biosynthesis.

copolymers of 3-hydroxybutyrate (HB) with 3-hydroxypentanoate. The latter is usually referred to by its trivial name, 3-hydroxyvalerate, or HV for convenience.

Chemical Synthesis

It should also be pointed out at this stage that all of the poly(3-hydroxyesters) described here can be synthesized chemically from the relevant substituted propiolactones. Indeed, an extensive literature exists on the polymerization of such lactones using in general aluminium or zinc alkyl catalysts with water as cocatalyst.[25–29] Some experimenters have even used chiral monomers to produce polymer with optical activity in the same sense, but to a lesser degree, as the biopolymer.[30,31]

Although academically interesting, the polymers produced in this way could never be commercially competitive with PHB by fermentation since the lactone monomers are expensive to make, very toxic and difficult to handle and yields are usually low. Nevertheless, the synthetic materials can be made to resemble the biopolymer very closely, even to the extent of being rapidly biodegradable by PHB-utilizing bacteria.[32]

Granule Structure

Before leaving the subject of PHB synthesis, something should be said about the morphology of the polymer granules laid down by bacteria in their cytoplasmic space. The fact that these are not just lumps of plastic is most readily seen in freeze fracture electron micrographs of whole cells. These are obtained by rapidly freezing a PHB-containing bacterial culture, fracturing it at low temperature and shadow-staining the freshly exposed surface with a heavy metal by evaporation from a source located at an acute angle.[33] The metal replicas of the surface can then be floated off, washed and examined by transmission electron microscopy. A typical result is shown in Fig. 4. The outline of the whole cell can be seen and inside this there are a number of shadowed 'craters' representing the holes left by granules that stayed with the other half of the cell during the fracturing process. However, the most striking features are the 'horns' of polymer that were pulled out from granules within the half of the cell examined.

The ultrastructure of PHB granules has been extensively studied[34–39] by this and other techniques such as X-ray analysis and electron microscopy of freeze-etched samples. A number of models for the granule structure have been proposed[36,39] and these have the common features of a central core composed of microfibrils like a ball of string surrounded by a lipid coat containing the polymerase enzymes. It is even reported that, under mild conditions, the granules can be separated from the rest of the cell and then fractionated by density gradient centrifugation to give one layer of granules with lipid coat and one of granules without their protective jacket.[40]

Although plastic deformation of quite brittle polymer particles such as polystyrene and polyacrylate latices during freeze-fracture is well documented[41] the magnitude of the effect for PHB is still surprising. The process is explained in terms of the energy dissipated during fracture causing massive local temperature rises to bring the polymer above its glass transition temperature. However, even above T_g, injection-moulded PHB is still a relatively brittle semicrystalline thermoplastic and the ductile behaviour evident in Fig. 4 is probably related to the special granule morphology rather than the intrinsic mechanical properties of the polymer above T_g.

FIG. 4. Transmission electron micrograph of shadowed carbon replica of freeze-fractured cells of *A. eutrophus* showing plastic deformation of PHB granules.

An extensive study of granule development during fermentation of *Alcaligenes eutrophus* in nitrogen limitation has just been completed and will be reported fully in the near future.[42] However, the major conclusions are that the number of granules per bacterium is defined in the very early stages of the fermentation. The granules then swell as the accumulation proceeds with the cell maintaining its ellipsoidal shape until the volume fraction of polymer reaches about 0·6 μm. At this point the bacteria are compelled to change shape to a more spherical geometry in order to accommodate more polymer and the rate of accumulation slows down. This is of more than scientific interest since it would appear that the limiting factor of PHB production is not loss of polymerase activity or cell viability but the packing density of granules in the cell. The bacteria literally cannot make more polymer or they would burst!

POLYMER PRODUCTION

Fermentation

The simplest way of producing PHB is in a two-phase batch fermentation. In the first stage an inoculum of bacteria is introduced into a sterile solution of trace-metal nutrients and a suitable carbon source such as glucose or acetic acid. The amount of carbon must be carefully calculated since it determines the dry weight of cells in the fermenter at the end of the first, cell growth phase. However, of more importance is the initial concentration of the essential nutrient on which the culture will be limited in the second, polymer accumulation phase. This may be nitrogen as ammonium ion, phosphorus as phosphate or atmospheric oxygen. The bacteria grow and multiply rapidly under the conditions of carbon and nutrient excess in the first stage until the initial supply of carbon is used and the essential nutrient is depleted. The cells now contain only 10–20% PHB and the major polymer accumulation phase is initiated by starting to feed more carbon, as glucose, acetate or ethanol for example. In the absence of the essential nutrient, cell growth is limited and the added carbon is converted almost quantitatively into PHB.

The production of copolymers using mixed carbon substrates is similar and has already been described. It is analogous to the free radical initiated copolymerization of two monomers and pseudoreactivity ratios could probably be assigned to the components of the mixed feed if desired.[43]

TABLE 1
MOLECULAR WEIGHT OF PHB EXTRACTED FROM DIFFERENT BACTERIAL STRAINS[a]

Bacteria	Carbon source	Extraction method	Weight-average mol. wt, M_w
Azotobacter vinerandii	Sucrose	Methylene chloride	1 500 000
Rhizobium meliloti	Glucose	Chloroform	1 000 000
Azotobacter chroococcum	Glucose	Chloroform	1 000 000
Azotobacter beijerinckii	Glucose	1,2-Dichloroethane	800 000
Alcaligenes eutrophus	Fructose	Chloroform	600 000
Methylobacterium B3–Bp	Methanol	1,2-Dichloroethane	300 000
Pseudomonas AM-1	Methanol	Chloroform	50 000
Bacillus cereus	Glucose	Hypochlorite	10 000
Rhodospirillum rubrum	Glucose	Chloroform	5 000

[a] Results are for typical batch fermentations. The actual values obtained will also depend upon the precise fermentation conditions and extraction method used.

The molecular weight of the final polymer, however, is somewhat more difficult to control. It is known that a given species of bacteria will reproducibly synthesize polymers of a fixed molecular weight under a defined set of growth conditions and that this characteristic molecular weight is different for different bacteria.[44] A few typical examples are given in Table 1. The mechanisms involved, however, are largely unknown.[45]

Solvent Extraction

Having produced the polymer in cells as a dilute aqueous fermentation broth, the next step is to isolate and purify it. The traditional method of achieving this is to evaporate most or all of the water by freeze- or spray-drying and then extract the PHB with a suitable organic solvent such as chloroform,[46] methylene chloride,[47] 1,2-dichloroethane,[48] pyridine[49] or propylene carbonate.[50,51] Alternatively, the polymer may be extracted into the organic phase directly from the fermentation broth if a solvent with very low water uptake such as 1,2-dichloroethane is used[52] and emulsion formation is avoided. The cell debris may then be removed by filtration in the former case or by flotation in the latter to recover a clean polymer solution.

Very pure PHB may be obtained using these solvent extraction techniques, particularly if the dried cells are prewashed with methanol to remove lipids that would otherwise coextract into the organic phase. However, the need to filter viscous PHB solutions to remove cell residues limits the concentration of polymer that can be processed, and even with the best solvent recovery system available, the operation is inevitably very expensive.

Non-solvent Extraction

The alternative is to leave the polymer as a solid and dissolve away the cell debris. This can be achieved by the use of sodium hypochlorite to degrade oxidatively and make soluble all of the non-PHB cell components.[53] However, the polymer itself is degraded to some extent by the hypochlorite, which is also difficult to remove completely and acts as a melt prodegradant.[54] Various mild digestive extraction processes have been used to isolate pure PHB granules[40,55,56] but these are extremely time-consuming and unsuitable for scale-up since they involve difficult or expensive processes such as sonication or French pressing to break open the cells followed by density gradient centrifugation to separate the polymer from the lighter cell debris. Recent developments in this field, however, have given much more encouraging results.[57,58]

ANALYSIS

Polymer Content of Cells

The most reliable method of determining the polymer content of a bacterial cell culture or partially purified digest is to freeze-dry the sample and extract gravimetrically the PHB into chloroform solution. Indeed this is the technique used to calibrate all other methods but it is rather time-consuming and several more rapid analytical procedures have been developed.

A classic method is to digest the PHB-containing cell culture with sodium hypochlorite and measure the turbidity of the suspension produced by release of the polymer granules.[53,59] Alternatively the polymer can be extracted into chloroform solution and its infrared absorbance at $1728\,cm^{-1}$ determined quantitatively.[60] Another spectroscopic method involves hydrolysing and dehydrating the polymer to crotonic acid with concentrated sulphuric acid and assaying this by conventional quantitative UV analysis.[61,62]

However, greater accuracy can be achieved using chromatographic methods. In this case the dried polymer composition is hydrolysed to the methyl ester and examined by gas chromatography,[63,64] or to the free acid for ion exchange liquid chromatography.[65] Both methods are reproducible and have the advantage that they can be used to determine the comonomer composition of PHB copolymers at the same time as measuring overall polymer content.

Comonomer Ratio

The hydrolysis–gas chromatography technique applied to copolymer analysis has already been described in the literature[20] and is accurate to within about ±2%. However, it contains a systematic error in that it is difficult to ensure complete conversion of the monomer units to their respective methyl esters and some dehydration to the corresponding α,β-unsaturated ester (e.g. hydroxybutyrate to crotonate) usually occurs. The liquid chromatography technique[65] suffers from similar problems but the hydrolysis to free acid using perchloric acid tends to be a cleaner reaction than methanolysis and it is our experience that comonomer ratios can be determined with an accuracy better than ±1%.

However, the hydrolysis step can be avoided completely if the comonomer ratio is measured by proton or ^{13}C NMR. Examples of typical spectra for hydroxybutyrate–hydroxyvalerate copolymer in deuterochloroform solution are given in Figs 5(a), (b) for ^{1}H and ^{13}C NMR respectively. The comonomer ratio can be determined in the former case by comparing the integrated peak areas of the methyl proton signals from the butyrate side group at 1·21 ppm with those from the ethyl side group of hydroxyvalerate at 0·81 ppm. The area of the signal from the two methylene protons of this ethyl group at 4·20 ppm should, of course, be two-thirds that of the signal from the three methyl protons and this provides a useful internal check on the quantitative interpretation of the spectrum.

Carbon-13 spectroscopy is a straightforward but rather expensive way of determining comonomer ratio in view of the very long data acquisition times required. However, it does have another use in that the spectrum also contains information on the sequence distribution of the comonomer units.

Comonomer Sequence Distribution

Under high resolution ^{13}C NMR, the chemical shifts of the carbonyl carbons of the poly(hydroxyalkanoates) which occur at around 169 ppm relative to TMS are sensitive to dyad sequences and to the

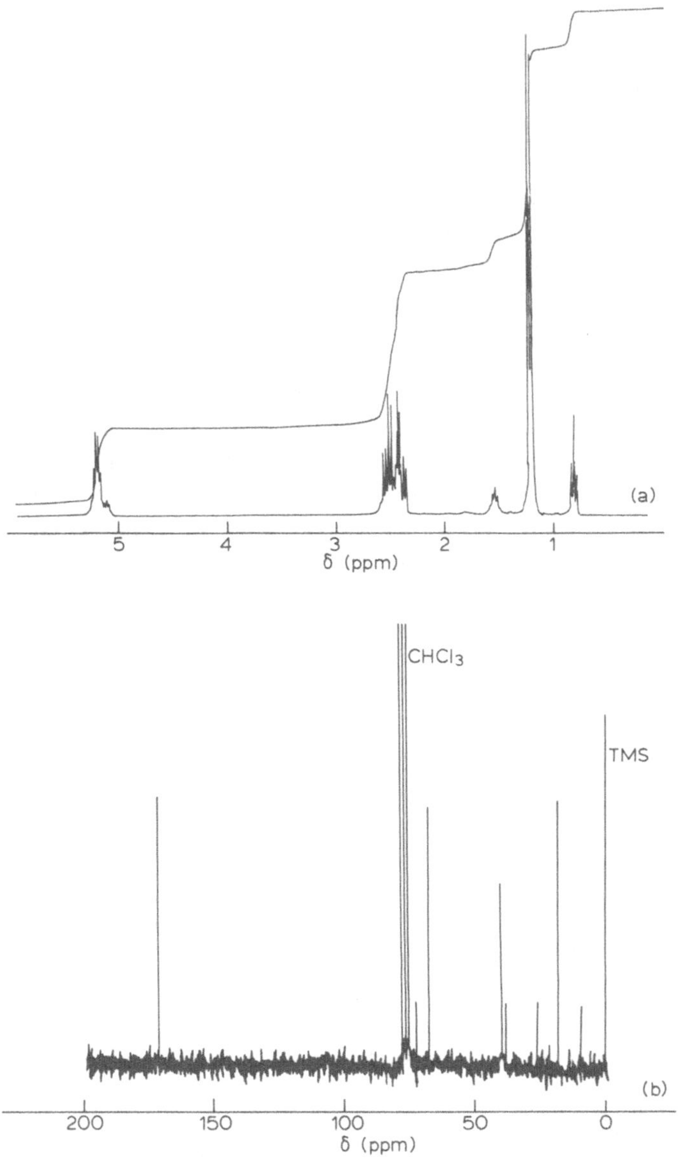

FIG. 5. (a) ^{1}H NMR spectrum, and (b) ^{13}C NMR spectrum, of an HB–HV copolymer, mole ratio 83:17.

stereochemistry at the adjacent carbon atoms. Thus, if the carbonyl signal from biologically produced poly((*R*)-3-hydroxybutyrate) is defined as having a chemical shift of zero, then chemically synthesized racemic poly((*RS*)-3-hydroxybutyrate) has two equal signals with shifts of 0·00 and 0·10 ppm. Similarly, racemic poly((*RS*)-3-hydroxyvalerate) gives two carbonyl signals at 0·38 and 0·50 ppm corresponding to the *meso* and racemic dyads respectively.[66]

The sample of biologically produced poly((*R*)-3-hydroxybutyrate-co-hydroxyvalerate) was found to give carbonyl resonances at 0·00, 0·18 and 0·37 ppm. Presumably the first signal arises from carbonyl groups in *meso*-butyrate dyads, the last from the equivalent *meso*-valerate units, with the new signal from either or both of the butyrate–valerate dyads.

Experimental results for four typical HB–HV copolymers are given in Table 2 together with the calculated intensities assuming random distribution of the comonomers. Clearly, these polymers, which were produced by the continuous addition of a mixed carbon substrate of fixed proportion during the second stage of a fermentation of *A. eutrophus*, are largely random copolymers with a hint of blockiness in the material of highest HV content, sample B.

TABLE 2
CARBON-13 SEQUENCING OF HB–HV COPOLYMERS

	From ^{1}H NMR results (mol %)		*From ^{13}C NMR results[a] (mol %)*		
Sample	*HB*(δ1·21)	*HV*(δ0·81)	*HB–HB*(δ0·00)	*HV–HB*(δ0·18) *HB–HV*	*HV–HV*(δ0·37)
A	94 ± 1	6 ± 1	83 ± 2(88 ± 2)	17 ± 2(12 ± 2)	0 ± 2(0·4)
B	70 ± 2	30 ± 2	55 ± 2(49 ± 2)	31 ± 2(42 ± 2)	14 ± 2(9 ± 1)
C	92 ± 1	8 ± 1	82 ± 2(85 ± 2)	16 ± 2(15 ± 2)	2 ± 2(0·6)
D	96·5 ± 1	3·5 ± 1	99 ± 2(93 ± 2)	5 ± 2(7 ± 2)	0 ± 2(0·1)

[a] Theory (based on ^{1}H NMR results) for random distribution in parentheses.

Crystallinity

PHB homopolymer is a completely stereoregular polyester and it is not surprising, therefore, that it is capable of achieving high levels of crystallinity. The crystal structure has been extensively studied by X-ray diffraction[54,67–72] and found to consist of polymer chains arranged as 2/1 helices in orthorhombic crystals with $P2_12_12_1$ space group. The fibre repeat (**c**) distance corresponding to two monomer units in one 360° turn is 5·96 Å with the other unit cell dimensions being **a** = 5·76 Å and **b** = 13·20 Å.[69] The polymer forms surprisingly thin lamellar crystals which are organized into spherulites when grown from the melt or arrow-shaped single crystals from dilute solution; these forms will be discussed in more detail later.

The mass fraction crystallinity of PHB can be obtained from X-ray diffraction studies,[73] density measurement or DSC thermal analysis.[74] At 20°C the amorphous density is 1·177 g cm^{-3} and the crystalline density 1·260 g cm^{-3}. The heat of fusion of PHB has been estimated at 146 J g^{-1}.[75] Whichever method is used, the mass fraction crystallinity of a typical PHB sample left to crystallize fully for several weeks at a temperature above its glass transition is in the range 60–90%.

The effect on melting behaviour of introducing hydroxyvalerate comonomer into the polymer is demonstrated in Fig. 6 (see also note added in proof). It should be noted that these results are for a series of HB–HV copolymers that were characterized by proton NMR to determine comonomer ratio and by DSC at 20°C min^{-1} to determine

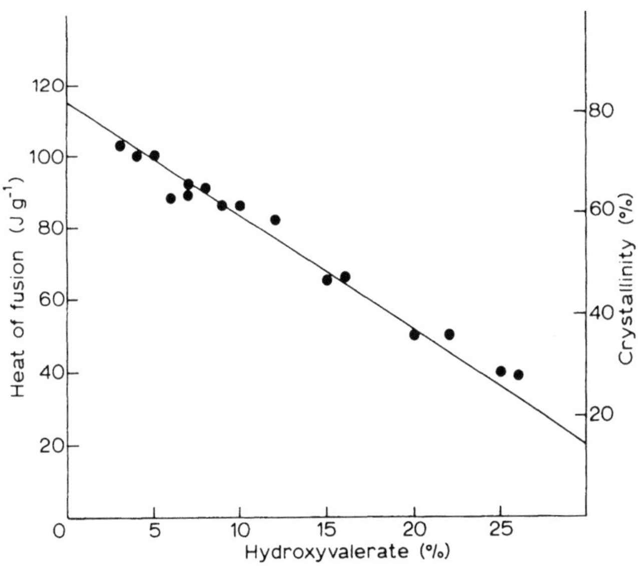

FIG. 6. Effect of HV comonomer content on the heat of fusion of PHB.

heat of fusion. No attempt was made to anneal the polymers or give them a defined thermal history, although all samples were obtained by precipitating from chloroform solution into methanol as a non-solvent and drying *in vacuo* overnight at room temperature. This is important because there is some evidence reported in the literature[76] that isomorphous replacement of the HV unit in HB crystals is possible and hence the degree of crystallinity of copolymers will depend upon annealing conditions. Certainly there is ample evidence that a whole range of poly(hydroxyalkanoates), including PHB and PHV, have isomorphous crystal structures.[71,77]

Thermal Transitions

Melting point and glass transition temperature are also affected by copolymer composition and the former is shown in Fig. 7 for the same series of copolymers under the same conditions as were described above. Again the observed melting point will depend on lamellar thickness[78] and the rate of lamellar thickening that occurs during the

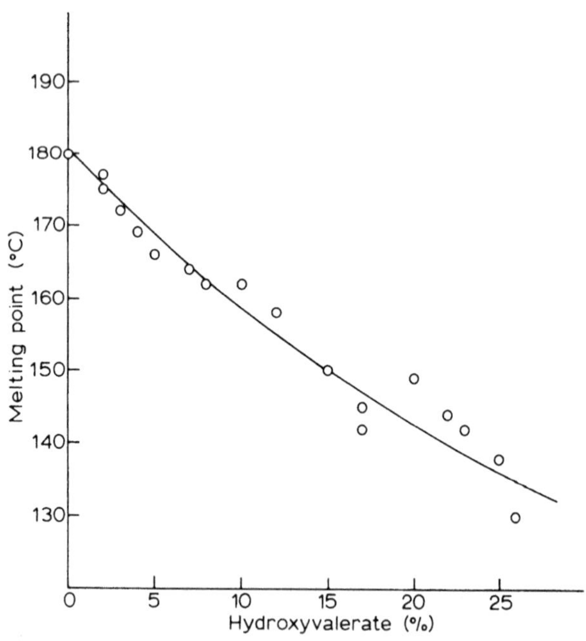

FIG. 7. Effect of HV comonomer content on the melting point of PHB.

melting experiment, and both of these will be influenced by potential isomorphous replacement of HV units in the HB crystals.

The glass transition temperature, T_g, is not significantly affected by copolymer composition since both HB and HV homopolymers have similar values of T_g at around 0°C. However, measured T_gs are dependent to a greater extent on the degree of crystallinity of the samples. For example, dynamic mechanical testing at low frequency of a quenched, amorphous sample of PHB will give a peak in tan δ at 0°C. If the same sample is annealed to crystallize it fully and then re-run, two effects become apparent. Firstly, the size of the tan δ signal is much reduced, as one might expect since only the amorphous phase contributes to T_g, and secondly the transition has moved to a much higher temperature at 25–30°C, reflecting the increased constraints placed on the amorphous polymer chains by the crystalline crosslinks. Dynamic mechanical analysis also reveals a smaller third-order transition at −80°C. A similar process is found in poly(ethylene terephthalate) and has been attributed to a relaxation in the crystalline phase.[79]

The final thermal transition in PHB is associated with degradation between 250 and 300°C.[80] Thermogravimetric analysis shows complete weight loss in a single step between these temperatures and corresponds to quantitative conversion of the polymer to crotonic acid.[81]

Molecular Weight

PHB is soluble in a number of common solvents and absolute molecular weights can be determined by light scattering, osmometry, or sedimentation analysis in chloroform,[82,84,86] trifluoroethanol,[82–86] 1,2-dichloroethane[82,84] or dichloroacetic acid.[82,84] Data from absolute measurements can then be used to determine Mark–Houwink–Sakurada parameters relating intrinsic viscosity to molecular weight; the values given in the literature are summarized in Table 3. This information can also be used to derive true molecular weights from gel permeation chromatography systems calibrated with polystyrene standards.[75] Whichever method is used, care must be taken to ensure that the polymer is not degraded during the sample preparation stage since it has been shown that dissolution of the polymer followed by precipitation can cause a severe decrease in molecular weight.[87] Moreover, a frequently used solvent for PHB is chloroform, which usually contains ethanol as a stabilizer. Refluxing the polymer for extended periods in this solvent to dissolve it can result in significant

TABLE 3
REPORTED VALUES OF THE MARK–HOUWINK–SAKURADA PARAMETERS RELATING INTRINSIC VISCOSITY TO MOLECULAR WEIGHT FOR PHB

$$[\eta] = K \cdot (M_w)^\alpha \ (\text{dl g}^{-1})$$

Solvent	*Temp.* (°C)	$K \times 10^4$	α	*Ref.*	*Year*
Chloroform	30	0·77	0·82	82	1970
	30	1·90	0·74	87	1975
	30	1·18	0·78	84	1976
	30	1·66	0·76	86	1979
Trifluoroethanol	30	2·51	0·74	83	1970
	30	1·25	0·80	84	1976
	25	2·22	0·76	85	1977
	25	1·75	0·78	86	1979
1,2-Dichloroethane	30	1·68	0·74	86	1979

esterolysis, particularly if acid impurities are present to catalyse the reaction.

SOLUTION PROPERTIES

Dilute Solution

The viscosity–molecular weight relationships for dilute PHB solutions have already been given in the previous section. The most unusual feature of these data is the very large excluded-volume effect in trifluoroethanol (TFE). One interpretation is that the polymer adopts a helical conformation in this solvent[83,88] but the most recent view is that the excluded-volume parameter, Z, is near the upper limit of the values predicted by modern theories of dilute polymer solutions.[86]

Investigations of this type are particularly arduous with PHB for two main reasons. Firstly, fractionation of the polymer to obtain monodisperse samples is very difficult although two solvent systems, dichloromethane with propylene glycol[84] and chloroform with acetone,[87] have been suggested. The second and more important factor is the lack of a theta solvent for PHB. It is interesting to note, therefore, that a theta solvent, butyl chloride at 13°C, has been reported for chemically synthesized poly((*RS*)-3-hydroxybutyrate).[86] Moreover, the molecular weight dependence of the intrinsic viscosity of this atactic polymer in TFE and chloroform is the same as that for the biopolymer. Thus, it should be possible for conclusions from characterization of the theta system to be extended to include dilute PHB solutions.

One effect shown exclusively by the isotactic biopolymer, however, is rotation of the plane of polarized light passing through it. Several optical rotary dispersion and circular dichroism studies on dilute solutions of PHB have been reported in the literature.[54,82,84,89] All agree that the rotation is nearly zero at 589 nm but increases to a strong positive value in the ultraviolet, reaching a maximum at about 240 nm. However, the earlier polarimetry data[82] were interpreted as suggesting a helical conformation in certain solvents, notably trifluoroethanol and 1,2-dichloroethane, and this was supported by an independent NMR study looking at lanthanide-induced shifts.[88] The later study did not confirm this and chose to interpret very similar data in terms of a highly expanded random coil model[84] which the same authors have since confirmed.[86]

Concentrated Solution

More concentrated solutions of PHB in methylene chloride and chloroform show viscosity effects typical of a high molecular weight polymer in a good solvent with no tendency towards liquid crystal formation or shear thinning. Examples of viscosity–shear rate data, obtained using a Haaka rotovisko system, for a series of PHB homopolymers (M_w 500 000) in chloroform solutions of increasing concentration at 50°C are given in Fig. 8.

Films of PHB and its copolymers produced by solution casting using conventional equipment[90] have a very fine spherulitic morphology because of the low crystallization temperature and concomitant high nucleation density and are extremely ductile. The relationship between morphology and mechanical properties will be discussed in more detail later.

Gels

Solutions of the polymer in poorer solvents such as propylene carbonate, butan-2-one and 1,2-dichloroethane (DCE) provide another means of product formation. These solutions gel on cooling and can be processed in that state. The standard high molecular weight biopolymers do not contain low molecular weight oligomers and gels formed from them are free-draining with most of the solvent being recoverable by squeezing.[91] Moreover, high-modulus and high-

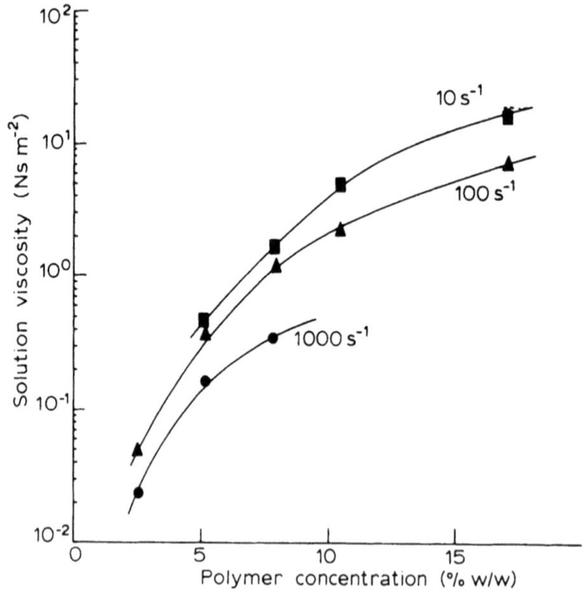

FIG. 8. Viscosity of PHB (M_w 500 000) in chloroform solution at 50°C as a function of polymer content at various shear rates.

strength fibres can be obtained by spinning and drawing suitable PHB gels in much the same way that polyethylene gels are ultradrawn.[92,93]

MELT PROPERTIES

Melt Stability

PHB has always been regarded as melt-unstable,[2] degrading almost quantitatively to crotonic acid at temperatures slightly above its melting point[81,94] by the β-elimination mechanism shown below. This reaction is identical to that involved in the pyrolysis of simple alkyl esters such as isopropyl acetate[95] and is similar to that seen in the decomposition of poly(vinyl acetate) at temperatures in the range 200–250°C.[96]

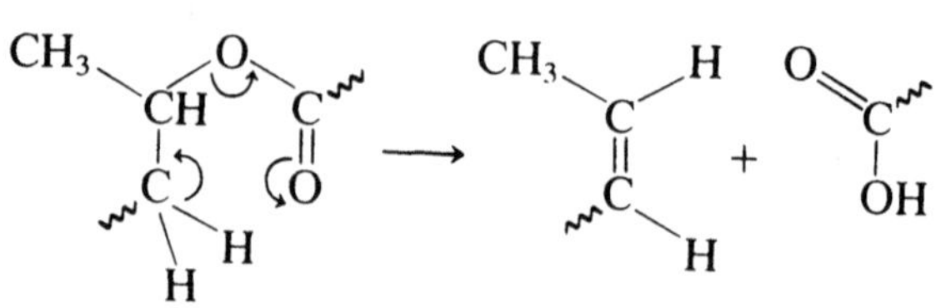

In fact all polymers, even polyethylene, are thermally unstable to some extent and will degrade slowly at processing temperatures. As a result, an industrial 'rule of thumb' has been established. In order to be regarded as injection-mouldable on a commercial scale, the melt flow index (MFI)[97] of a polymer must not increase by more than a factor of two during a 10 min period at the processing temperature. More stringent thermal stability requirements are applied to extrusion and blown-film processing, but let us see how PHB fares in this MFI doubling time (T_d) test.

Figure 9 shows the results of a series of MFI tests on high molecular

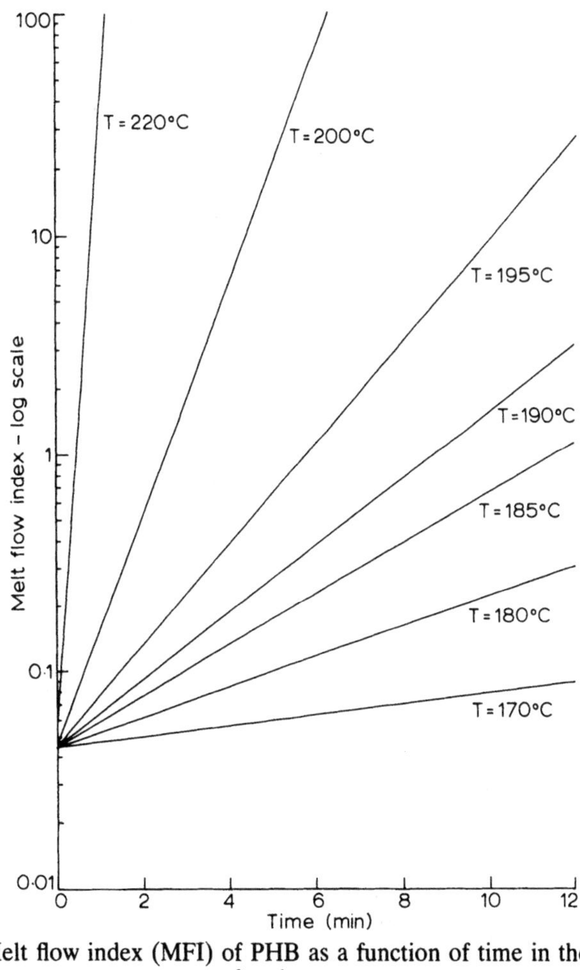

FIG. 9. Melt flow index (MFI) of PHB as a function of time in the melt for a range of melt temperatures.

weight ($M_w \sim 500\,000$) PHB homopolymer over a range of temperatures. Polymer (3g) was charged into the barrel of a standard MFI grader fitted with a 2·2 mm diameter × 8 mm length die and a 2 kg load applied to the piston after a 4 min warm-up period. Polymer extruding from the die was then cut off at 1 min intervals until the barrel was empty and each piece of extrudate weighed. MFI is defined as the weight of polymer extruding under the standard conditions described over a 10 min period. This, of course, assumes a more-or-less constant extrusion rate, which is not the case for a thermally unstable polymer. Therefore, the results plotted in Fig. 9 are the weight of polymer extruding in a given 1 min period *multiplied by ten* (to obtain an 'instantaneous' MFI) against the average residence time of that particular sample in the grader barrel.

The T_d of the homopolymer at a typical melt processing temperature of 190°C is about 2 min, which is well short of the 10 min target. However, reducing the melt temperature to below 170°C allows the biopolymer to meet this arbitrary standard. Unfortunately this temperature is actually a few degrees *below* the melting point of PHB and the results were obtained by melting the polymer in the grader at 190°C for the first 3 min of the warm-up period and then allowing the barrel to cool to 170°C. The polymer does not, of course, recrystallize until far larger undercoolings are achieved.

Similar experiments can also be performed with the HB–HV copolymers and, because of their lower melting points (Fig. 7), we do not have to resort to the above complication to obtain T_d data below 170°C. Typical results for the homopolymer and two copolymers are summarized in Fig. 10 as a plot of log T_d against melt temperature. Clearly low-melting HB–HV copolymers processed at or below 165°C meet the industrial rule of thumb on melt stability and should be regarded as commercially mouldable.

In fact all of the biopolymers are injection-mouldable in practice and can even be blown into film quite successfully provided that care is taken to minimize melt temperatures and residence times. It is also possible to adopt inverted temperature profiles along extruders such that the polymer is first melted at a relatively high temperature and then conveyed to the die through much cooler zones. This further reduces thermal degradation.

In conclusion, the HB–HV biopolymers are around or just below the acceptable limits of melt stability. The dominating degradation reaction is a β-elimination reaction, the kinetics of which have been

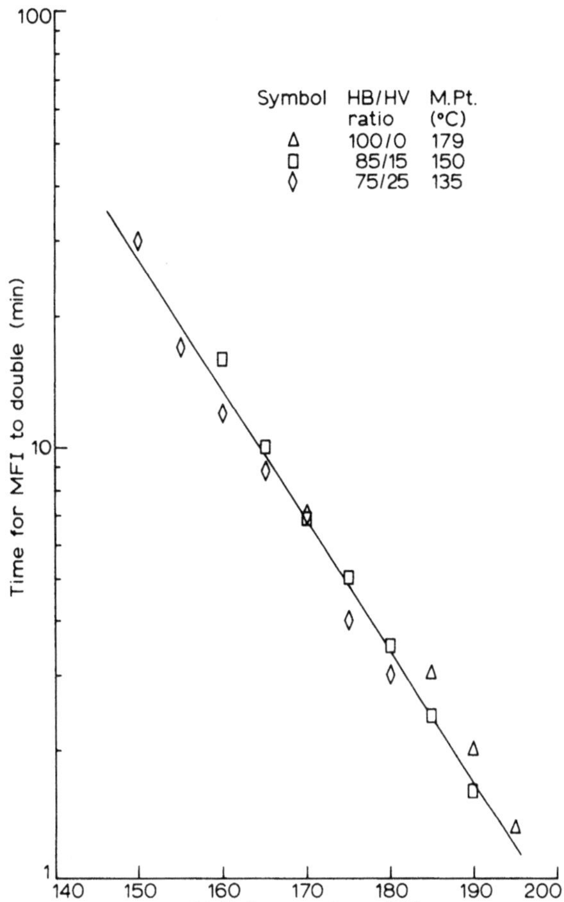

FIG. 10. Effect of melt temperature on the time required for the MFI of a range of HB–HV copolymers to increase by a factor of two (T_d).

studied in some detail and fitted to a random scission model.[98] The attempts to inhibit this reaction and stabilize the polymers have been reviewed recently[99] but all standard stabilizing systems such as acid acceptors, free-radical traps, etc., were found to be ineffective. The only method of modifying the process appears to involve the free-radical-initiated repolymerization of the polymer fragments using multifunctional unsaturated crosslinking agents.[100] A typical system would be dicumyl peroxide with triallyl cyanurate. Great care must be taken, however, to ensure that the rate of the repolymerization reaction does not exceed that of the depolymerization since this inevitably results in complete crosslinking of the polymer.

Thermal Decomposition

Although crotonic acid, or 2-pentenoic acid from HV comonomer, is the main decomposition product from the biopolymers at low temperature, other degradation products appear under high-temperature pyrolysis conditions.[80] For PHB homopolymer these include propene and carbon dioxide from the breakdown of crotonic acid itself, isocrotonic acid from the obvious reorganization reaction and β-butyrolactone as a short-lived intermediate that spontaneously decomposes by *O*-alkyl scission to propene and carbon dioxide or by *O*-acyl scission to acetaldehyde and ketene.[101] Similar reactions can be proposed for the comonomer decomposition and some pyrolysis experiments have already been reported.[102]

Melt Rheology

There has been little or no published work on the melt rheology of PHB, presumably because of problems associated with its melt instability. However, some information on melt viscosity can be derived from melt flow index data such as that given in Fig. 9, using the empirical relationship[103]

$$\log_{10} \mathrm{MFI} = \mathrm{constant} - \log_{10} \eta_0 \tag{1}$$

where η_0 = melt viscosity at zero shear. For polyethylene an MFI of 1 under the standard conditions of die size, piston and load weight (total 2·16 kg), etc., corresponds to a melt viscosity of $1{\cdot}5 \times 10^4\ \mathrm{N\,S\,m^{-2}}$ and hence the constant can be calculated to be 4·176.[104] The density difference between polyethylene and PHB means that this constant will not be the same for both polymers since MFI is a gravimetric measure of flow rate and viscosity is concerned with volumetric flow rates. However, the relationship does give a useful order-of-magnitude estimate of melt viscosity at zero shear from MFI data.

MORPHOLOGY

Solution-Grown Crystals

It is possible to grow single crystals of PHB from solution and examine them by X-ray diffraction and electron microscopy,[78] although the

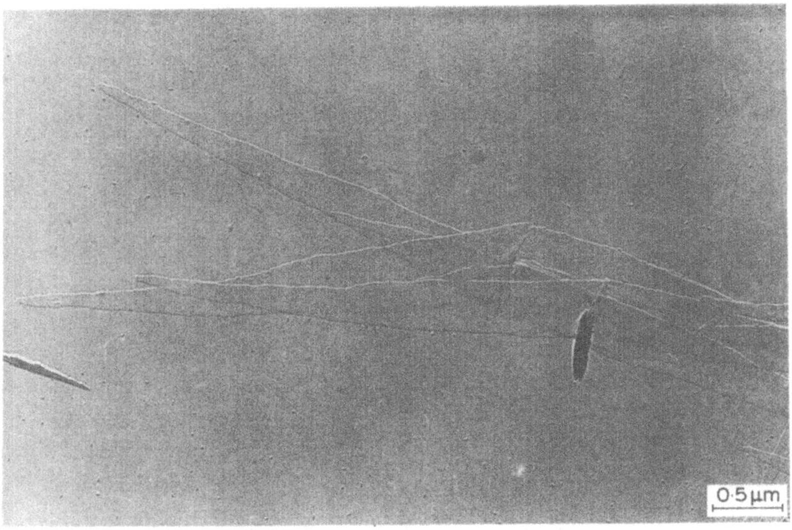

FIG. 11. Electron micrograph of PHB single crystals grown from a dilute solution in propylene carbonate.

polymer is very susceptible to electron beam damage. The crystals have an elongated arrow-like shape with quite rough edges and there are indications that they originally have a three-dimensional tent-like structure which is collapsed as they settle on to a solid surface.[75] This is illustrated in Fig. 11, an electron micrograph of PHB single crystals grown from a dilute solution in propylene carbonate.

Selected area diffraction from part of a crystal indicates that the crystallographic **a**-axis runs along the direction of each crystal and examination of fracture modes suggests that the direction of chain folding is also parallel to the long axes of the crystal edges, i.e. along the [100] planes.[75] This actually poses interesting questions about the spatial arrangement of adjacent chains and their interconnection via chainfolds since we know from X-ray studies that the two PHB chains in each unit cell are antiparallel,[67–72] as illustrated in Fig. 12. If the direction of chain folding in single crystals is to be predominantly along the **a**-axis and adjacent chains in this direction are parallel then, for an adjacent re-entry model, the chains must preserve a memory of their last fold direction and fold alternately at 45° and −45° to the **a**-axis.

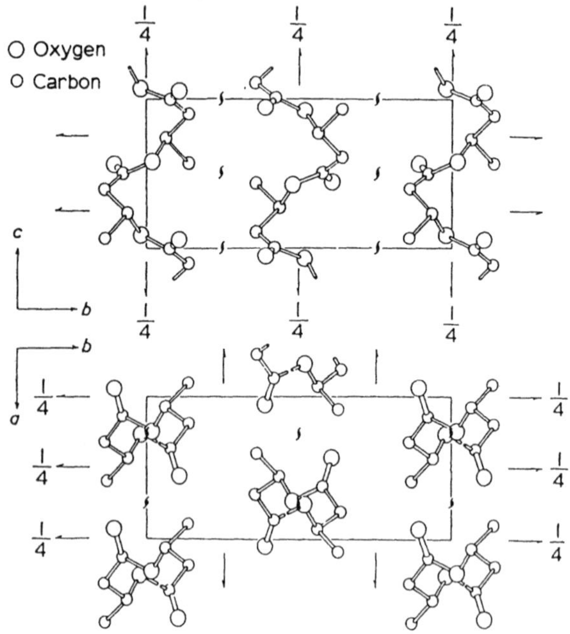

FIG. 12. Crystal structure of PHB (**a** = 5·76 Å, **b** = 13·20 Å, **c** = 5·96 Å).

Moreover, the crystals are very thin, between 4 and 6 nm depending on crystallization temperature.[75] This represents only six to ten crystallographic repeat units of two monomers each and, since the sample is known to be 80% crystalline, leaves just three or four monomer units in the amorphous phase to perform a complete 180° turn and allow the chain to re-enter the crystal. Clearly there is little room in the amorphous fold to allow anything but adjacent re-entry and this leaves the question of why the chains apparently zig-zag along the **a**-axis instead of continuing to fold linearly at 45° to it.

Melt-Grown Spherulites

When PHB is crystallized from the melt it forms large banded spherulites. Because of its biological origin and the extensive purification process used to separate it from cell debris, the polymer does not contain inorganic catalyst residues or other impurities which could act as heterogeneous nucleation centres. As a result it is very easy to obtain samples of PHB with low nucleation density that form massive spherulites on cooling from the melt. With a modicum of care, spherulites of several millimetres diameter that are visible to the naked eye can be grown. These are obviously very easy to study and make PHB an ideal material for the investigation of spherulite morphology.

For example, it is possible to obtain selected-area X-ray diffraction patterns from parts of a spherulite, and deduce the orientation of the crystal axes within it. In fact the crystallographic **a**-axis is radial in the spherulite with the other two axes rotating about it to produce the typical banded appearance.[75] The band spacing varies with the crystallization temperature but as yet there are no mathematical models relating band spacing to the fundamental parameters of nucleation and crystal growth.

The thickness of the lamellar crystals in PHB spherulites was also found to vary with crystallization temperature but is uniformly low, typically around 6–7 nm. Representing fewer than 10 repeat units per fold stem and in view of the high degree of crystallinity, this again leaves little room for looseness in the fold structure. A large number of folds must be made within the small volume fraction of amorphous phase available.

An interesting phenomenon was noted by the authors of the above study[75] when they plotted the polymer melting point, which is of course dependent on lamellar thickness as a function of crystallization temperature.[78] A curious but quite reproducible peak in melting point appeared at a crystallization temperature of around 60°C in this normally smooth curve. No explanation for this behaviour was offered but it gives rise to an unusual effect in that if a sample of PHB is first partially crystallized at a temperature away from the peak (say 90°C) and subsequently fully crystallized at this peak, then on reheating the core of the spherulites melts first to leave an annular ring of higher-melting polymer. It is interesting to reflect on the morphological implications of this observation.

EFFECT OF MORPHOLOGY ON MECHANICAL PROPERTIES

Spherulite Defects

The advantage of the coarse spherulite texture of melt-cast PHB is that it makes morphological investigations relatively easy. Its disadvantage is that the resulting structure is extremely brittle. Indeed, when pure polymer is crystallized from the melt on a microscope slide, the resulting large spherulites are frequently cracked and little extra

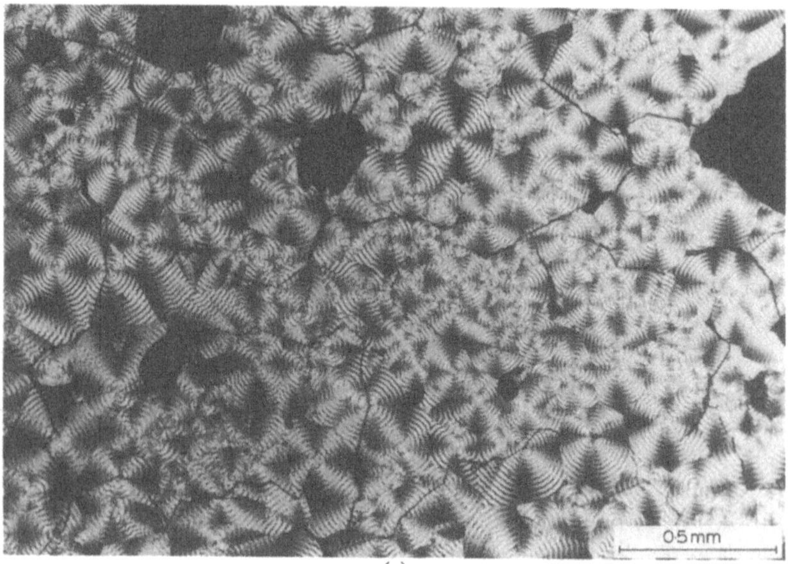

(a)

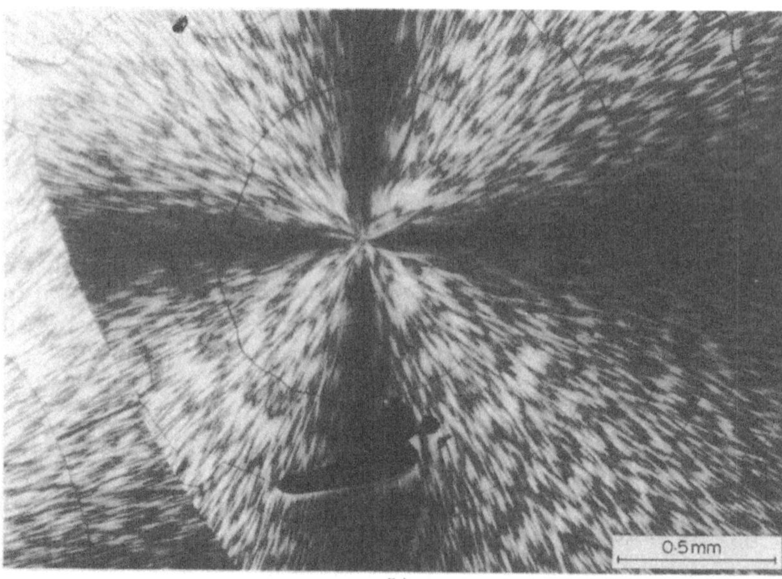

(b)

FIG. 13. Optical micrographs taken between cross-polars of thin films of PHB, (a) crystallized at 60°C, and (b) crystallized at 130°C. Note radial cracks in (a) and circumferential cracks in (b).

strain is required to open up these cracks and produce brittle failure of the sample.

Interestingly, in a detailed study of crack formation in PHB spherulites, it was reported that the cracks run radially in spherulites grown at temperatures below about 120°C and circumferentially in those produced at higher temperatures,[105] as illustrated in Fig. 13. Moreover, in the latter case the first cracks only appeared when the spherulites had grown to at least 35 μm in diameter. Thus, one way of avoiding spherulite crack formation in PHB is to crystallize the polymer at a high temperature, say 130°C, to avoid the radial defects and ensure a high nucleation density so that none of the spherulites obtains the critical diameter for circumferential crack formation. The latter may be achieved by 'self-seeding', which will be described in more detail later, and it is encouraging to note that significant improvements in the ductility of melt-cast PHB sheets were reported for samples prepared under these conditions.

Cold-Rolling

An alternative method of improving the ductility of brittle PHB sheets is to cold-roll them.[105,106] The precise morphological changes that occur when the polymer is squeezed in the nip of calender rolls are still being studied but it is apparent from optical microscopy that the spherulite cracks are 'healed' by this process. In fact the spherulites are progressively distorted to ellipsoids by the rolling process until the original spherulite boundaries become indistinguishable and the hair-line cracks, which originally open up under very low strain, gradually disappear even under much higher degrees of applied strain. The effect on mechanical properties can be seen in Table 4.

Spherulite Size

The effect of spherulite size on the mechanical properties of PHB is immediately apparent from the comparison of solution-cast and melt-cast films given in Table 5. Both films have spherulitic textures but the comparatively low crystallization temperature of the solution-cast material ensures that its nucleation density is maximized and the spherulite size reduced by two orders of magnitude with respect to the melt-cast sample. Clearly the smaller spherulite size produces major benefits in improved ductility.

However, there is always concern in such comparisons that residual solvent acts as a plasticizer in the former case. To avoid this charge,

TABLE 4

EFFECT OF COLD-ROLLING ON MECHANICAL PROPERTIES OF COMPRESSION-MOULDED PHB PLAQUES

	Tensile properties			
Elongation by rolling (%)	*Modulus at 0·5% strain (GPa)*	*Tensile strength (MPa)*	*Extension to break (%)*	*Tensile impact strength* ($kJ\,m^{-2}$)
0	2·80	28	7	45
10	2·16	25	23	53
18	1·79	32	180	77
45	1·46	30	180	240
60	1·33	50	200	450
110	1·23	46	250	550
300	1·20	45	300	1000

Table 5 also contains data on a melt-fabricated sample that has been 'self-seeded' to produce a higher nucleation density and consequently a smaller spherulite size. Again the resulting improvement in ductility is apparent.

'Self-seeding' is observed when a polymer is only partially melted

TABLE 5

EFFECT OF SPHERULITE SIZE ON MECHANICAL PROPERTIES OF PHB

Parameter	*Solution-cast plaque*	*Unnucleated moulding*	*Self-seeded moulding*	*Nucleated moulding*
Melt-casting temp. (°C)	25	195	185	195
Average spherulite diameter (μm)	1	200	30	10
Young's modulus (GPa)	2·8	3·5	3·2	3·2
Notched Izod impact strength ($J\,m^{-1}$)	—	20	60	65
Falling weight impact strength (J)	17	0·5	2·7	3·5
Extension to break (%)	20[a]	2	6	8

[a] Yield point.

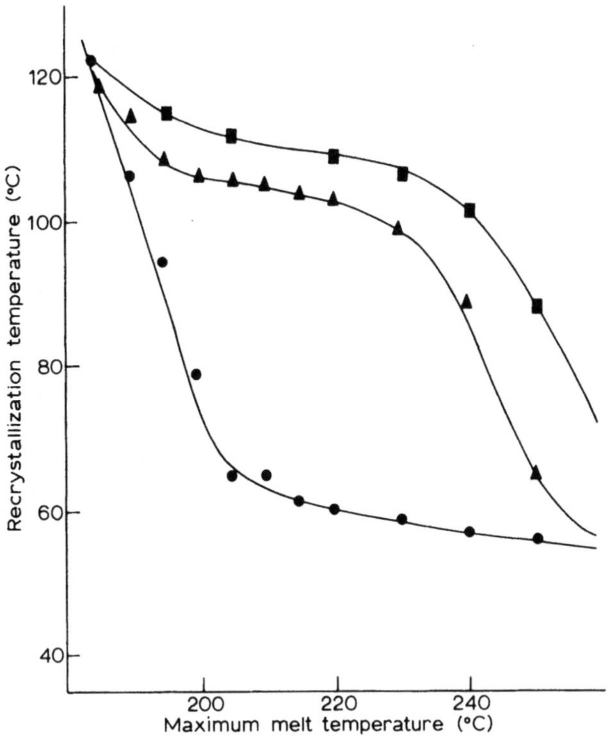

FIG. 14. Recrystallization temperature of unnucleated and nucleated PHB on cooling from the melt at 20°C min^{-1} as a function of melt temperature. (●), PHB; (▲), PHB + 1% talc; (■), PHB + 1% Boron Nitride.

before cooling and recrystallizing. The residual crystallites in the melt act as additional, ready-formed nucleation centres for the rest of the polymer and result in the observed reduction in spherulite diameter. The effect is most clearly demonstrated using differential scanning calorimetry (DSC). In the results given in Fig. 14, the extent of self-seeding is controlled by heating the polymer to increasing maximum melt temperatures (T_{max}) before cooling at a controlled rate and monitoring the nucleation effect from the temperature of the recrystallization; the better the nucleation effect the sooner recrystallization occurs on cooling, resulting in a higher recrystallization temperature, T_c. The self-seeding effect for PHB homopolymer can be seen to diminish slowly as the melt temperature increases from 180 to 200°C and all of the polymer is finally melted.

It is interesting to note the magnitude of the undercooling necessary to recrystallize PHB in the absence of self-seeding—almost 120°C! Similar effects can be seen with other polymers but the maximum undercoolings measured are usually only 60 or 70°C. The massive temperature range over which nucleation and crystallization can be observed in PHB is one of the major reasons for academic interest in the material as a model polymer for morphological and crystallization kinetics studies.

Of more commercial interest is the improvement in mechanical properties resulting from enhanced nucleation. For this reason a range of nucleating agents for the polymer have been identified. Many of these work with other aliphatic polyesters,[107] and finitely ground talc is a typical example. Their mode of action has been described as 'enhanced self-seeding' because the nucleating species is not the inorganic particle itself but polymer tightly bound to its surface.[108] Thus if talc is added directly to a PHB melt and the mixture cooled, no nucleation effect is observed. However, some polymer crystallizes on to the talc particles and is bound strongly to them so that on reheating they do not melt with the unbound polymer and are available to act as nucleation centres when cooled. The DSC method described earlier has been used to illustrate this in Fig. 14 and the effect of nucleating agents on mechanical properties is shown in the final column of Table 5.

Copolymerization

The effect on morphology of introducing comonomers into the PHB chain has not been studied in any great detail to date. The only literature report relates to an examination of a low molecular weight mixture of copolymers.[109] The crystallographic unit cell dimensions of this heteropolymer were found to correspond closely to those reported for racemic PHV[77] with the same orthorhombic $P2_12_12_1$ space group. Solution-grown single crystals of the copolymer, however, were square in shape, in distinct contrast to the spear-shaped PHB crystals. It would be interesting to study the transition of one type into the other as the comonomer ratio was varied.

Comonomer ratio certainly has a profound effect on mechanical properties of the biopolymers. This is illustrated in Table 6 for a range of injection-moulded random HB–HV copolymers. These samples contained a small amount of talc as a nucleating agent and were annealed at 60°C for 24 h before mechanical testing at room tempera-

TABLE 6
THERMAL AND MECHANICAL PROPERTIES OF RANDOM HB–HV COPOLYMERS

HB:HV ratio	*Melting point (°C)*	*Glass transition*[a] *(°C)*	*Heat distortion temperature (°C)*	*Young's modulus (GPa)*	*Tensile strength (MPa)*	*Notched Izod impact strength*[b] *(J m*$^{-1}$*)*
100:0	179	10	157	3·5	40	50
97:3	170	8	140	2·9	38	60
91:9	162	6	125	1·9	37	95
86:14	150	4	112	1·5	35	120
80:20	145	−1	99	1·2	32	200
75:25	137	−6	92	0·7	30	400

[a] Position of E'' peak at 5 Hz.
[b] With 1 mm radius notch.

ture. As expected, increasing the HV comonomer content produced a reduction in stiffness (modulus) and increase in toughness (impact strength).

The increase in toughness as a result of increasing the HV comonomer content was very apparent but it was also somewhat limited. This is because the glass transition temperature of the copolymer was not significantly reduced and, although the proportion of amorphous phase was increased, its ability to dissipate energy after low-temperature impact was not enhanced. Dynamic mechanical analysis has been used to illustrate this[110] and Table 6 shows the limited effect of comonomer content on the temperature of the glass transition process. Melting point is, of course, more affected by the comonomer; this is reflected in the large reductions in heat distortion temperature for the copolymers which are also included in Table 6.

CRYSTALLIZATION KINETICS

Nucleation

Five types of nucleation process have been identified for PHB in a recent review.[108] These are: homogeneous nucleation, occurring infrequently at random sites when the polymer is cooled from a melt at or above 210°C; self-seeding, which has already been described as nucleation by unmelted crystallites when the maximum temperature of the melt does not exceed the melting point by more than a few degrees; enhanced self-nucleation, which is an extension of self-seeding in that, when the maximum melt temperature is limited to only 1°C above melting point, the original spherulites simply reappear on cooling rather than going through the normal radial growth process; nucleation by impurities such as talc in the enhanced self-seeding manner described earlier as the talc is coated with bound crystallites having a higher-than-usual melting point; and finally genuine epitaxial nucleation by saccharin, which will initiate nucleation when added to supercooled melts with spherulite growth initiated preferentially at certain faces of the saccharin crystal.

Three of these cases—homogeneous nucleation, nucleation by talc and nucleation by saccharin—were evaluated quantitatively and the nucleation rates, measured by hot-stage microscopy, fitted to eqn (2).

$$I = I_0 \exp \frac{-U^*}{R(T - T_\infty)} \cdot \exp \frac{-C(T_m^0)^2}{\Delta H^2 \cdot \Delta T^2 \cdot kT} \qquad (2)$$

The first exponential term represents the transport of molecules to the growing nucleus with U^* the activation energy for this process and $(T - T_\infty)$ the temperature difference between the crystallization temperature T and the temperature at which backbone motions substantially cease. For most polymers T_∞ is about 50°C below the glass transition temperature. The second exponential term represents the work required to form a critical nucleus[111] where T_m^0 is the melting point of an infinitely thick lamellar crystal, ΔT is the supercooling and ΔH the heat of fusion. The term C contains various interfacial energy terms and depends upon the precise mechanism of the nucleation process. For homogeneous nucleation

$$C = 32\sigma^2 \cdot \sigma_e$$

where σ is the side free energy and σ_e the free energy of the fold surface.

Using values for the various parameters derived from other studies,[75] the experimental data for pure PHB could be fitted to eqn (2). The ultimate melting point, T_m^0 was calculated to be 197°C to give the best fit. However, data for the talc-nucleated samples could only be reconciled with eqn (2) if a higher value of T_m^0, 222°C, was used. This confirms the theory that the nucleating species in this case is not talc itself but polymer bound to the surface of the inorganic particles, the lower entropy of the bound polymer resulting in its higher melting point.

The saccharin nucleation data were not consistent with eqn (2) for any value of T_m^0 when using the expression for C derived for homogeneous nucleation. However, assuming that nucleation occurs epitaxially at the saccharin/polymer interface, the correct expression for C should be approximately[108]

$$C = 16\sigma^2 \,.\, \sigma_e$$

This provided excellent agreement with the experimental data when the original value of T_m^0 for the pure polymer was used.

In fact the only discrepancy between the theoretical prediction of eqn (2) and the experimental data was found to be in the pre-exponential factor I_0. This was over 20 orders of magnitude lower than typical values for other polymers[108] and, although no explanation for the massive discrepancy has been proposed, it must reflect the difficulty in nucleating PHB. It also confirms our earlier observation about the poor nucleation behaviour of PHB: that the polymer must be cooled from the melt to a much greater extent than is usual for other materials before nucleation is initiated.

There are no obvious reasons for this behaviour in the thermodynamic parameters for PHB, which are compared with those for several other polymers in Table 7. Even the undercooling required to obtain the maximum spherulite growth rate, T at G_{max}, is not excessively large. Thus the problem clearly centres around the nucleation process rather than the subsequent crystal growth. It may be that the use of a better nucleating agent would bring PHB more into line with other polymers. From the reported values of maximum nucleation rate,[108] saccharin ($I_{max} = 600$ events $s^{-1}\,mm^{-3}$) would appear to be a better nucleating agent than talc ($I_{max} = 32$ events $s^{-1}\,mm^{-3}$), which is still an order-of-magnitude improvement on the pure polymer ($I_{max} = 3{\cdot}0$ events $s^{-1}\,mm^{-3}$).

Unfortunately, in a practical sense, saccharin is *not* a nucleating agent for PHB. It actually dissolves in, or co-melts with, the polymer and will function only if added as a crystalline solid to a supercooled melt. The effect was discovered serendipitously because saccharin, with its sharp melting point at 183·19°C, was being used as a temperature calibrant for hot-stage microscopy of PHB. Inevitably, some of the saccharin dropped into the polymer melt and the scientist involved was alert enough to note the resulting nucleation process.

TABLE 7
THERMODYNAMIC PARAMETERS FOR SOME COMMON POLYMERS INCLUDING PHB

Polymer	T_m° (°C)	T_g (°C)	ΔH_f ($MJ\,m^{-3}$)	ΔS_f ($MJ\,m^{-3}\,K^{-1}$)	$T_m - T_c$ at G_{max} (°C)	G_{max} ($\mu m\,min^{-1}$)	U^* ($kJ\,mol^{-1}$)	$\sigma . \sigma_e$ ($J^2\,m^{-4}$)	σ_e ($J\,m^{-2}$)
Poly(ethylene terephthalate)	270	70	182	0·33	90	10	—	—	—
Polystyrene (isotactic)	240	100	99	0·19	60	0·2	6·55	$2·7 \times 10^{-4}$	0·035
Poly(hexamethylene adipamide)	232	62	233	0·46	40	1200	6·00	$9·4 \times 10^{-4}$	0·118
PHB	197	0	185	0·39	90	300	10·25	$11·16 \times 10^{-4}$	0·038
Polypropylene	186	−8	198	0·43	75	20	—	—	—
Polyethylene	141	−40	292	0·70	30	5000	6·3	$13·8 \times 10^{-4}$	0·098
Polyoxymethylene	183	−80	368	0·81	40	400	6·3	$13·8 \times 10^{-4}$	0·061

Spherulite Growth

The radial growth kinetics of PHB spherulite formation have been evaluated in terms of Hoffman's theoretical relationship:[75]

$$G = G_0 \exp \frac{-U^*}{R(T - T_\infty)} \, . \, \exp \frac{-K\sigma\sigma_e T_m^0}{\Delta H \, . \, \Delta T \, . \, kT} \qquad (3)$$

where the terms have the same meaning as in eqn (2) with the addition of the constant K, which is a function of the crystal growth mechanism. Hoffman describes three distinct phases depending upon the rate of nucleus formation at a new face of the lamellar crystal compared with the rate of spreading from a nucleus along the face.[112] This ratio is determined by the crystallization temperature. Thus at low supercoolings the rate of nucleus formation is very low and, once formed, polymer will spread from this nucleus right along the face before another nucleus can form on that particular face. This is called Regime I and the constant K in eqn (3) has the value 2. In Regime II, at slightly greater supercoolings, the two rates are comparable and several nuclei will form and spread across the crystal face simultaneously. The value of K is then 4. Eventually, nuclei are formed at every site on the face, precluding any spreading, and the value of K under this Regime III again adopts a value of 2.

Plotting $\ln G + U^*/R(T - T_\infty)$ against $1/T \, . \, \Delta T$ from eqn (3) should give a straight line with slope $-K\sigma\sigma_e T_m^0/k \, \Delta H$ and intercept $\ln G_0$. For PHB this graph was found to have an inflection corresponding to a crystallization temperature of 130°C with the slopes of the linear curves before and after this being in the ratio 2:1.[75] This is strong evidence for a transition from Regime II to Regime III kinetics at 130°C. It is interesting to note that the radial to circumferential spherulite cracking effect described earlier also occurs at the same temperature. The Regime II–Regime III transition is of course predicted for all crystalline polymers, but in most cases it is impossible to observe experimentally because the crystallization process occurs too quickly at high supercoolings.

The effect on spherulite growth kinetics of introducing HV comonomer units into the HB chain has not been quantitatively studied to date, although it is clear that the crystallization rate is reduced by increasing the HV content. Similarly the nucleation processes in HB–HV copolymers have not been extensively examined but these polymers do appear to be nucleated by the same materials that are effective in PHB.

ORIENTATION

Uniaxial

Since PHB is a polyester it might have been expected that it could be drawn and oriented into fibres quite easily using similar techniques to those developed for poly(ethylene terephthalate). This is not the case. PET is normally melt-spun into a fibre which is then quenched and wound up as an amorphous, glassy material.[113] In a separate stage it is then drawn over a hot pin at a temperature above its T_g and crystallized in this oriented state to produce the familiar strong polyester fibres. This type of stress-induced crystallization from the quenched amorphous state does not seem to be possible with PHB. Indeed, the application of strain to a supercooled PHB melt can inhibit recrystallization rather than promote it.

This may be demonstrated by extruding two identical pieces of pure, high molecular weight PHB homopolymer lace from the melt into air at ambient conditions using, for example, a simple MFI grader. If one piece is left to crystallize isotropically as a control, while the other is stretched immediately by 50–100%, then the latter will remain elastic for several minutes after the former has fully crystallized. Both samples, however, eventually harden to the same isotropically crystalline and very brittle condition.

The results of a more quantitative investigation into the cold-drawing behaviour of amorphous PHB are presented in Table 8. Tensile bars were cut from compression-moulded plaques that had been quenched into liquid nitrogen and then tested under a variety of strain rates using an Instron machine with a temperature-controlled sample cabinet. At temperatures below T_g, i.e. below about −5°C, the samples failed by brittle fracture at very low extension as expected for a glassy material. Around T_g itself, in the range −2 to +7°C, the samples yielded at a tensile stress of about 10 MPa and then drew out under very low stress (1–2 MPa) to more than 1000% extension. However, these drawn samples were extremely elastic and recovered almost to their original dimensions on release of the stress. Attempts to heat-set the extended tapes usually failed since the samples tended to fail by necking rupture. Finally, at higher temperatures, the material became rather cheesy and failed at very low loads after only modest extension.

Thus amorphous PHB is extremely difficult to crystallize in the oriented state under an applied strain. In addition, the fully crystalline

TABLE 8

EFFECT OF TEMPERATURE AND STRAIN RATE ON MECHANICAL PROPERTIES OF QUENCHED AMORPHOUS PHB

Temp. (°C)	*Strain rate (min^{-1})*	*Modulus at 0·5% strain (GPa)*	*Tensile stress* Yield (MPa)	Break (MPa)	*Strain* Yield (%)	Break (%)
−20	1	1·600	—	40	—	3·5
−10	1	1·200	—	35	—	4·5
0	1	0·910	20	10	5	1140
0	5	0·833	—	25	—	10
+5	1	0·227	—	4·2	—	250
+5	5	0·357	8·7	3·1	5	1240
+7	10	0·250	7·75	2·3	4·5	1100
+7	25	0·200	11·40	1·6	—	1500
+10	1	0·043	—	2·2	—	100[a]
+15	1	—	—	0·8	—	200[a]
+23	1	—	—	0·2	—	130[a]

[a] Necking rupture.

homopolymer, in contrast to polyolefins for example, will not draw under any conditions of temperature and strain rate and uniformly undergoes brittle failure. The only exceptions to this are the observations described earlier, that well nucleated PHB crystallized at temperatures above 130°C to avoid radial cracks in the spherulitic structure is ductile near the melting point[105] and that cold-rolling can induce ductility in brittle crystalline samples.[106] However, the former effect has proved difficult to reproduce consistently and seems to depend upon using a very high molecular weight polymer.[114]

Fortunately, there is one other method of producing drawn fibres of PHB from the melt. Returning to the experiment with the extruded polymer lace: if instead of straining it immediately it is first allowed to crystallize partially under isotropic conditions, then at some critical level of crystallinity the fibre may be cold-drawn at a neck up to eight times its length to produce a reasonably strong oriented yarn.[115] This is effectively an in-line spin–draw operation with a partial crystallization stage between the two processes. Drawn fibres of both PHB polymer and HB–HV copolymers can be obtained in this way but the crucial parameters of temperature and duration of the crystallization stage will depend upon the nucleation and crystallization kinetics of the particular material, the former being influenced by the presence of

TABLE 9
MECHANICAL PROPERTIES OF SPUN-DRAWN PHB AND HB–HV COPOLYMER FIBRES

Material code	*Draw ratio*	*Denier (g/9000 m)*	*Tenacity (g/den)*	*Extension to break (%)*
PHB	1·0	1381	0·24	6
	4·8	253	0·98	90
	5·6	170	1·35	65
	6·4	150	1·65	60
	7·2	147	1·82	50
	8·0	142	2·01	35
	9·0	135	2·43	32
P7/21	1·0	940	0·20	10
(10% HV)	5·0	190	1·70	80
	8·5	108	2·30	50
P7/23	1·0	1650	0·20	7·5
(3% HV)	5·0	301	2·00	58
	6·0	265	2·20	51
	7·0	198	2·40	46

nucleating agents and the latter varying with comonomer ratio. Clearly, if the material is not crystalline enough then it will deform elastically during the drawing stage, and if it is too crystalline the material will merely undergo brittle failure. Mechanical properties of some typical spun–drawn PHB and the HB–HV copolymer fibres are given in Table 9.

Solution Spinning

A similar technique can be employed when dry-spinning PHB from solution. In fact there is somewhat greater latitude in the drawing conditions for this process since the spherulitic morphology of solution-cast PHB is much finer and intrinsically more ductile than that of melt-cast polymer. Moreover, residual solvent in the thread line can act as a temporary plasticizer to aid the drawing process.

The main difficulty with dry spinning PHB is in obtaining a sufficiently concentrated and viscous dope that can form a continuous fibre when pushed through a spinneret and maintain this stable condition despite the full weight of the thread line tending to draw it down prematurely into individual droplets. PHB homopolymer tends to crystallize from solution before the critical viscosity is reached, whereas the more soluble HB–HV copolymers provide a somewhat lower viscosity for the same polymer concentration. Nevertheless, dry spinning is a process option for these materials. It is certainly possible to cast PHB on to a substrate from solution to obtain isotropic films which may then be cold-rolled to improve ductility if required.

Wet spinning also overcomes the problems associated with the very high solution viscosities required for dry spinning. However, a suitable coagulant has yet to be found for the PHB systems. Spinning a PHB in chloroform solution into methanol, for example, results in too rapid precipitation of the polymer and the production of extremely brittle, voided fibres.

Gel spinning would appear to be the most appropriate solvent-based spinning technology for PHB. Using the methods developed for polyethylene[92,93] it is possible to spin and draw to very high extension ratios fibres of PHB obtained from gels of the polymer in 1,2-dichloroethane. There are no problems with solution viscosity or coagulants and the gels can be prepared quite simply by dissolving the polymer under reflux to the appropriate concentration and then chilling in a refrigerator to effect gelation.[91] The low-boiling solvent is relatively easy to remove from the finished product and, provided a high molecular weight polymer is used, very high stiffness and strength fibres can be produced. Typical properties for highly drawn gel-spun PHB would be a modulus in the region of 8 GPa and tenacity around 0·9 GPa.

Fibre Properties

Unfortunately, the properties of gel-spun fibres are not typical of those produced by melt or solution spinning. As can be seen from Table 10, the properties of conventionally produced PHB fibres are lower by a factor of four or five than the figures quoted for the gel-spun material. In particular, the tenacity of highly drawn, melt-spun PHB is unusually low at around 0·2–0·3 GPa compared with other fibre-forming polymers. However, the most puzzling feature of the mechanical strength of oriented PHB fibre is that it increases only slowly with draw ratio. Thus, isotropic, undrawn polypropylene has a similar tensile strength to isotropic PHB. However, the tenacity of polypropylene fibre increases by a factor of over 20 when it is drawn to seven times its original length.[116] In contrast PHB fibres drawn similarly are only a maximum of 10 times stronger than the undrawn polymer.

TABLE 10
MECHANICAL PROPERTIES OF TYPICAL ORIENTED FIBRES

Sample	*Overall draw ratio*	*Modulus (g/dtex)*	*Tenacity (g/dtex)*	*Extension to break (%)*
Rolled and drawn PHB	7	20	2·2	34
Melt-spun and drawn PHB	9	21	2·4	32
Gel-spun PHB	7	90	9·0	16
Poly(ethylene terephthalate)	5	99–117	5·4–7·7	6–14
Polypropylene	7	81	4·5–8·6	15–25
Nylon-6,6	7	41–72	5·4–8·6	12–26
Poly(glycolic acid)	7	—	5·0–8·0	15–35

The results of gel spinning indicate that it is possible to spin stronger fibres of PHB, but this observation is in itself anomalous. The ultimate or crystal modulus of PHB has been measured by examining the deformation of the crystal structure under applied stress using X-ray techniques. Assuming that the macroscopic stress is transferred efficiently to the lamellar crystals, and the measurements were performed below T_g to ensure this, then the observed change in lattice spacing indicates that the crystal modulus of PHB is in the range 8–9 GPa,[117] which is about the same as the macroscopic room-temperature modulus of the gel-spun polymer! Thus the modulus of the amorphous phase in this material must also be equal to that of the crystalline component—a remarkable observation for a polymer above its quoted T_g. In fact, dynamic mechanical analysis revealed virtually no β-relaxation in this sample, so the model of a low-modulus crystal phase embedded in an amorphous phase of similar stiffness is at least possible.

Clearly much more work on the structure–property relationships for PHB and HB–HV copolymer fibres is required before we can come to a complete understanding of the observed phenomena.

Biaxial Orientation

With the current lack of understanding of uniaxial orientation process for PHB apparent in the previous section, it would be surprising if the biaxial orientation mechanisms of the biopolymer had been fully characterized. Solution-cast foils, particularly if they contain a small

TABLE 11
MECHANICAL AND GAS BARRIER PROPERTIES OF POLYMER FILMS

Polymer	Draw ratio	Tensile properties: Young's modulus (GPa)	Tensile properties: Stress at break (MPa)	Tensile properties: Extension to break (%)	Oxygen permeability[a] ($ml\ m^{-2}\ atm^{-1}\ day^{-1}$)
PHB homopolymer	4 × 4	4·0	100	75	45
83:17 HB:HV copolymer	5 × 5	1·9	54	50	60
Poly(ethylene terephthalate)	4 × 4	4·4	200	120	70
Polypropylene (uncoated)	7 × 7	2·0	160	140	1700
83:17 HB:HV copolymer containing 20% plasticizer	—	0·39	21	50	150
Polyethylene (low-density)	—	0·3	10	700	7000

[a] Normalized for 25 μm thick film.

amount of residual solvent as plasticizer, are relatively easy to draw biaxially at elevated temperatures (40–80°C below melting point) under simultaneously applied transverse strains in a machine such as a T.M. Long film stretcher or an inflating bubble device. This process is also made easier and more reproducible if the initial foil is first subjected to a compressive step in a press or calender rolls to anneal voids and microscopic irregularities in the sample.[106]

The mechanical properties of biaxially drawn PHB are intermediate between those of poly(ethylene terephthalate) and polypropylene as shown in Table 11. The biopolymer films are mechanically strong and tough, and they are transparent with good clarity. Moreover, their gas barrier properties, particularly to oxygen, are excellent and much superior to those of uncoated polyolefin films. In fact PHB and its copolymers would appear to have a promising future as packaging materials in both oriented and unoriented form.

ADDITIVES

Nucleants

The role of nucleating agents in PHB has already been described in great detail but, as they are potentially the most useful additives in both homopolymer and copolymer processing, attention is once again drawn to the results in Fig. 14.

Plasticizers and Polyblends

The 3-hydroxyvalerate and other biochemically acceptable 3-hydroxyalkanoate comonomers are, of course, internal plasticizers for PHB and have the expected effects in terms of degree of crystallinity, melting point and mechanical properties as shown in Figs 6 and 7 and Table 6 respectively. However, the glass transition temperature of the biopolymer is not significantly reduced by *in vivo* polymerization with the available comonomers so that the low-temperature properties of all the materials are roughly similar. In contrast, external plasticizers can have a dramatic effect on T_g with concomitant improvements in ductility and crystallization kinetics at the expense of lower softening points.

As a polyester, PHB can partake in many of the hydrogen-bonding type of specific interactions with other functional additives that lead to partial miscibility and compatibility. For example, the miscibility of polyesters with chlorinated polymers, polyamides, polycarbonates, cellulose derivatives and other functional polymers is well documented,[118–122] and PHB is no exception to this general observation.[123] However, these interactions are dominated by the tendency to self-crystallize with exclusion of the additive to the amorphous phase. For example, an 80/20 melt compounded and injection-moulded sample of PVC/PHB polyblend appears initially to be exceptionally tough with the PHB acting as a polymeric plasticizer. The presence of the PVC retards but does not stop crystallization of the PHB at room temperature and the material eventually becomes brittle. Under extreme circumstances, the PHB phase can actually achieve almost 100% crystallinity within the blend, as determined by X-ray analysis and DSC. Thus, plasticized formulations and polyblends involving PHB itself are limited to relatively low levels of additive because only the minor amorphous phase of the biopolymer is involved in the interaction. Even so, some plasticizers have been proposed for PHB.[124]

In general, the more polar the plasticizer the greater its impact on the polymer and compounds featuring phosphate, sulphonamide, chlorosulphonate, chloro, amino and ester groups are commonly effective.

Impact Modifiers

Impact modifiers are a special case of polyblend where the crucial parameter is the degree of adhesion between the brittle PHB matrix and the rubbery additive. The same rules of polymer–polymer miscibility apply and materials such as chlorinated polyethylene, polychloroprene[125] and poly(butyl acrylate) have been evaluated. Superficially, these additives do improve the toughness of PHB injection mouldings. However, more detailed analysis using fracture mechanics reveals that this is a result of a reduction in the yield stress rather than a genuine increase in the material's toughness as judged by the critical stress intensity factor, K_{1c}. For example, the K_{1c} of PHB homopolymer is 3·0 MN $m^{-3/2}$ at 23°C, which is slightly higher than that for polypropylene. However, this can be reduced by a factor of two by the incorporation of an 'impact modifier' such as poly(butyl acrylate).

Fillers

Inorganic particulate fillers such as calcium carbonate and china clay can be compounded with PHB and extruded or moulded. The mechanical properties of the resultant products, however, provide no particular surprises, being stiffer but more brittle than the base polymer. Even so, calcium hydroxyapatite filled PHB is currently being evaluated as a potential bone substitute in maxillofacial surgery, for example, and in fracture fixation plates in much the same way as the recent development of apatite filled polyethylene.[127] The main advantages of PHB over polyethylene (PE) in these applications are its biocompatibility and biodegradability, which will be discussed in more detail later.

In purely mechanical terms, chopped glass fibre reinforced PHB offers a good balance of stiffness and toughness, as can be seen from the comparison with glass filled PET and polypropylene given in Table 12. The material is stiffer than equivalent grades of PP and tougher than those of PET. However, its thermal resistance in terms of heat distortion temperature and Vicat softening point are significantly inferior to those of the glass filled engineering plastics. Just to complete the picture, Table 12 also contains some details of continuous glass and carbon fibre reinforced PHB. These composites reflect the properties of the reinforcing fibre, with the biopolymer acting merely as a glue.

Miscellaneous Additives

A whole range of miscellaneous additives may be compounded with PHB to provide specific effects. These include lubricants, whiteners,

TABLE 12
MECHANICAL AND THERMAL PROPERTIES OF FIBRE REINFORCED POLYMERS

Polymer	*Fibre*	*Fibre loading (wt%)*	*Flexural modulus (GPa)*	*Tensile break stress (MPa)*	*Instrumented falling-weight impact strength (J)*	*Vicat softening pt (°C)*
PHB	Chopped glass	30	9·0	60	3·0	156
		40	11·7	55	3·7	160
		50	14·3	42	1·9	156
PET	Chopped glass	30	10·0	150	0·7	225
		45	13·8	193	1·0	226
PP	Chopped glass	20	4·9	72	1·0	158
		30	6·8	86	1·6	164
PHB	Continuous glass	55	37·1	—	—	—
PHB	Continuous carbon fibre	55	92·3	685	—	—

brighteners, antioxidants, UV stabilizers and processing aids. The normal range of commercially available effect chemicals for polymers can be used with PHB, although specific materials should first be evaluated to ensure that they are not prodegradants for the polymer. Antioxidants, however, are not usually required since accelerated weathering tests indicate that PHB is quite resistant to oxidation, being much superior to polypropylene in this respect. A 250 μm film of PHB, for example, embrittled after 38 days in a Xenotest compared with only six days for an equivalent sample of unstabilized polypropylene.

BIODEGRADABILITY

Environmental

PHB and its copolymers with other β-substituted, 3-hydroxyalkanoates are synthesized by microorganisms as carbon and energy reserve materials in response to a sudden deficiency of an essential nutrient such as nitrogen, phosphorus, sulphur or oxygen. Under these conditions, growth is not possible and excess carbon is stored as a polymer so that the osmotic balance of the cell is not jeopardized by large amounts of monomer. This exercise would be futile if the bacteria did not have access to the storage product when the deficiency is lifted. It is not surprising, therefore, that many different types of microorganism produce enzymes capable of depolymerizing aliphatic polyesters.[128–138]

Under optimum conditions the rate of the enzymic degradation of PHB is extremely fast.[128] For example, if a batch fermentation of *Alcaligenes eutrophus* is phosphate-limited and a small amount of phosphoric acid is then introduced at the end of the usual PHB accumulation phase, the amount of polymer in the cells can fall from 60–70% of cell dry weight to only 10–15% within a few hours. Extracted polymer in the form of intact granules is less susceptible to enzymic attack, but the process is still quite rapid.[132–134] In contrast, extracted polymer that has lost its biological granule morphology as a result of solution or melt processing is only attacked relatively slowly over periods of days and weeks rather than minutes and hours.[129–131]

These PHB-specific depolymerase enzyme systems are believed to be located in the lipid membrane surrounding the polymer granules and to consist of two distinct enzymes.[132–135] The first enzyme degrades the high molecular weight polymer to a dimer and the second completes the depolymerization from dimer to monomer. It has also been proposed that these enzymes are the same ones that are involved in the synthesis of the polymer, but operating in reverse mode.[139] This would certainly help to explain the rapid depolymerization of PHB in cells in the absence of a nutrient limitation since the bacteria would only have to switch the function of the existing polymerase enzymes rather than begin synthesizing a new set of degrading enzymes.

Some microorganisms, however, clearly possess extracellular rather than membrane-bound depolymerases. Fungi, for example, are incapable of accumulating PHB, but many strains have the ability to use it as a substrate, as can be seen in Fig. 15. This scanning electron micrograph shows a mycelial fungus growing on the surface of a compression-moulded PHB plaque. Extracellular enzymes from the microorganisms degrade and solubilize the polymer in the immediate vicinity of the cell so that the resulting nutrient-rich solution can be absorbed. Bacteria can also degrade PHB in this way, as is evident from Fig. 16, which shows a bacterial colony growing on an agar plate with a plaque of polymer in the centre as the sole carbon source. Moreover, the extracellular depolymerases are not limited to attacking PHB and the other biologically produced poly(3-hydroxyalkanoates) and have been found to degrade a whole range of synthetic aliphatic polyesters.[137,138] The specificity of enzymes from bacteria are also reported to be completely different from those of the equivalent fungal

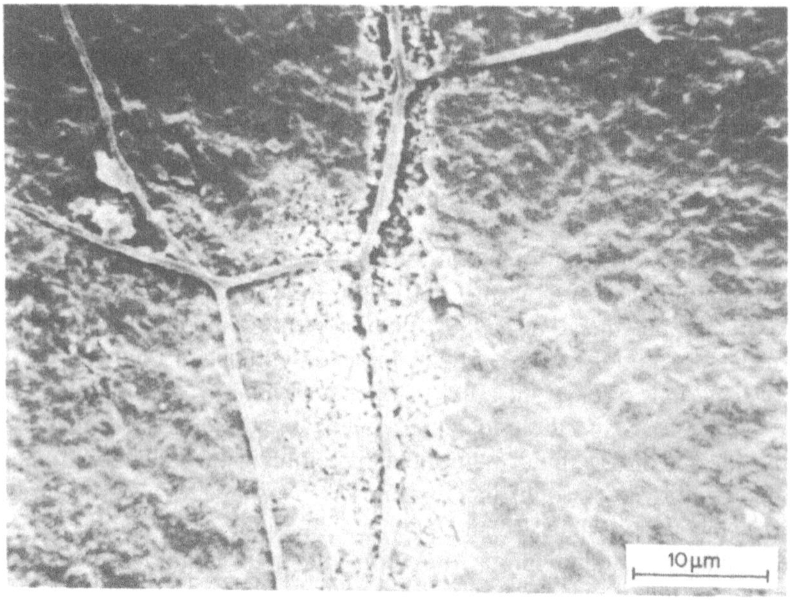

FIG. 15. Scanning electron micrograph of a PHB film surface showing colonization by a fungus.

depolymerases. For example, poly(propiolactone) is completely degraded by the former and not touched at all by the latter, whereas the reverse is true for polymers of adipic acid and aliphatic glycols.

The factors influencing the rate of microbiological attack on a polymer film or moulding are many and varied. They include material parameters such as the chemical composition of the polymer, its molecular weight, degree of crystallinity and level of orientation, and factors relating to microbial growth. The latter begin with the presence of some water in contact with the polymer as a prerequisite and continue to include pH, temperature, presence or absence of oxygen (aerobic or anaerobic conditions), ionic strength and the availability of other nutrients essential for growth. In addition there is a third group of conditions affecting microbial attack that are less easy to define. These relate to the mechanism of biodegradation which, as we have seen, is by colonization of the polymer surface by bacteria and fungi. The way in which microorganisms stick to this surface is, therefore, important and this will be influenced by surface tension, the surface texture and porosity, the degree of agitation in the local environment

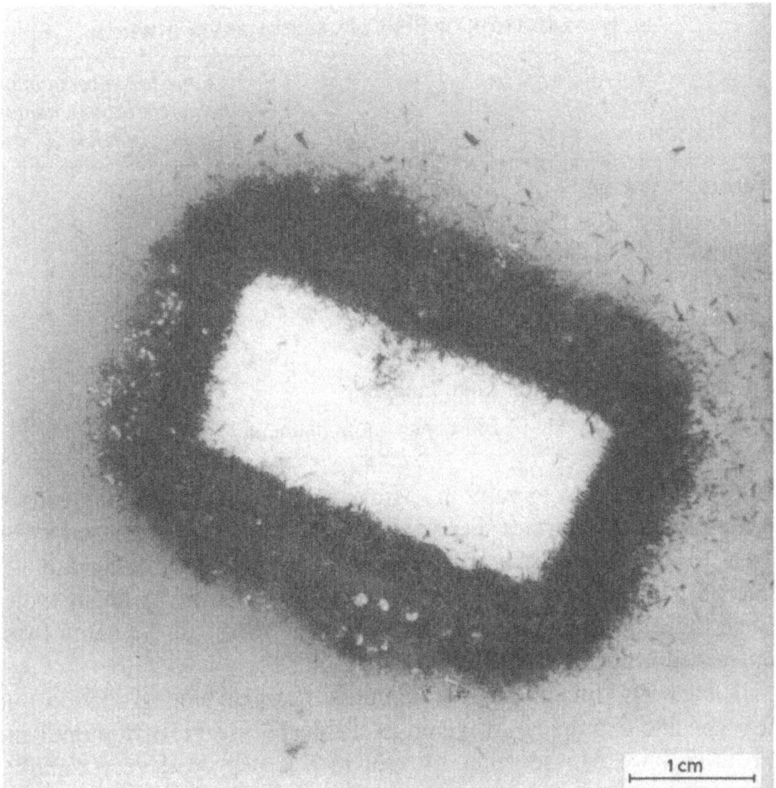

FIG. 16. Bacterial colony growing on an agar plate with a plaque of PHB as the sole carbon source.

which could cause nascent colonies to be washed away and, of course, the surface area available for colonization.

In a practical sense many of these parameters will change during the biodegradation of a particular film or moulding in the environment. For example, when a PHB film is buried in well-watered soil, the first step in the biodegradation process is a modification of the film surface from hydrophobic to hydrophilic. This wetting-out procedure usually takes about one week and is presumably caused by surface hydrolysis to increase the number of hydroxyl and carboxylic acid groups available or by the intervention of microbially produced surfactants. Bacterial and fungal colonization can then begin but the rate of growth will depend upon the daily fluctuations of temperature and the changes

TABLE 13
BIODEGRADATION OF PHB IN VARIOUS ENVIRONMENTS

Environment	*Sample form*	*Time for substantially complete biodegradation (weeks)*
Anaerobic sewage	1 mm moulding	6
	75 μm film	1
Well-watered soil at 25°C	1 mm moulding	75
	75 μm film	12
Estuarine sediment at 20°C	1 mm moulding	40
	75 μm film	8
Bovine rumen	1 mm moulding	25
In vivo	200 denier monofilament	250

in pH resulting from rainfall. Moreover, as the attack proceeds the initially smooth surface becomes pitted, exposing a greater surface area to the bacteria and enhancing the biodegradation rate. The microorganisms may not even be given the chance to finish the job since there is some evidence that thin PHB films can be eaten by soil mites and other insect pests!

It is clearly difficult to give definitive rates of biodegradation but a few specific examples are given in Table 13 for PHB homopolymer. From this we can see that the polymer is degraded very rapidly in anaerobic sewage and progressively more slowly, in warm soils, aerobic sewage, estuarine sediments and sea water.

Medical

PHB has also been reported to be biodegradable *in vivo* as a subcutaneous or intramuscular implant.[140] In addition, the ultimate biodegradation product, (*R*)-3-hydroxybutanoic acid, is a normal metabolite in human blood[141] and the polymer itself exhibits good biocompatibility with no cytotoxic response. PHB shows negligible oral toxicity since the LD_{50} is greater than 5000 mg kg^{-1} and polymer in cells has actually been evaluated as a nutritional supplement in the diet of growing pigs.[142] However, healthy human tissue, fortunately for us, does not contain PHB-utilizing bacteria and fungi, so how is the polymer being biodegraded under these circumstances?

The most common biodegradable, or to be more accurate resorbable, polymer used in surgery is poly(glycolic acid) and this material has been shown to degrade *in vivo* at more or less the same rate as it hydrolyses *in vitro*.[143] PHB is also hydrolysed in water with the normal universal acid–base catalysis for esters. The rate of degradation is quite rapid at high pH but in neutral buffer at body temperature the hydrolysis proceeds very slowly. However, the kinetics do not appear to follow first-order behaviour[144] since the reciprocal molecular weight does not decrease linearly with time. In fact the plot of the logarithm of molecular weight versus time is more nearly linear.

Moreover, initial experiments using a standard protocol[145] suggest that the rate of degradation of PHB *in vivo* is significantly faster than the *in vitro* hydrolysis rate at the same temperature and pH. The obvious interpretation is that non-specific esterase and lysozyme enzymes secreted by the body's immune system catalyse the process. Indeed it is relatively easy to demonstrate that the polymer is attacked by hog's liver esterase, for example.

This variation of the degradation rate of PHB *in vivo* with the activity of the body's immune system may help to explain some of the discrepancies in the literature reports of PHB biodegradation. These range from the very rapid biodeterioration of implanted films,[140] through a modest but measurable resorption of subcutaneously located microcapsules[146] to virtually undetectable weight loss of fibre-monofilaments over an 18-month period.

Veterinary

The final areas in which PHB biodegradation is of interest are the specialized cellulose-degrading locations of animals, such as the rumen of cattle and sheep and the colons of goats. PHB is degraded in these environments and, as one would expect, the rate is intermediate between slow hydrolysis *in vivo* and the more rapid attack of bacteria and fungi in soil or sewage.

Representative examples from specific experiments have again been included in Table 13.

Miscellaneous Properties

The susceptibility of PHB to attack in aqueous environments is part of the wider consideration of the polymer's general chemical resistance. This is summarized in Table 14 and is represented as the time taken for an injection-moulded PHB homopolymer bar to fracture when immersed at room temperature in the particular solvent and placed under a constant 1% strain. As expected, the resistance to acid and

TABLE 14
CHEMICAL RESISTANCE OF PHB

Chemical	*Uptake after total immersion at 23°C for 1 month (%)*	*Time to fracture with ca 1% strain at 23°C(h)*	*Assessment of resistance*
Tap water	0·13	>336	Good
1M-H_2SO_4	0·46	4·15	Poor
2M-HCl	0·41	2·6	Poor
2M-NaOH	0·77[a]	0·02	Poor
Methanol	2·08	140	Moderate
Ethyl methyl ketone	5·41	27·6	Moderate
Genklene	0·83	>457	Good
Motor oil	0	>505	Good
Toluene	0·05	7·4	Poor
Chloroform	—	2·8	Poor
1,2-dichloroethane	—	<0·01	Poor

[a] Sample broke up when removed from NaOH.

base is relatively poor whilst at neutral pH the samples survive quite well, with only 0·13% water uptake after one month. The polymer is also attacked by the chlorinated solvents in which it is soluble, such as chloroform and 1,2-dichloroethane, but not by the other chlorinated solvents commonly used in dry cleaning, e.g. 1,1,1-trichloroethane, which PHB resists better than many non-solvents, including the lower alcohols and ketones. Resistance to the higher alkanes used in motor oil is also quite good but in general the chemical resistance of PHB and its biologically produced copolymers must be regarded as poor.

A more positive aspect of the chemical functionality of PHB is that the polymer can act as a processing acid for other polymers such as PVC and polyacrylonitrile. Both PHB homopolymer and the HB–HV copolymers reduce the gelation time and improve the surface finish of extrudates when blended with unplasticized PVC at the 1% level prior to processing. They function in much the same way as high molecular weight polyacrylates and chlorinated polyethylene, which are already used commercially in this application. In addition, the power required to melt-extrude polyacrylonitrile-based resins, such as the Barex range from Lonza AG, is much reduced by the addition of 1% PHB to the polymer.

The dielectric properties of PHB are typical of plastic insulators. Its dielectric constant (relative permittivity) is around 3 and the volume resistivity is average at $10^{16}\ \Omega$ cm at room temperature. Loss curves are also fairly typical of a clean polyester with a low-temperature transition at −100°C corresponding to a crystallinity-independent relaxation of the ester group, a major relaxation at around room temperature caused by the glass–rubber transition and a high-temperature runaway process above 100°C.

Of more interest are the piezoelectric properties of the polymer. Many biological polymers exhibit piezoelectric effects and the phenomenon appears to have physiological significance in the stimulation of bone growth.[147] PHB behaves in a similar manner to poly(γ-methyl-L-glutamate) in that the piezoelectric response is generated by the application of a shear stress to oriented polymer crystallites.[148] This is a direct result of crystal symmetry considerations and distinguishes these biological polymers from the most-studied piezoelectric material, poly(vinylidene fluoride), which has different symmetry and is active under tensile and hydrostatic stresses.[149]

To determine the piezoelectric strain coefficient, d_{14}, and the stress coefficient, ε_{14}, for PHB under shear deformations is experimentally very difficult. There is no point in shearing an isotropic block of polymer, since the resulting changes in polarization at individual crystal lamellae will cancel to give zero macroscopic effect, and it is almost impossible to shear uniaxially oriented fibres and tapes in the correct deformation mode. The simplest solution to this problem is to cut rectangular samples from uniaxially drawn film at 45° to the orientation direction; subjecting these to a tensile extension results in a pseudo-shear deformation of the oriented crystallites and the generation of an electrostatic charge on the surface of the film. If conducting electrodes are first evaporated on to the sample the magnitude of the charge may be determined using an electroscope and, knowing the value of the applied strain, a piezoelectric strain coefficient may be calculated. The relationship of this experimentally determined constant to the fundamental ε_{14} parameter, which is defined in terms of the change in polarization resulting from a *shear* deformation, is uncertain but the technique does provide a practical way of measuring the piezoelectric properties of polymers with this particular crystal symmetry.

Experiments of the type described have been reported for PHB and the results analysed in terms of a two-phase model.[150] The overall piezoelectric stress coefficient was assumed to be composed of additive contributions from the crystalline phase and the oriented amorphous phase according to eqn (4):

$$d'_{14} = d_c F_c X_c + d_a F_a (1 - X_c) \tag{4}$$

where d_c and d_a are the piezoelectric stress coefficients for the crystalline and amorphous phases respectively, F_c and F_a are the appropriate orientation functions and X_c is the volume fraction crystallinity. The authors assumed F_c and F_a to be approximately equal and plotted their experimental results of d'_{14} against X_c to obtain values of d_c and d_a at the limits.

The most surprising feature of this study is that d_c and d_a were found to have opposite signs, the former being around $+5\,pC\,N^{-1}$ and the latter $-2\,pC\,N^{-1}$ at 0°C. The reason for this strange behaviour is not fully understood but it hints at the fact that the polymer chains, or at least the associated dipoles, rotate in different directions in the amorphous and crystalline phases when subjected to a shear stress.

Many assumptions have been made in the derivation of this model. For example, apart from the implications involved in using a macroscopic tensile stress to induce a shear deformation at the crystallites there is an implicit assumption that the applied stress is transferred uniformly to both amorphous and crystalline phases. This may be true at temperatures below T_g but is most unlikely to hold when the amorphous phase becomes rubbery. Nevertheless, the piezoelectric data may be providing some clues about the fundamental reasons for the difficulty in promoting strain-induced crystallization in PHB.

Chiral Synthesis from PHB

The alkyl esters of 3-hydroxybutanoic acid are a particularly useful group of compounds in organic synthesis because they are available in both chiral forms. Ethyl (*S*)-3-hydroxybutanoate, for example, may be prepared by the reduction of ethyl acetoacetate on the *re* side of the keto group with fermenting yeast[151,152] and leads to an enantiomeric purity of about 85% which may be enriched further via a crystalline dinitrobenzoate.[153] Synthesis of the *R* form was a little more difficult but could be achieved with 83–88% purity by the reduction of the acetoacetate with hydrogen in the presence of Raney nickel which had been treated with tartaric acid.[154]

Referring to Scheme 1, preparation of ethyl(*R*)-3-hydroxybutanoate (**2a**) in 100% optical purity is now somewhat easier since it may be achieved by the esterolysis of PHB (**1**) in the presence of a suitable

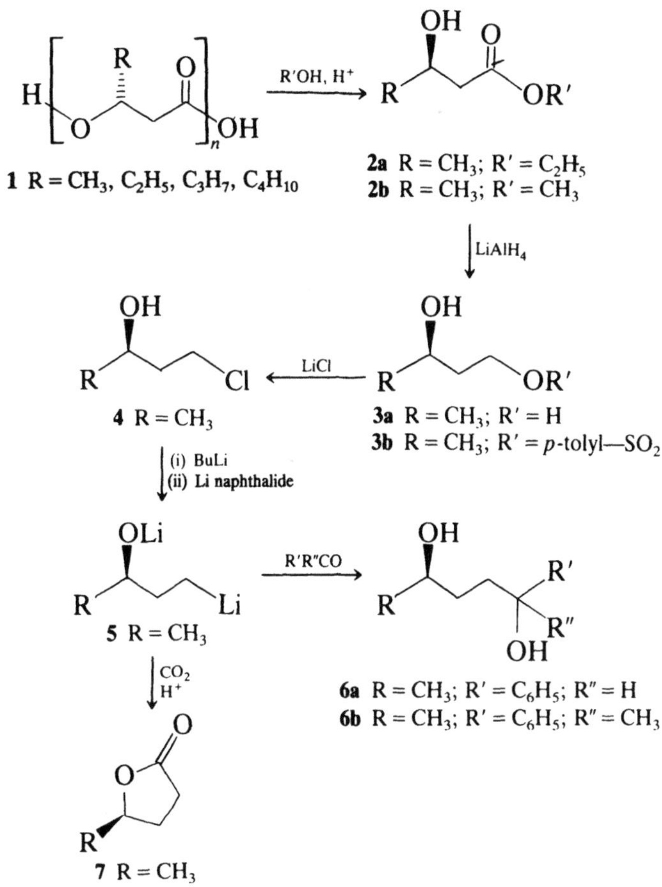

SCHEME 1.

catalyst, such as *p*-toluenesulphonic acid, with recovery of the product by fractional vacuum distillation.[155,156] Alternatively, the glycol **3a** can be obtained by the reduction of PHB with lithium aluminium hydride in tetrahydrofuran and used as a starting material for synthesis.[155]

Several other basic chiral building blocks can be derived by simple organic chemistry transformations of these PHB derivatives. For example, (*R*)-1-chloro-3-butanol (**4**) may be obtained from the glycol **3a** via the tosyl derivative (**3b**) by nucleophilic substitution with lithium chloride.[157] The product can then be subjected to sequential metallation with butyllithium and then lithium naphthalide in THF at −78°C to give stable solutions of the highly reactive *C*-metallated alkoxide (**5**). This material reacts with aldehydes and ketones to give 1,4-diols such as **6a** from benzaldehyde or **6b** from acetophenone and with carbon dioxide to produce the lactone (**7**).

All of these reactions may be repeated starting with methyl (2*R*,3*R*)-3-hydroxy-2-methylbutanoate (**9a**) which is obtained, as shown in Scheme 2, by the double deprotonation of methyl 3-hydroxybutanoate (**2b**) using lithium di-isopropylamide (LDA) to give the alkoxide-enolate (**8**) and alkylation with a methyl halide in THF at −78°C.[158,159] After reduction with lithium aluminium hydride, tosylation of the alcohol group and nucleophilic attack by lithium chloride, the (2*R*,3*R*)-1-chloro-2-methyl-3-butanol (**10**) is available for the sequential metallation reactions and conversion to 1,4-diols or lactones.[153]

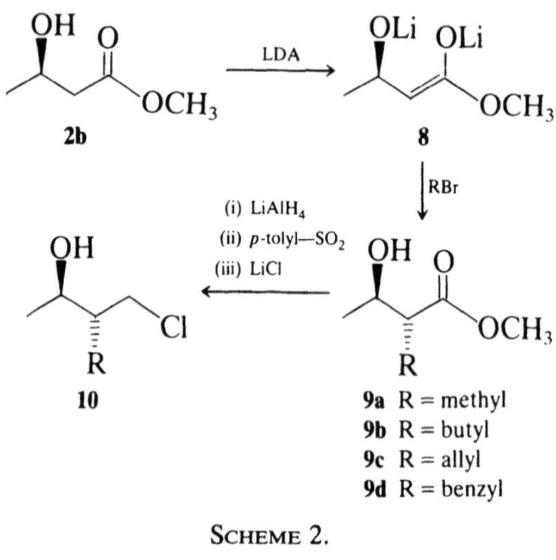

SCHEME 2.

Examples of the use of these chiral building blocks in the synthesis of drugs and natural products abound in the literature. They include the synthesis of grahamimycin A_1,[160] the macrocyclic component of the antibiotic elaiophylin,[161] fungicides norpyrenophorin, pyrenophorin and vermiculin,[162] phorcantholide I and J, which are defensive secretions of the eucarypt longicorn beetle,[163] citronellol[164] and a pheromone of the western corn rootworm.[165] Clearly the monomeric chiral derivatives of PHB have a role to play in enantiomer-specific synthesis.

Uses

PHB and a range of HB–HV copolymers are currently available in test quantities from Marlborough Biopolymers Ltd of Stockton-on-Tees, a subsidiary of ICI. The polymers, however, are still in the development stage and, although a great deal is known about their physical properties and suitability in a range of applications, their eventual commercial uses will depend upon their position on the price–volume exclusion curve for plastic materials. Thus, at one end of the spectrum there are low-volume, exotic applications in medicine and chiral synthesis. In the mid-range there are speciality uses in paramedical disposables, veterinary medicine and horticulture. Finally, should the polymers be as cheap as polyethylene, there are bulk applications in packaging films, commodity disposables and raffia-type woven sacking.

Examples of medical applications include controlled drug release,[145,166–167] surgical sutures,[140,168] lubricant powders for surgical gloves,[169] bone plates[127,170] and wound care.[171,172] All of these potential uses, however, depend upon the biocompatibility and slow resorption of PHB *in vivo*.

The utility of the optically active (*R*)-3-hydroxybutyrate monomer, or indeed its oligomers,[173–176] as a carbon source in intravenous infusion solutions[141] and as reagents in chiral drug synthesis have already been discussed. However, another application that could exploit the chiral nature of the polymer is in the chromatographic separation of optical isomers. The principle of using chiral polymers as the stationary phase in both gas chromatography (GC) and high-performance liquid chromatography (HPLC) has already been established[177,178] and specific examples, such as the resolution of the enantiomeric alcohols *exo*- and *endo*-2-norbornanol using poly(propylene glycol) from (*S*)-2-methyloxirane as a coating in GC soft glass capillaries, have appeared in the literature.[179]

Applications of intermediate volume centre around the biopolymers as disposable films. Their good oxygen-barrier properties open up opportunities in packaging but most interest has been shown in their use as moisture barrier films or coverstock in disposable items such as diapers, sanitary napkins, incontinence products, and ostomy devices.[180]

Some speciality agricultural and horticultural applications, for example in the sustained release of plant growth regulators or pesticides,[181] also represent large prospective markets for PHB. Of particular interest here are laminates of PHB with paper to provide low-cost waterproof, but completely biodegradable, seed tapes. Lamination may be achieved using standard film-based techniques or via a biodegradable lacquer formulation based on PHB.[182] More speculative applications include biodegradable hard-water detergents[183] and, as the acyl derivatives, shortening aids in bread products.[184]

However, if we forget about the biodegradability and other special properties of PHB and regard the polymer as a conventional commodity thermoplastic with the advantage of being a renewable resource *not* derived from oil, then many bulk applications become possible. Fungicides and bactericides could be incorporated into the polymer to inhibit biodegradation and the whole range of film, fibre and moulded plastic products made from it. In particular, the biopolymers could be viewed as replacements for polypropylene in many applications or as a processing aid in PVC.

Many of these applictions have been reviewed recently in more detail elsewhere.[185] However, it is clear that the list of possible uses has not been exhausted and that the PHB-based biopolymers represent one of the most versatile ranges of materials that can be derived from renewable sources rather than from fossil fuels, upon which we currently rely so heavily.

CONCLUSIONS

Microbially produced poly(3-hydroxyalkanoates) are scientifically fascinating materials and are likely to be of increasing commercial importance. They present many challenges to the scientist, from trying to understand how, and even why, they are produced by bacteria to comprehending the subtleties involved in their melt extrusion, nucleation, crystallization and orientation behaviour. Much more research is required before these biopolymers can become household names like polythene, but it is hoped that this chapter will stimulate further interest in this unique range of environmentally degradable and socially useful thermoplastics.

REFERENCES

1. Howells, E. R., *Chem. Ind.*, 1982, 508–51.
2. Lemoigne, M., *Ann. Inst. Past.*, 1925, **39**, 144–73.

3. Dawes, E. A. and Senior, P. J., *Adv. Microbial. Phys.*, 1973, **10,** 138–266.
4. Lepidi, A. A., Nuti, M. P. and Bertoldi, M., *Agr. Ital. (Pisa)*, 1972, **72,** 166–84.
5. Wong, P. P. and Evans, H. J., *Plant Physiol.*, 1971, **47,** 750–5.
6. Stokes, J. L. and Parson, W. L., *Can. J. Microbiol.*, 1968, **14,** 785–9.
7. Gottschalk, G., *Angew. Chem.*, 1962, **74,** 342–7.
8. Matin, A., Veldhuis, C., Stegeman, V. and Veenhuis, M., *J. Gen. Microbiol.*, 1979, **112,** 349–55.
9. Lundgren, D. G. and Bott, K. F., *J. Bacteriol.*, 1963, **86,** 462–73.
10. Stevenson, L. H. and Socolofsky, M. D., *J. Microbiol. Serol.*, 1973, **39,** 341–50.
11. Reusch, R. N. and Sadoff, H. L., *J. Bacteriol.*, 1983, **156,** 778–88.
12. Merrick, J. M., *Photosynth. Bact.*, R. K. Clayton and W. R. Sistrom, Eds, 1978, Plenum, New York, pp. 199–219.
13. Ward, A. C., Rowley, B. I. and Dawes, E. A., *J. Gen. Microbiol.*, 1977, **102,** 61–8.
14. Gottschalk, G., *Arch. Mikrobiol.*, 1964, **47,** 236–50.
15. Siegel, R. S. and Ollis, D. F., *Biotech. Bioeng.*, 1984, **26,** 764–70.
16. Skinner, K. J., *Chem. Eng. News,* 18 August 1975, 22–41.
17. Tanaka, Y., Saito, T., Fukui, T., Tanio, T. and Tomita, K., *Eur. J. Biochem.*, 1981, **118,** 177–82.
18. Wallen, L. L., *Amer. Chem. Soc., Div. Environ. Chem. Prepr.*, 1973, **13,** 30–1.
19. Wallen, L. L. and Rohwedder, W. K., *Env. Sci. Tech.*, 1974, **8,** 576–9.
20. Findlay, R. H. and White, D. C., *Appl. Env. Microbiol.*, 1983, **45,** 71–8.
21. Capon, R. J., Dunlop, R. W., Ghialberti, E. and Jefferies, P. J., *Phytochemistry,* 1983, **22,** 1181–84.
22. Holmes, P. A., Wright, L. F. and Collins, S. H. (Imperial Chemical Industries plc), Eur. Patent Appln, EP 52459, 1981.
23. Holmes, P. A., Wright, L. F. and Collins, S. H., (Imperial Chemical Industries plc), Eur. Patent Appln, EP 69497, 1983.
24. De Smet, M. J., Eggink, E., Witholt, B., Kingma, J. and Wynbert, H., *J. Bacteriol.*, 1983, **154,** 870–8.
25. Agostini, D. E., Lando, J. B. and Shelton, J. R., *J. Polym. Sci., Part A1,* 1971, **9,** 2775–87.
26. Tani, H., Yamashita, S. and Teranishi, K., *Polymer J.*, 1972, **3,** 417–18.
27. Vergara, J. and Figini, R. V., *Makromol. Chem.*, 1977, **178,** 267–70.
28. Iida, M., Araki, T., Teranishi, K. and Tani, H., *Macromolecules,* 1977, **10,** 275–84.
29. Yasuda, T., Aida, T. and Inoue, S., *Macromolecules,* 1983, **16,** 1792–6.
30. Agostini, D. E., Lando, J. B. and Shelton, J. R., *J. Polym. Sci., Part A1,* 1971, **9,** 2789–99.

31. TERANISHI, K., IIDA, M., ARAKI, T., YAMASHITA, S. and TANI, H., *Macromolecules,* 1974, **7,** 421–7.
32. ARAKI, T. and HAYASE, S., *J. Polym. Sci., Polym. Chem. Ed.,* 1979, **17,** 1877–81.
33. BULLIVANT, S., *Advanced Techniques in Biological Electron Microscopy,* J. Koehler, Ed., 1973, Springer, Berlin, p. 67.
34. LUNDGREN, D. G., PFISTER, R. M. and MERRICK, J. M., *J. Gen. Microbiol.,* 1964, **34,** 441–6.
35. MERRICK, J. M., LUNDGREN, D. G. and PFISTER, R. M., *J. Bacteriol.,* 1965, **89,** 234–9.
36. ELLAR, D. J., LUNDGREN, D. G., OKAMURA, K. and MARCHESSAULT, R. H., *J. Mol. Biol.,* 1968, **35,** 489–502.
37. JENSON, T. E. and SICKO, L. M., *J. Bacteriol.,* 1971, **106,** 683–6.
38. GRIEBEL, R. and MERRICK, J. M., *J. Bacteriol.,* 1971, **108,** 782–9.
39. ROBARDS, A. W. and DUNLOP, W. F., *J. Bacteriol.,* 1973, **114,** 1271–80.
40. NICKERSON, K. W., *Appl. Env. Microbiol.,* 1982, **43,** 1208–9.
41. SLEYTR, U. B. and ROBARDS, A. W., *J. Microscopy,* 1977, **110,** 1–25.
42. KITTLETY, R. S. and STAGEMAN, J. F. (Imperial Chemical Industries plc), private communication, 1984.
43. ALFREY, T., BOHRER, J. J. and MARK, H., *High Polymers,* Vol. VIII, *Copolymerization,* 1952, Interscience, New York.
44. LUNDGREN, D. G., ALPER, R., SCHNAITMAN, C., and MARCHESSAULT, R. H., *J. Bacteriol.,* 1965, **89,** 245–51.
45. NUTI, M. P., DE BERTOLDI, M. and LEPIDI, A. A., *Can. J. Microbiol.,* 1972, **18,** 1257–61.
46. HOLMES, P. A., WRIGHT, L. F., ALDERSON, B. and SENIOR, P. J., (Imperial Chemical Industries plc), Eur. Patent Appln EP 15123, 1980.
47. BAPTIST, J. N., (W. R. Grace and Co.), US Patent 3 044 942, 1962.
48. SOLVAY ET CIE, Eur. Patent Appln EP 14490, 1979.
49. BAPTIST, J. N., (W. R. Grace and Co.), US Patent 3 036 959, 1962.
50. LAFFERTY, R. M. and HEINZLE, E., *Chem. Rundschau,* 1977, **30,** 15–16.
51. LAFFERTY, R. M., (Agroferm AG), US Patent 4 101 533, 1977.
52. HOLMES, P. A. and WRIGHT, L. F., (Imperial Chemical Industries plc), Eur. Patent Appln. EP 36669, 1981.
53. WILLIAMSON, D. H. and WILKINSON, J. F., *J. Gen. Microbiol.,* 1958, **19,** 198–209.
54. ALPER, R., LUNDGREN, D. G., MARCHESSAULT, R. H. and COTE, W. A., *Biopolymers,* 1963, **1,** 545–56.
55. GRIEBEL, R., SMITH, Z. and MERRICK, J. M., *Biochemistry,* 1968, **7,** 3676–81.
56. BARBER, L. C., *Diss. Abstr. Int. B.,* 1973, **34**(6), 2445; Univ. Microfilms, Ann Arbor, Mich. 73–29 266.
57. HOLMES, P. A., (Imperial Chemical Industries plc), Eur. Patent Appln. EP 46335, 1982.
58. HOLMES, P. A. and LIM, G. B., (Imperial Chemical Industries plc), Eur. Patent Appln 8411670, 1984.
59. POINDEXTER, J. S. and ELEY, L. F., *J. Microbiol. Methods,* 1983, **1,** 1.

60. JUETTNER, R. R., LAFFERTY, R. M. and KNACKMUSS, H. J., *Eur. J. Appl. Microbiol.*, 1975, **1**, 233–7.
61. SLEPECKY, R. A. and LAW, J. H., *Anal. Chem.*, 1960, **32**, 1697–9.
62. WARD, A. C. and DAWES, E. A., *Anal. Biochem.*, 1973, **52**, 607–13.
63. BRAUNEGG, G., SONNLEITNER, B. and LAFFERTY, R. M., *Eur. J. Appl. Microbiol. Biotech.*, 1978, **6**, 29–37.
64. APOSTOLIDES, Z. and POTGIETER, D. J. J., *Eur. J. Appl. Microbiol. Biotech.*, 1981, **13**, 62–3.
65. KARR, D. B., WATERS, J. K. and EMERICH, D. W., *Appl. Env. Microbiol.*, 1983, **46**, 1339–44.
66. IIDA, M., HAYASE, S. and ARAKI, T., *Macromolecules*, 1978, **11**, 490–3.
67. OKAMURA, K. and MARCHESSAULT, R. H., *Conform. Biopolymers, Pap. Int. Symp., Madras*, 1967, **2**, 709–20.
68. MARCHESSAULT, R. H. and CORNIBERT, J., *Polymer Prepr.*, 1971, **12**, 185–91.
69. CORNIBERT, J. and MARCHESSAULT, R. H., *J. Mol. Biol.*, 1972, **71**, 735–56.
70. YOKOUCHI, M., CHATANI, Y., TADOKORO, H., TERANISHI, K. and TANI, H., *Polymer*, 1973, **14**, 267–72.
71. CORNIBERT, J. and MARCHESSAULT, R. H., *Macromolecules*; 1975, **8**, 296–305.
72. TADOKORO, H., *Makromol. Chem., Suppl. 4*, 1981, 129–41.
73. RULAND, W., *Acta Crystallogr.*, 1961, **14**, 1180–85.
74. GRAY, A. P., *Inst. News*, 1970, **20/2**, 1–3.
75. BARHAM, P. J., KELLER, A., OTUN, E. L. and HOLMES, P. A., *J. Mat. Sci.*, 1984, **19**, 2781–94.
76. BLUHM, T. L., HAMER, G. K. and MARCHESSAULT, R. H., *Int. Union of Crystallography Meeting, Hamburg*, 1984.
77. PUNDSACK, A. L. and BLUHM, T. L., *J. Mat. Sci.*, 1981, **16**, 545–7
78. MARCHESSAULT, R. H., COULOMBE, S., MORIKAWA, H., OKAMURA, K. and REVOL, J.-F., *Can. J. Chem.*, 1981, **59**, 38–44.
79. ILLERS, K. H. and BREUER, H., *J. Colloid Sci.*, 1963, **81**, 1–31.
80. GRASSIE, N., MURRAY, E. J. and HOLMES, P. A., *Polym. Deg. Stab.*, 1984, **6**, 47–61.
81. BAPTIST, J. N. and WERBER, F. X., *Soc. Plast. Eng. Trans.*, 1964, **4**, 245–50.
82. MARCHESSAULT, R. H., OKAMURA, K. and SU, C. J., *Macromolecules*, 1970, **3**, 735–40.
83. MARCHESSAULT, R. H., CORNIBERT, J., BENOIT, H. and WEIL, G., *Macromolecules*, 1970, **3**, 741–6.
84. AKITA, S., EINAGA, Y., MIYAKI, Y. and FUJITA, H., *Macromolecules*, 1976, **9**, 774–80.
85. MIYAKI, Y., EINAGA, Y., HIROSYE, T. and FUJITA, H., *Macromolecules*, 1977, **10**, 1356–64.
86. HIROSYE, T., EINAGA, Y. and FUJITA, H., *Polymer J.*, 1979, **11**, 819–26.
87. DE MOLA, A. H., MARX-FIGINI, M. and FIGINI, R. V., *Makromol. Chem.*, 1975, **176**, 2655–67.

88. Delsarte, J. and Weill, G., *Macromolecules,* 1974, **7,** 343–8.
89. Delsarte, J. and Weill, G., *Macromolecules,* 1974, **7,** 450–3.
90. Couzens, E. G., *Film Casting Process,* 1965, Bexfords Ltd, Manningtree, Essex.
91. Barham, P. J. and Selwood, A., (Imperial Chemical Industries plc), Eur. Patent Appln EP 24810, 1981.
92. Smith, P. and Lemstra, P. J., *J. Mat. Sci.,* 1980, **15,** 505–14.
93. Barham, P. J. and Keller, A., *J. Mat. Sci.* 1985, **20,** 2281–2302.
94. Macrae, R. M. and Wilkinson, J. F., *J. Gen. Microbiol.,* 1958, **19,** 210–22.
95. Taylor, R., *J. Chem. Soc. Perkin II,* 1975, 1025–9.
96. Grassie, N., *Trans. Faraday Soc.,* 1952, **48,** 379–87.
97. ASTM D1238, 1970.
98. Grassie, N., Murray, E. J. and Holmes, P. A., *Polym. Deg. Stab.,* 1984, **6,** 95–103.
99. Holmes, P. A., Henman, T. J. and Billingham, N. C., *in Developments in Polymer Degradation—7,* N. Grassie (ed.), 1987, Elsevier Applied Science, London, pp. 81–121.
100. Henman, T. J. and Holmes, P. A., (Imperial Chemical Industries plc) Eur. Patent Appln EP 52946, 1982.
101. Grassie, N., Murray, E. J. and Holmes, P. A., *Polym. Deg. Stab.,* 1984, **6,** 127–34.
102. Morikawa, H. and Marchessault, R. H., *Can. J. Chem.,* 1981, **59,** 2306–13.
103. Boenig, H. V., *Polyolefins,* 1966, Elsevier, Amsterdam, Chap. 8, p. 262.
104. Van Krevelen, D. W., *Properties of Polymers,* 1976, Elsevier, Amsterdam.
105. Barham, P. J. and Keller, A., *J. Polym. Sci., Polym. Phys. Ed.,* 1986, **24,** 69–77.
106. Holmes, P. A. and Barham, P. J., (Imperial Chemical Industries plc) US Patent 4 427 614, 1984.
107. Binsbergen, F. L., (Shell International Research), British Patent 1 139 528, 1967.
108. Barham, P. J., *J. Mat. Sci.,* 1984, **19,** 3926–34.
109. Marchessault, R. H., Morikawa, H., Revol, J.-F. and Bluhm, T. L., *Macromolecules,* 1984, **17,** 1882–4.
110. Owen, A. J., (University of Regensburg), private communication, 1984.
111. Price, F. P., in *Nucleation,* A. C. Zettlemayer (Ed.), 1969, Marcel Dekker, New York.
112. Hoffman, J. D., *Polymer,* 1983, **24,** 3–26.
113. Moncrieff, R. W., *Man-Made Fibres,* 1975, Newnes–Butterworths, London.
114. Keller, A., (University of Bristol) and Ward, I. M., (University of Leeds), private communication, 1984.
115. Holmes, P. A., (Imperial Chemical Industries plc), Eur. Patent Appln 8224683, 1982.
116. Samuels, R. J., *Structured Polymer Properties,* 1974, Wiley–Interscience, New York.

117. Ward, I. M., Davies, G. R. and Unwin, P., (University of Leeds), private communication, 1984.
118. Ziska, J. J., Barlow, J. W. and Paul, D. R., *Polymer,* 1981, **22,** 918–23.
119. Cruz, C. A., Barlow, J. W. and Paul, D. R., *Macromolecules,* 1979, **12,** 726–31.
120. Paul, D. R. and Newman, S., *Polymer Blends,* Vols I and II, 1978, Academic Press, New York.
121. Olabisi, O., Robeson, L. M. and Shaw, M. T., *Polymer–Polymer Miscibility,* 1979, Academic Press, New York.
122. Aubin, M. and Prud'homme, R. E., *Macromolecules,* 1980, **13,** 365–9.
123. Holmes, P. A., Newton, A. B. and Wilmouth, F. M., (Imperial Chemical Industries plc), Eur. Patent Appln 52460, 1982.
124. Baptist, J. N., (W. R. Grace and Co.), US Patent 3 182 036, 1964.
125. Hummel, K. and Siapkas, S. *Kautsch. Gummi Kunst.,* 1981, **34,** 549–51.
126. Bucknall, C. B., *Toughened Plastics,* 1977, Applied Science, London.
127. Bonfield, W., Grynpas, M. D., Tully, A. E., Bowman, J. and Abram, J., *Biomaterials,* 1981, **2,** 185–6.
128. Hippe, H., *Arch. Mikrobiol.,* 1967, **56,** 248–77 and 278–99.
129. Stinson, M. W., and Merrick, J. M., *J. Bacteriol.,* 1974, **119,** 152–61.
130. Malik, K. A., and Claus, D., *J. Appl. Bacteriol.,* 1978, **45,** 143–6.
131. Tanio, T., Fukui, T., Shirakura, Y., Saito, T., Tomita, K., Kaiho, T. and Masamune, S., *Eur. J. Biochem.,* 1982, **124,** 71–7.
132. Gavard, R., Dahinger, A., Hauttecoeur, B. and Reynaud, C., *C. R. Hebd. Seances Acad. Sci., Ser. D,* 1966, **263,** 1273–5.
133. Gavard, R., Reynaud, C., Hauttecoeur, B. and Dahinger, A., *C. R. Hebd. Seances Acad. Sci., Ser. D,* 1967, **265,** 1557–9.
134. Lusty, C. J. and Doudoroff, M., *Proc. Natl. Acad. Sci. USA,* 1966, **56,** 960–6.
135. Merrick, J. M. and Yu, C. I., *Biochemistry,* 1966, **5,** 3563–8.
136. Tanaka, H., Tonomura, K. and Kamibayashi, A., *Biseibutsu Kogyo Gijutsu Kenkyusho Kenkyu Hokoku,* 1976, **48,** 67–74 and 75–9.
137. Tanaka, H., Tonomura, K. and Kamibayashi, A., *Nippon Nogei Kagaku Kaishi,* 1976, **50,** 431–6.
138. Tokiwa, Y., Ando, T. and Suzuki, T., *Hakko Zasshi,* 1976, **54,** 603–8.
139. Fukui, T., Akio, Y., Mamoru, M., Shunji, H., Terumi, S., Hiroko, N. and Kenkichi, T., *Arch. Microbiol.,* 1976, **110,** 149–56.
140. Baptist, J. N., (W. R. Grace and Co.), US Patent 3 111 496, 1966.
141. Pawan, G. L. S. and Semple, S. J. G., *Lancet,* 1983, August, 15–17.
142. Brune, H. and Niemann, E., *Z. Tierphysiol.,* 1977, **38,** 81–93.
143. Frazza, E. and Schmitt, E. E., *J. Biomed. Mat. Res.,* 1971, **1,** 43–58.
144. Kuhn, W., *Chem. Ber.,* 1930, **63,** 1503–9.
145. Lawrie, P., *Brit. J. Surgery,* 1959, **96,** 634–7.
146. Korsatko, W., Wabnegg, B., Tillian, H. M., Braunegg, G. and Lafferty, R. M., *Pharm. Ind.,* 1983, **45,** 1004–7.
147. Fukada, E., *Quart. Rev. Biophys.,* 1983, **16,** 59–87.
148. Fukada, E., *Adv. Biophys.,* 1974, **6,** 121–55.

149. MURAYAMA, N., NAKAMURA, R., OBARA, H. and SEGAWA, M., *Ultrasonics*, 1976, **14,** 15–23.
150. ANDO, Y. and FUKADA, E., *J. Polym. Sci., Polym. Phys. Ed.*, 1984, **22,** 1821–34.
151. LEMIEUX, R. V. and GIGUERE, J., *Can. J. Chem.*, 1951, **29,** 678–90.
152. DEOL, B. S., RIDLEY, D. D. and SIMPSON, G. W., *Aust. J. Chem.*, 1976, **29,** 2459–67.
153. HUNGERBUHLER, E., SEEBACH, D. and WASMUTH, D., *Helv. Chim. Acta,* 1981, **64,** 1467–87.
154. HARADA, T. and LZUMI, Y., *Chem. Lett.*, 1978, 1195–6.
155. SEEBACH, D. and ZYGER, M., *Helv. Chim. Acta,* 1982, **65,** 495–503.
156. SUGAII, T., FUJITA, M. and MORI, D., *J. Chem. Soc. Japan,* 1983, 1315–21.
157. NAJERA, C., YUS, M. and SEEBACH, D., *Helv. Chim. Acta,* 1984, **67,** 289–300.
158. FRATER, G., *Helv. Chim. Acta,* 1979, **62,** 2825–8.
159. SEEBACH, D. and WASMUTH, D., *Helv. Chim. Acta,* 1980, **63,** 197–200.
160. SEIDEL, W. and SEEBACH, D., *Tetrahedron Lett.*, 1982, **23,** 159–60.
161. SUTTER, M. A. and SEEBACH, D. *Liebigs Ann. Chem.*, 1983, 939–49.
162. SEURIG, B. and SEEBACH, D., *Liebigs Ann. Chem.*, 1978, 2044–73.
163. KITAHARA, T., KOSEKI, K. and MORI, K., *Agric. Biol. Chem.*, 1983, **47,** 389–93.
164. MORI, K., *Tetrahedron,* 1977, **33,** 289–94.
165. MORI, K. and WATANABE, H., *Tetrahedron,* 1984, **40,** 299–303.
166. KORSATKO, W., WABNEGG, B., BRAUNEGG, G., LAFFERTY, R. M. and STREMPFL, F., *Pharm. Ind.*, 1983, **45,** 525–7.
167. BISSERY, M. C., VALERIOTE, F. and THIES, C., *Microspheres and Drug Therapy,* E. Tomlinson and S. S. Davies (eds), 1984, Elsevier, Amsterdam, pp. 217–27.
168. BAPTIST, J. N. and ZIEGLER, J. B., (W. R. Grace and Co.), US Patent 3 225 766, 1965.
169. RAMSEY, W. B. and DELAPP, D. F., (American Cyanamid Co.), US Patent 3 781 349, 1973; *Idem,* US Patent 3 846 382, 1974.
170. VAINIONPAA, S., *Fixation Properties and Disappearance of Biodegradable Material in Bone Tissue,* Ph.D. Thesis, University of Tampere, Finland, 1986.
171. POLISTINA, R. A., (American Cyanamid Co.), US Patent 3 875 937, 1975.
172. COSEY, D. J. (American Cyanamid Co.), Fr. Demande 2 265 412, 1975.
173. MARCHESSAULT, R. H. and FAURE, A. J., *ACS Div. Polym. Chem. Polym. Prepr.*, 1974, **15,** 87–8.
174. HAUTTECOEUR, B., JOLIVET, M. and RAYMOND, G., *C.R. Hebd. Seances Acad. Sci., Ser. D,* 1975, **280,** 2801–4, 2899–902.
175. VANLAUTEM, N. and GILAIN, J., (Solway et Cie) Eur. Patent Appln EP 43620, 1982.
176. COULOMBE, S., SCHAUWECKER, P., MARCHESSAULT, R. H. and HAUTTECOEUR, B., *Macromolecules,* 1978, **11,** 279–81.

177. JONES, R. A., *An Introduction to Gas–Liquid Chromatography*, 1970, Academic Press, New York.
178. PIRKLE, W. H., *J. Amer. Chem. Soc.*, 1981, **103,** 3964–6.
179. KIRMSE, W. and SIEGFRIED, R., *J. Amer. Chem. Soc.*, 1983, **105,** 950–6.
180. POTTS, J. E., (Union Carbide Corp.), US Patent 4 372 311, 1983.
181. WHITE, B. G., BARTLETT, D. W., HOLMES, P. A. and POWELL, K. A., British Patent Appln 8221567, 1982.
182. DHEIN, R., REUTER, K., RUDOLPH, H. and PFITZNER, J., US Patent 4 324 880, 1982.
183. HELBA, E.-A. I. and WILLIAMS, A. L., (Mobil Oil Corp), US Patent 3 770 643, 1973.
184. STOCKMANN, H. H., RAY-CHAUDHURI, D. K. and BOETTGER, R. M., (National Starch and Chemical Corp.), US Patent 3 579 549, 1971.
185. HOLMES, P. A., *Phys. Tech.*, 1985, **16,** 32–6.

Note Added in Proof

The evidence for isodimorphism in the poly(3-hydroxybutyrate-co-3-hydroxyvalerate) system, with each crystalline phase accommodating the repeating unit of the other monomer with little distortion of the lattice parameters, is now well established (see for example R. H. Marchessault *et al.*, *Macromolecules,* 1986, **19,** 2860–2876). Thus, all of the P(HB-co-HV) copolymers described here are capable of crystallizing to a high degree. The observed reduction in the heat of melting determined by DSC with increasing HV content is due to a progressive decrease in the molar heat of fusion as the composition of the crystalline phase changes with incorporation of HV units rather than a decrease in the degree of crystallinity. The rate of crystallization, however, is severely retarded by the presence of HV monomer in the copolymer chain.

Chapter 2

PERMANGANIC ETCHING AND ITS INSIGHTS INTO CRYSTALLINE POLYMERS

D. C. BASSETT

J. J. Thomson Physical Laboratory, University of Reading, UK

SUMMARY

The macroscopic properties of crystalline polymers are controlled in large measure by their physical microstructure. Permanganic etching allows systematic study of real microstructures. It is a spectrum of techniques whereby potassium permanganate dissolved in strong acids reveals lamellar arrangements and structural details for an increasing range of crystalline polymers. With this method, one can now examine representative interior lamellar organizations of most, if not all, commercially important crystalline polymers at high resolution with the electron microscope. Diffraction studies are also possible. This chapter begins with an explanation of the importance of microstructure and the use of permanganic etching in relation to other morphological techniques. The wide potential of the method is illustrated with examples of achievements in particular cases. Analysis of specimens can be deepened, giving more precise and quantitative conclusions, if textural differences are amplified by controlled annealing and recrystallization treatments. In these ways, permanganic etching coupled with a complementary morphological approach, provides a powerful and unprecedented means of identifying and quantifying the lamellar textures underlying the properties of crystalline polymers.

INTRODUCTION

Permanganic etching is a technique, introduced from this laboratory, which allows systematic study with the electron microscope of the internal organization of many crystalline polymers to resolutions of 5 nm and better.[1,2] This method has, at long last, made it possible to acquire detailed knowledge of the lamellar texture and organization, within a range of different polymers and morphologies. This is essential information for a fundamental understanding of physical-structure/property relationships in these important materials. The purpose of this chapter is to detail current practice of the technique (which must be varied to suit each sample and polymer) and to provide a concise review of its applications to date.

The Importance of Microstructure

The importance of physical texture is that it determines how the macroscopic behaviour of a sample relates to the underlying microscopic property. The microstructure of a material, together with inherent molecular characteristics, determines its properties. It is because of changes in physical texture that different ways of processing the same polymer produce different properties. Nowadays we have become accustomed to the preparation of high-performance polymer fibres, including polyethylene, whose Young's modulus can exceed that of steel (210 GPa) and approach theoretical values for the crystal lattice (~300 GPa), i.e. some two to three orders of magnitude greater than for the conventionally processed polymer. Such improvements are consequences of suitably modified internal microstructures, in this case the replacement of a series lamellar/interlamellar composite (whose relatively high compliance is due to the soft interlamellar regions) by an oriented parallel morphology, i.e. one in which the soft regions are stiffened by crystal bridges (Fig. 1). If we are to have a fundamentally based understanding of physical properties it is, therefore, essential that we know the underlying microstructure.

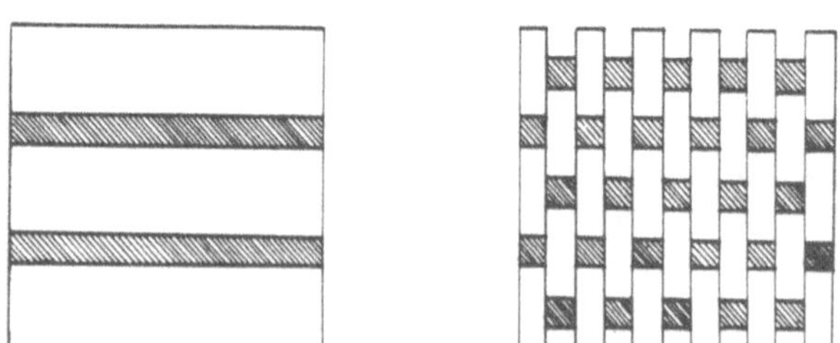

FIG. 1. Sketch of the distinction between series (left) and parallel (right) arrangements of lamellar/interlamellar composites with the same crystallinity.

Real, as opposed to idealized, organization is best assessed by microscopy. The eye-and-brain combination is able to recognize patterns in an image which can be of considerable complexity whereas in modelling one naturally begins with the simplest constructions. Once identified, these features can be measured (often most conveniently by methods involving diffraction or reciprocal space which readily yield average values over the whole of a sample) and so provide the basis for quantitative science. The essential first step, however, is correctly to identify the pattern of organization. One of the revelations of permanganic etching has been to show how drastically over-simplified model textures have been in comparison with reality.

Spherulites and Lamellae

In crystalline polymers it has long been known from optical microscopy that there is a pronounced tendency to organize into spherulites. The name means 'little spheres' and during growth from an unperturbed melt it is often the case that individual spherulites do indeed have a spherical envelope at low resolution. Once they have impinged, however, this can no longer be true, and the envelope of an individual spherulite will then be polyhedral with flat or, more generally, curved boundaries. Equally, in the early stages of growth, embryonic spherulites commonly pass through stages of development of a radiating polycrystalline habit from a simple crystalline precursor. Such development often looks sheaf-like, whence the widespread use of the term 'sheaf' in this context, although nowadays we know that such 'sheaves' often do not possess cylindrical symmetry but are rather the projected appearance of less symmetric lamellar objects.

When spherulites were first reported in polymers (by Bunn and Alcock[3] in polyethylene) it was also found, from the sign of optical birefringence and subsequently by X-ray diffraction, that the long molecules of the polymer crystallized tangentially, or nearly so, to spherulites. This was difficult to understand in terms of the then-prevalent fringed-micelle model of polymer crystallinity. It was, however, an immediate corollary of the discovery of chainfolded polymer lamellae that if such lamellae grew radially outwards in spherulites, then their constituent molecules would have a tangential or near-tangential orientation depending on whether or not chains were exactly normal within lamellae.[4,5] What has been controversial is whether such lamellae really were typical of melt-crystallized systems.[6] As will be evident from the work described in this chapter, permanganic etching followed by electron microscopy has shown unequivocally that the melt-crystallized polymers so far studied, including systems with crystallinities of order 30%, are indeed lamellar, thereby confirming the conclusion reached painstakingly over two decades from combinations of many indirect experiments and non-microscopic methods.

The presence of lamellae within crystalline polymers has major consequences for macroscopic properties. This is because, for example, modulus, strength, thermal conductivity, thermal expansion etc., depend on the relative placement of lamellae and their neighbours. When there is a continuous crystalline skeleton, such as parallel lamellar arrangements provide (Fig. 1), then these macroscopic properties will approximate to that of the crystalline component. Conversely the normal series lamellar arrangement in as-crystallized polymers produces properties reflecting more the interlamellar or 'amorphous' component of the composite morphology.[7] This is a very different situation from that presupposed in the fringed-micelle model whose individual crystallites were necessarily surrounded by essentially equivalent amorphous environments. The properties of such organization would be restricted to those of a series composite, with values depending on (usually proportional to) the degree of crystallinity. With lamellar organization, on the other hand, the influence of the degree of crystallinity is modest in comparison with the changes which can be effected by modifying the morphology.

The problems of being able to study representative internal organization in melt-crystallized polymers and particularly lamellar organization, have, however, taken a long time to solve. Permanganic etching now provides one solution to this problem. It is able to reveal lamellae, their relative arrangement and the placement of different lamellar populations together with indications of the characters (length, branch content and possibly tacticity) of associated molecules. This is essential information for a fundamental understanding of properties, but it still needs to be supplemented by knowledge of the nature of interlamellar connections and organization (such as entanglements) outside lamellar interiors to provide the basis of precise calculations of properties.

Electron Microscopy of Polymers

There is a hierarchy of organization within crystalline polymers, most of it at dimensions requiring the resolution of the electron microscope.[4] At low resolutions the optical microscope usually indicates internal texture seemingly tantalizingly just below its resolution in the sub-micron range. In addition, low-angle X-ray or neutron diffraction indicates periodicities, usually interpreted as lamellar/interlamellar composites, of order tens of nanometres. The underlying morphology in both of these cases can now be studied systematically following permanganic etching. Other recent advances[8–10] have allowed lattice resolution, and especially the study of lattice defects within crystallites which may be of particular importance for electrical conduction.[11] (It should not be forgotten, however, that lattice defects were successfully resolved and studied as long ago as 1958 using moire methods.[12]) There is unfortunately no current prospect of going beyond this and resolving order in the disordered component.

If one approaches the problem of examining crystalline polymers in the electron microscope with adequate resolution from first principles, there is first of all the choice of instrument: transmission or scanning. Scanning (SEM) instruments are less demanding in specimen preparation but, until the very recent introduction of high-resolution instruments, they have not generally been able to resolve polymer lamellae; also, they are restricted to the examination of surfaces. They are widely used on such problems as the location and internal texture of spherulites but their limitations need to be remembered. There are cases in the literature where the lack of resolution of SEM has led to misinterpretations of the texture which are evident when individual lamellae are resolved in a transmission instrument. One needs also to be aware of the possibility of bias in the surface being examined. It is common to study interior morphologies via fracture surfaces, but these have been exposed by a crack whose propagation is likely not to have followed a random path. For example, we now know that in linear polyethylene fracture exposes {200} surfaces with high preference.[13] Other views, especially the important and diagnostic view down **b**, the spherulitic radius, are almost always absent. In addition, if there are two or more different lamellar populations within a sample resulting, say, from crystallization first isothermally then on cooling, it can be the case that thinner lamellae are much under-recorded (and in some cases appear to be completely absent) in fracture surfaces because of ductile fracture.[14] Images of fracture surfaces must, therefore, be regarded with due caution. The problem of obtaining a specimen showing representative internal morphology remains, and for the best resolution one would use the transmission instrument. Here again, there are two routes: examination of a specimen directly in transmission, or indirect study via surface replication.

Transmission electron microscopy requires thin specimens both to transmit electrons and to minimize the proportion suffering energy loss and thus a change in wavelength leading to chromatic aberration. For metals or inorganics it is usually possible to thin specimens chemically down to thicknesses of 100 nm or less, producing a transparent specimen and contrasting image. For biological specimens the normal route is to microtome specimens to produce sections, say, 50 nm thick. Being formed mostly of carbon and other low-mass elements, these will tend to lack contrast and it is common to enhance and stabilize this by staining.

For polymers, sectioning is the most general way to produce thin specimens, although much work has also been carried out on specially prepared thin films—whose reliability as guides to general features needs then to be established—because of the difficulties involved in making sections of good quality. This is still much of an art but the chances of success tend to be improved by sectioning below the brittle/ductile transition, using a very sharp knife with attention to its angle of address and the speed of cutting. Difficult specimens can sometimes be made tractable by reducing their size (to below 1 mm) and especially by embedding such a small specimen in a proprietary electron microscope epoxy resin. Even when good sections have been prepared, they still tend to have inherently low contrast, and their usefulness is severely limited by radiation damage.

It is radiation damage above all which has restricted polymer science from benefiting comparably with other areas of materials science from the advent of modern electron microscopes 30 years ago. Nevertheless, much was learned by rapid working at low magnifications, most notably the inference that solution-grown lamellae contained molecules in chainfolded conformation.[15] In due course this led to the realization that the low Young's moduli of traditional polymer fibres were due to the interruption of strong and stiff covalent bond sequences by backfolding.[16] The reduction or elimination of chainfolding then became a major aim in developing new fibres. In newer materials such as Kevlar and high-modulus polyethylenes this has been substantially realized. However, knowledge of lamellar organization within spherulites and bulk polymers remained rudimentary.

A wealth of photographs in Geil's book[17] left no doubt that lamellae were present in a very wide range of synthetic crystalline polymers but these were illustrations either of external or of fracture surfaces. There was always the justifiable doubt that what was observed was special, specifically lower molecular mass material which could either have migrated to an external surface or guided the propagation of a crack to give non-representative views. Both these situations do occur but the application of permanganic etching has removed any lingering doubt that melt-crystallized polymers are profusely lamellar. For a long time this conclusion was resisted in certain quarters. Even accepting the existence of lamellae, the question of to what extent these are regularly chainfolded has been highly controversial.[6] At present the consensus seems to be that there is a spectrum of ordering, improving at higher crystallization temperatures. Recent work from this laboratory showing sectorization in melt-crystallized isotactic polystyrene provides a powerful argument for the occurrence of regular chainfolding at low supercoolings.[18]

As has been mentioned earlier, macroscopic properties are controlled by microstructure. For melt-crystallized polymers this means primarily the relative dispositions of lamellae and interlamellar material, the extent of fractional crystallization and the nature of interlamellar connections. Knowledge of the first of these, for representative textures, can now be derived following permanganic etching for many polymers but it did not come from the direct approach of transmission electron microscopy of thin sections because of radiation damage. This was not just a question of the destruction of crystallinity giving a limited lifetime in which to make observations, but was much more a consequence of radiation damage causing mass transport in the specimen leading to unstable images of low inherent resolution. Detailed case-histories of such effects have been reviewed by Grubb.[19] With hindsight, electron diffraction studies on solution-grown polymer crystals can be appreciated to have benefited greatly from lamellae being in contact with a carbon substrate.[20] This inhibited mass transport in monolayers (though not in multilayers whose farther layers were not constrained) and reduced the difficulties to those of a limited lifetime. Moreover, inherent detail was comparatively sharp and even when crystals had been rendered amorphous in the electron beam, they retained their original shapes, thereby providing much relevant detail. Unfortunately similar behaviour does not occur for melt-crystallized polymers. Their thin sections start featureless, then develop contrast which, at best, is indirectly related to the original morphology.[21] Such problems have made the electron microscopy of melt-crystallized polymers both difficult and comparatively unrewarding for a long period. Then, some ten years ago, two new techniques of specimen preparation transformed the possibilities of characterizing lamellar organization in typical melt-crystallized polymer morphologies.

Chlorosulphonation

Strong contrast in chlorosulphonated polyethylene was reported by de Korosy and Zeigerson[22] in 1967 but the development of chlorosulphonation as a technique for the electron microscopy of melt-crystallized polyethylene was pioneered by Kanig.[23,24] Ostensibly this is a staining method which introduces heavier atoms, principally chlorine, sulphur and sometimes uranium into interlamellar regions. The value of the method derives, however, especially from the stability it brings to the image. In marked contrast to untreated sections, images of chlorosulphonated polyethylene are sharp and stable in the electron beam with little or no mass transport evident. Chlorosulphonated polyethylene is easier to section, but it is no longer the original polyethylene. There are considerable reductions in crystallinity, but what remains has a melting point raised by ~15 K.[25] The progressive conversion can be monitored by thermal analysis which reveals that there is a discontinuous change from the original to the chlorosulphonated state. This change also causes strong blackening, thereby allowing visual inspection of the extent of conversion from the surface of a sample. Different polyethylenes require different times of treatment. Longer is required for linear polymer and thick lamellae (higher crystallization temperatures) than for thin lamellae and branched polymer. A typical treatment for linear polyethylene would be to immerse the polymer (of ~1 mm thickness) in 60% chlorosulphonic acid for 3 h or more. This time would be reduced at room temperature for low-density material. Very recently Phillips has chlorosulphonated specimens at temperatures within the melting endotherm as a means of identifying particular melting species.[26]

Chlorosulphonated polyethylene is examined as thin sections in transmission. The contrast observed varies considerably with lamellar orientation and the ratio of section to lamellar thickness (which should be about unity for optimal interpretation).[25] It is most striking when lamellae are parallel to the electron beam and then shows black interlamellar regions separating light lamellar interiors. Departure from this orientation rapidly reduces the contrast as the sheets of staining atoms are looked through obliquely rather than edge-on. The substantial areas of relatively uniform (pale) contrast frequently observed in chlorosulphonated polyethylenes arise from such tilted orientations. This can be confirmed in particular cases by applying reverse tilts to the specimen with a suitable stage to obtain the familiar striped contrast.[25] In certain circumstances the degree of tilting can be assessed by inspection. This occurs when the lamellar thickness is comparable with the section thickness. This is not generally the case for the ~20 nm thick lamellae typically present in polyethylene but has been demonstrated for those an order of magnitude thicker produced by crystallization treatments at high pressure.[25] The sensitivity to orientation can also be turned to advantage to identify those regions in which lamellae are parallel to the electron beam. A notable example is in banded spherulites where the sequence of rings is clearly evident. Nevertheless, the apparent packing of lamellae within a band is very open with consistent gaps. Individual lamellae themselves show little or no signs of twisting about the radius of a spherulite and tilting experiments designed to test whether a progressive rotation of the normal of an individual lamella did spiral about the spherulitic radius produced a negative result.[27] Such evidence shows that banded spherulites consist of interleaving arrays of lamellae in which the gross twisting around the radius required from optical studies is achieved by the progression of arrays, rather than from twisting within individual lamellae as has often been postulated. More recent experiments involving permanganic etching show that the twisting is associated with systematic arrays of screw dislocations in which successive layers of the associated spiral terrace are not in contact.[28]

The above is one example of the complementary nature of chlorosulphonation and permanganic etching. The former, with its strong emphasis on a particular lamellar orientation, is akin to dark-field microscopy. The latter, on this analogy, is like a bright-field technique. In polyethylene, the only polymer for which both methods are readily applicable, each has confirmed the textures shown by the other.[13] There is no doubt, therefore, that the morphologies now revealed in bulk polyethylene are genuine.† When permanganic

† Notwithstanding the total qualitative agreement between polyethylene textures recorded by the two techniques, there is a quantitative disagreement in the measure of lamellar thickness. That recorded in chlorosulphonated sections is always lower than the distance in permanganically etched surfaces

(continued)

etching is applied to other polymers the textures observed, though not identical with polyethylene, are of similar kinds and are generally consistent with other textural indicators. One is confident, therefore, that these too, are genuine microstructure.

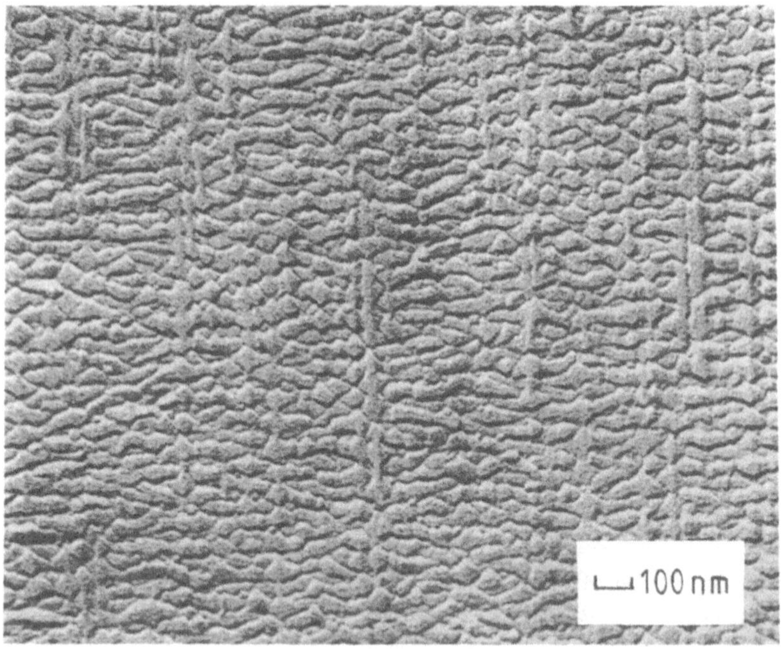

(a)

FIG. 2. Electron micrographs of high-modulus, high-strength polyethylene fibres.[29] The samples were prepared for electron microscopy by (a) permanganic etch and replication, or (b) chlorosulphonation and sectioning. The morphologies revealed are essentially the same, but the spacing appears smaller in the chlorosulphonated section.[30] However, lightly chlorosulphonated sections have an advantage that they can be used for electron diffraction in addition to imaging.

---

of the same sample. Thus, for a 26 000 mass fraction with a repetition parallel to the chain axis for lamellae crystallized at 130·6°C measured as 96 ± 2 nm (by nitration and subsequent gel permeation chromatography), etched surfaces revealed thicknesses of ~100 nm whilst the figure for chlorosulphonated sections was only ~60 nm.[13] A comparison at more typical lamellar thicknesses is shown in Fig. 2; the lamellar repetition is greater for etched than for stained specimens. This discrepancy remains to be fully resolved.

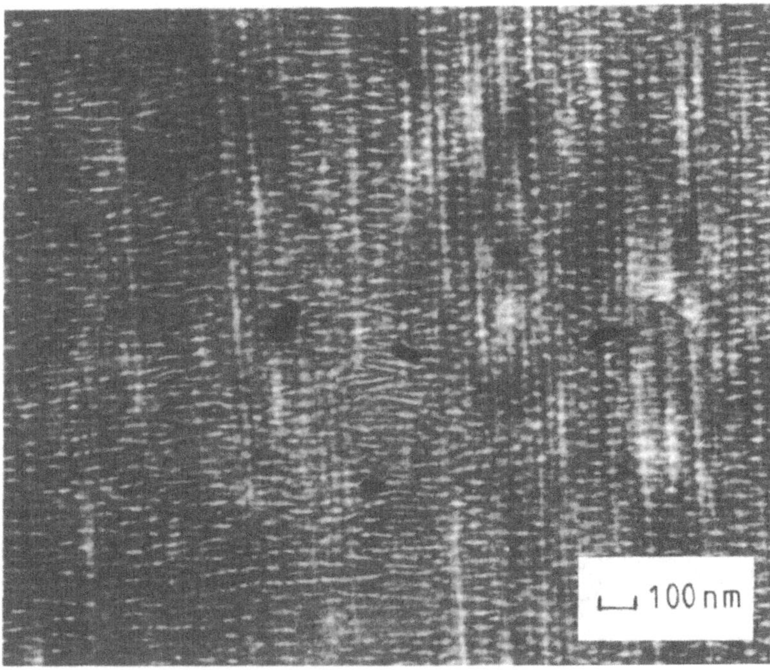

(b)
FIG. 2—*contd.*

THE TECHNIQUE OF PERMANGANIC ETCHING

The background to the introduction of permanganic etching lay in the knowledge that fuming nitric acid rendered polyethylene friable, into a condition in which it could be separated into individual lamellar fragments.[31] Such friability is undesirable for replication but it suggested the potential of milder oxidizing treatments. The impetus for testing these came from the realization that the best alternative method (fracture surface replication) was very limited in its applicability to the detailed study of lamellar deformation even in the most favourable case, namely anabaric or pressure-crystallized polyethylene.[32] Specifically, the fracture-surface striation defining the chain axis (Fig. 5) tended to obscure other detail. Another method was needed if this problem was to be elucidated further. A series of oxidants was, therefore, tried as etchants and tested on fracture surfaces of pressure-crystallized polyethylene; removal of the **c**-axis striation was a test for their efficacy. In the event, the reagents tended not to be very effective and conditions were adjusted for maximum action. So it was that the combination of potassium permanganate dissolved in concentrated sulphuric acid came to be used at high concentrations (7%) and high temperatures (100 °C).[1] This pairing was much superior to all others investigated and remains the basis of current practice, although nowadays treatment is at room temperature, concentrations are usually an order of magnitude less and orthophosphoric acid and water have been introduced into the spectrum of reagents used.[2] It is best to adjust the proportions of etchant for individual types of specimen rather than to rely on a universal recipe. Nevertheless, a very adequate reagent for polyethylene and isotactic polypropylene is 0·7% w/v potassium permanganate in a 2:1 mixture of concentrated sulphuric acid and orthophosphoric acid. The incorporation of water (to give a specific brown colour, presumably an oxidation state) produces a more aggressive reagent though a less uniform reaction, i.e. a rougher etched surface. This is advantageous in the study of isotactic polystyrene.[33]

Etching is stopped by a prescribed washing procedure. Our practice remains that of using 5 cm^3 of reagent per specimen, cooling this to 0°C, decanting, then washing successively in cold dilute sulphuric acid, hydrogen peroxide, water and acetone.[1] Others have devised their own procedures.[34,35] Even here there is need for caution. Different operators following the same prescription have produced different results because of differing times in the cold acid prior to adding hydrogen peroxide. In the case of PEEK (polyetheretherarylketone), too rapid a cessation causes obscuration of lamellar detail which slower treatments will avoid.[36] The difference appears to be that in the former case a little recrystallization occurs at the surface, but not in the latter.[37]

There is also the question of artefacts. It was reported initially[1] that the detail revealed by the etching solutions of potassium permanganate in sulphuric acid alone was not entirely genuine. A few regions, generally of some microns across, were present in which novel hillock-like structures, without parallel in thin optical sections of samples, appeared to be superposed on lamellar features. In the early work these were avoided and surveys of genuine lamellar morphologies in polyethylene were completed.[38–40] These were carried out at fixed 2% concentration but it became appreciated that the finer details, especially in the complex textures (e.g. that of monoclinic polypropylene) or following deformation, benefited from less concentrated solutions. Unpublished investigations of the effect of different concentrations of permanganate showed that under particular conditions, the proportion of artefacts became very large. The problems could, however, be minimized by not etching for too long. At the same time there were reports in the literature which showed that in some investigations of microstructure in polyethylene cables, artefacts had been mistaken for real morphology.[41] At this stage a new investigation found that the artefacts were a consequence of crystals being deposited on the specimen surfaces in the later stages of etching with unsurprising preference for nucleating along surface scratches, etc.[2] They can be removed for examination and are pink, i.e. presumably a manganese salt, although actual chemical composition has yet to be established. These crystals would be expected to hinder attack, consistently with the development of hillocks on the surface superimposed on genuine detail. They are, however, removed in subsequent washing procedures so that only the legacy of a changed surface profile attests to their having intervened during etching.[2]

With common sense and careful monitoring† of the progress of etching appropriate to a given sample, artefacts in the $KMnO_4/H_2SO_4$ system need never be numerous or a serious disadvantage. However, with etching that is too prolonged the situation can deteriorate, as the work of Naylor and Phillips demonstrates.[42] These authors also etched using an ultrasonic bath (which roughly doubles the rate of attack) but removed far more material (20–40 μm) than has been our practice. This has the disadvantage of losing the relationship between the initial surface and the etched one which depends upon removing less than typical lamellar widths.[1] Selection of the initial surface (particularly fracture or cut) brings added subtlety to etching which can be used to advantage. Our practice has always been to remove ~1 μm or less for the above reason; current procedures remove 1 μm from linear polyethylene in ~1 h.[43]

The problem of artefacts was removed by the introduction of orthophosphoric acid into the etching mixture.[2] This has proved to be a substantial advance. Not only are there no artefacts but there is greater delicacy of etching. As a direct consequence, systematic studies on monoclinic polypropylene[44–46] and on deformed

† Our practice is to use Nomarski interference contrast optics for this purpose.

polyethylene[47] have been carried out which would scarcely have been possible with previous formulations.

Such studies and also work on PEEK require a very high quality of permanganic etching. Our experience is that it is best to optimize treatments for a particular sample and polymer by careful monitoring. Prescriptive treatments in advance are less successful and should be treated as no more than a guide from which to begin systematic adjustments of etchant composition for a given specimen. Published work on comparative etching conditions has shown up the saturation effects of prolonged etching.[34,42] In our work we do not approach this phenomenon but remove polymer at a linear rate.[43]

The list of polymers known to respond satisfactorily to permanganic etching is now long and continually growing. It consists of linear and branched polyethylene, four isotactic polyolefins (polypropylene, polystyrene, poly(4-methylpentene-1) and poly(butene-1)), related atactic polymers, poly(vinylidene fluoride) (hereafter denoted PVF_2), PEEK, and poly(ethylene terephthalate) (PET), together with various copolymers and others such as ethylene propylene rubbers and ethylene-propylene-diene (EPDM) terpolymer.

The selectivity of permanganic etching is astonishing by comparison with what had gone before and even from simple expectation of how an oxidizing reaction might proceed. The basic finding that a crystalline polymer is removed in integral lamellar stages is in accord with experience of both nitric acid[31] and gaseous oxidation[48] of polyethylene. It is consistent with a reaction path of breaking covalent bonds at fold surfaces followed by a much faster lateral removal of fold stems. In this way one would also leave the $\{hk0\}$ surfaces, as are evident for thick lamellae (~100 nm) in polyethylene. It is also to be expected that different components of a blend[1] and different crystal structures of the same polymer (as in polypropylene[49] and PVF_2) would be removed at different rates and thus to different depths within a specimen. This would simply be a reflection of differing chemical potentials, as would the preferential attack along the core of screw dislocations in parallel with many of the early papers on dislocations in inorganic materials. However, the selectivity went further. From the first, thinner lamellae were often found to be removed more than thicker ones, leading to what could be spectacular differences of relief.[4] *A priori,* one would scarcely have expected so big a difference even making allowance for an association of thin lamellae with shorter molecules, both being concerned in the later stages of crystallization.[13,39] As experience accumulated it became clear that such differential etching, as we shall term it, did not always occur but was rather unpredictable.

It is only recently that insight has been gained into the causes of this erratic behaviour with consequent gains in the understanding of how etching proceeds. This has come from the work of Freedman *et al.*, who attempted to develop quantitative measures of the permanganic etching of polyethylene.[43] Their aim was to construct a triple correlation of electron microscopy, showing where polymer had been removed, with thermal analysis, giving the proportions and lamellar thicknesses removed, and with gel permeation chromotography, identifying the molecular species involved. It was necessary to work with thin (~5 μm) sections for which the material removed is a sufficiently large proportion of the whole. Individual sections can be examined with the sensitivity of modern thermal analysis equipment.

Detailed investigation was made of polyethylene samples crystallized in two stages, approximately half at 128°C and the remaining half on quenching[43] (Fig. 3). Two corresponding peaks in the melting endotherms were used to monitor the various stages of treatment by measurement of their changing proportions and magnitudes. In the first place it was found that the size of the lower melting peak (i.e. quenched population) was substantially less than in the bulk for sections cut at room temperature and both peaks were shifted ~2 K to lower temperatures. However, endotherms similar to the bulk polymer but with both peak positions lowered by ~1 K were obtained when sections were cut at dry-ice temperatures. Alternatively, the lower peak could be restored by annealing sections at 131°C, i.e. between the two melting peaks of the bulk polymer. Such observations reveal a discrimination between the two lamellar populations due to selective deformation during microtomy at room temperature. This interesting phenomenon illustrates two facets of the interrelation between morphology and properties. One sees not only how the two populations give the sample two melting points but also that they respond differently to deformation. The properties are varying within the sample according to textural variations on the sub-micron scale. Still more interesting is the response of the sections to etching. Those which were cut cold showed a linear decrease of mass with time but with the lower-melting population of lamellae eaten away much the faster of the two. According to a simple model the lateral surfaces were removed at 3·3 and 1·6 Å s^{-1} for the quenched and isothermally

FIG. 3. Two populations of lamellae in a linear polyethylene (Rigidex 140/60) crystallized for 3 h at 128°C, then quenched into cold water. The thicker layers formed at 128°C and have established a framework within which the thin lamellae formed on quenching. Replica of etched cut surface.

FIG. 4. Two populations of lamellae in a linear low-density polyethylene (7·5 C_2H_5/1000 C atoms) which was crystallized for 9 min at 118°C, then quenched into dry ice/acetone. The thin lamellae comprising the matrix contain a higher proportion of branches than do the thick dominants. Replica of etched cut surface.

crystallized populations respectively. It is this difference in rates which is responsible for thinner lamellae being eated away to greater depths. However, when sections which had been annealed at 131°C were etched, the data for the two populations were barely distinguishable at 1·8 and 1·6 $A\,s^{-1}$ respectively. There is now little or no difference in etching between the two populations although their lamellar thicknesses and the associated molecular weights both still differ by factors of ~2. One is, therefore, led to conclude that it is neither the lamellar thickness nor the molecular weight which is responsible for differential etching between the populations but the differing deformation each has suffered during the cutting to prepare the specimens. What is being revealed is a secondary effect related to the morphology itself rather than an inherent property of the material.

The insensitivity of etching rate to the lamellar thickness confirms the inference from the shallow grooves observed to be etched between the basal surfaces of adjacent lamellae, namely that the rate-controlling step in etching is scission of chains at these surfaces which is slow compared with the rate of removal of residual stems. Further experiments have been conducted upon linear low-density polyethylene with 7·5 ethyl branches/1000 main chain carbon atoms.[50] In this case the two populations (grown at 118°C and on quenching) (Fig. 4) etched at 5·9 and 6·5 $A\,s^{-1}$ after preparation by cold cutting. But after annealing at 120°C the comparable figures were 5·0 and 6·5 $A\,s^{-1}$. In this instance we are now observing an inherent difference between the two populations which differ in lamellar thickness and branch content (with 9·7 branches/1000C atoms in the quenched population and 4·2 branches/1000C atoms in the isothermally crystallized one) and in molecular mass: ($1{\cdot}6 \times 10^4$ and $4{\cdot}4 \times 10^4$ respectively). The increase in etching rates results, however, from the more open structure of the interfacial regions which allows more rapid scission of molecular chains.

APPLICATIONS OF PERMANGANIC ETCHING

The potential applications of permanganic etching are very wide, as wide as the relationship of microstructure to properties.[4] As befits a major new technique of broad relevance, the applications are increasing rapidly. It is not our intention to attempt to list all of these, for any such attempt would fail and be incomplete before publication. Rather, salient themes will be identified from the point of view of how studies using permanganic etching have confirmed and consolidated previous understandings, with especial emphasis on those areas in which novel morphologies and conclusions have resulted. It will be convenient, nevertheless, to allow these themes to emerge from a discussion of morphologies in different systems.

Anabaric Polyethylenes

The description 'anabaric' is given to those polyethylenes which have crystallized or been annealed in the high-pressure phase of polyethylene[51] which was first predicted,[52,53] then discovered,[51] in our earlier work. Investigation of crystallization and annealing processes at high pressure (typically 2–6 kbar; 0·2–0·6 GPa) and the deformation of the resulting materials was the principal theme of research in this laboratory when the efficacy of permanganic etching for disclosing lamellar morphology was discovered. Indeed, it was the difficulty of investigating deformation in such materials, relying on fracture-surface replication to reveal lamellar morphology, which led to a search for alternative procedures. Knowledge of the morphology of anabaric polyethylenes had already been built up painstakingly by other means, some direct, some indirect. Permanganic etching not only confirmed this earlier understanding, including direct evidence in instances where only indirect had existed, but also exposed totally new features.[4,54]

When applied to fracture surfaces of anabaric polyethylene (Fig. 5), permanganic etching reproduces the size and perspective of thick (>100 nm) lamellae seen in the initial surface. In fact it nearly is the same surface because only ~1 μm or less is removed in typical treatments: this dimension is much less than typical lamellar widths, so that the lamellae exposed in the etched surface are mostly those of the original surface.[1] (This conclusion applies generally.) What is new is that all lamellae are revealed after etching whereas in fracture surfaces thinner lamellae (less than ~30 nm)—known to be present from peaks in the melting endotherm—are at best under-recorded, but usually are not clearly seen.[55] It may be that these thinner lamellae, being tougher because they have more chainfolds per molecule,† deflect the propagating crack. Alternatively, more chainfolding may give ductile rather

† It was shown long ago that the initial description of extended-chain crystals given to lamellae in pressure-crystallized polyethylene was wrong.[56] Full molecular extension is a comparative rarity; in general molecules are folded.[57] Whether a material is brittle or ductile depends on whether the ratio of molecular length to crystal thickness exceeds a ratio of ~10.[58]

FIG. 5. Unetched fracture surface of a linear polyethylene ($\bar{M}_m = 2{\cdot}9 \times 10^4$; $\bar{M}_n = 2{\cdot}3 \times 10^4$) which was crystallized at 0·5 GPa while cooling at 0·25 K min^{-1}. Many kink bands are identified by changes in the direction of the **c**-axis striation. Surface replica.

than brittle fracture and a confused image. A comparable effect is known to occur in very high molecular mass ($\sim 10^6$) polyethylenes. Replicas of their fracture surfaces contain many threads pulled from the rough surface at room temperature, with considerable obscuration of detail.[57] With shorter molecules conditions can be reached where only certain lamellar populations in a sample (namely those with more chainfolds per molecule) give this effect. With such materials permanganic etching produces a clear image and has made it possible to conduct systematic studies of the deformation of high molecular mass polyethylenes.[49] On the debit side, etching also removes fracture-surface striations parallel to **c**, so that information on the local orientation of the chain axis in lamellae is lost. However, with suitable replication techniques, it is often possible to recover some of this when some regions of the surface are pulled off the sample with the replica and the **c**-axis striation is again revealed.[38]

Fracture surfaces have always been used to study pressure-crystallized linear polyethylene but invariably showed the $\{hk0\}$ prism faces of lamellae (Fig. 5), in fact mostly $\{200\}$. Lamellar shapes were rarely, if ever, presented. With permanganic etching, all views of all lamellae are obtainable by etching cut surfaces, i.e. those exposed by cutting a sample open (Fig. 6). In this way one gains information not previously available.

The lamellae themselves are seen to be comparatively irregular. They often have one or perhaps two corners of hexagonal (~120°) appearance but, under typical crystallization conditions, the shapes are more or less irregular with a tendency to be extended in one dimension.[54] No doubt this reflects the liquid crystalline nature of the high-pressure phase. At the same time one observes internal boundaries within lamellae whose different response to etching is informative. Optical studies in the diamond-anvil cell had reported on a subdivision of lamellae parallel to their basal plane during cooling,[59] i.e. associated with the hexagonal-to-orthorhombic transition. Such internal boundaries are rare but can be detected and identified by their resistance to etching. They are not penetrated by the etchant and grooves do not form along them, in marked contrast to the external surfaces of lamellae. More common are internal boundaries parallel to the chain axis which can also be seen to develop in the diamond-anvil cell.[54] It has been suggested that these represent alternative orthorhombic orientations of unit cells derived from the common initial two-dimensional hexagonal structure of the liquid crystalline phase.[54,60]

After the transition to the orthorhombic structure, molecular chains are inclined to basal surfaces, whereas they are exactly normal in the high-pressure phase. The interconversion must involve shear, and fine parallel sets of lines are frequently found on the basal surfaces of etched lamellae (Fig. 16). These, it is suggested, are the traces of the shear planes on the basal surfaces. From parallel orthorhombic overgrowths of known orientation it appears that the lines are along the orthorhombic **b**-axis.[54]

Finally, etched samples provide detailed evidence for the texture, scale and location of species fractionated during the high-pressure crystallization. Early studies had demonstrated (by solvent extraction and molecular mass measurement) that low peaks in the melting

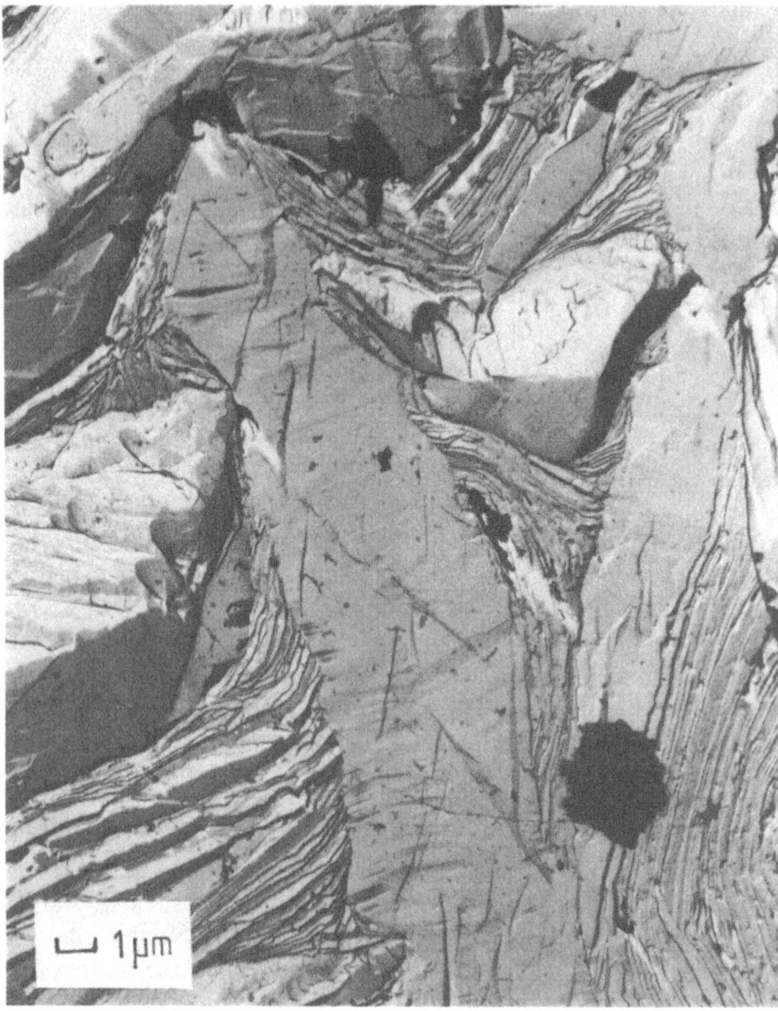

FIG. 6. The same sample as in Fig. 5 but now displayed in an etched cut surface. Various lamellar perspectives are presented, as are groups of thin lamellae which are not seen in Fig. 5. Surface replica.

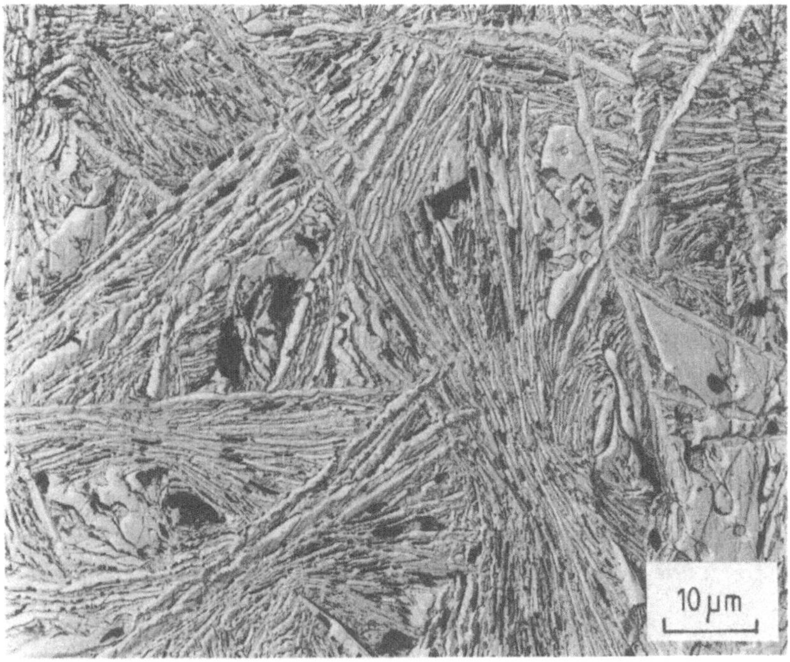

FIG. 7. Linear polyethylene (Rigidex 9) crystallized at 0·5 GPa while cooling at ~1 K min^{-1}. Note the splaying development of thick lamellae similar to that of melt-crystallized polymers with much thinner lamellae. Between the very thick lamellae are thinner layers comprising shorter molecules. Replica of an etched cut surface.

endotherms of anabaric polyethylenes were due to shorter molecules.[61] Fracture surfaces did not reveal the location of the corresponding populations inside a sample, but permanganic etching does. At least two types of site are identifiable. In one case there are pockets of thin lamellae characteristic of the last regions to crystallize (Fig. 6). In the other there can be extensive linear (strictly planar) crystallization between individual diverging thick lamellae[54] (Figs 6, 7). Both observations anticipate evidence for fractionation which has since become available for spherulitic crystallization from the melt in typical conditions for many polymers.[33,38–40,62]

Fractional Crystallization

Fractional crystallization is a principal reason for the systematic variation of microstructure and hence properties from point to point in a solid polymer.[4] Studies using permanganic etching have not only confirmed that this is a widespread phenomenon but have also constantly shown that real microstructures are systematically ordered but more complex than had previously been supposed.

Recognition of two different kinds of fractional crystallization followed[63] soon after the discovery of polymer single crystals grown from solution. At high temperatures, shorter molecules are unable to precipitate in reasonable times and only do so on cooling. Crystals which are filtered from their mother liquor at the crystallization temperature show a single X-ray long period and when collected form flexible mats. Those crystallized on cooling also show a single, though lower, long period but tend to form brittle, chalky aggregates. The molecules crystallizing isothermally are longer than those precipitating on cooling. Without hot filtration, two X-ray long periods result from the successive crystallizations.

Fractionation also occurs during the process of isothermal crystallization. In the first instance this was suggested by the finding that only the exterior regions of isothermally-grown monolayers would dissolve at suitable constant temperatures while their interiors, of the same thickness, were stable indefinitely.[63] The association of a successive reduction in molecular weight with increasing outward distance in the crystal was confirmed by measuring the length of molecules dissolved progressively from the edge towards the centre.[64] Similar fractionation effects have now been established to occur for polymeric crystallization from the melt.

Fractional crystallization from the melt is shown particularly clearly by polyethylene crystallized at high pressure.[54] Homopolymers so processed usually show multiple melting peaks associated with different populations of lamellae. These are related, in turn, to molecular mass differences, as may be demonstrated in two ways. If low-mass polymer is dissolved out prior to processing, then the lower-melting peaks do not occur.[61] Conversely, if polymer giving the lower peaks is dissolved out after processing, then measurements show that it consists of shorter molecules; molecular masses of as little as ~4000 comprise the material of the lowest peaks.[55] The cause in this instance is the inability of short molecules to form the high-pressure liquid crystalline phase so that when they eventually solidify on cooling it is directly into the orthorhombic structure.[56] Closely parallel behaviour is found on annealing, thereby providing detailed support for the thesis that the reorganization it produces is due to partly melting then recrystallizing the more unstable regions.[14,61] A similar pattern of segregation, with the same interpretation, occurs when polyethylene is annealed *in vacuo*.[65]

It is in the location of the segregated species, and in the wider problem of establishing the stages of growth, that permanganic etching has given much new insight. One needs to recognize particular components within the microstructure according to different morphological characteristics. The simplest case is when there are differences in lamellar thickness. Indeed, it is often advantageous in solving a problem if such differences can be engineered, for example by quenching at an appropriate stage of crystallization.

The greatest differences arise, once again, in the (anabaric) crystallization of polyethylene at high pressure. In this, the first application of permanganic etching, regions of thin lamellae were immediately identified, whereas these were rarely seen in fracture surfaces, to which previous research had been almost entirely restricted. For this system the basic division is between crystallization from the melt into the high-pressure phase giving thick lamellae (usually >200 nm thick) and into the orthorhombic when lamellae are usually <50 nm thick (Fig. 8). Superimposed on this divide are variations based on pressure, crystallization temperature and molecular mass which make the detailed texture especially informative on the details of crystallization processes.[54]

Two patterns are common. One is the conglomeration of very thin lamellae in roughly equi-axed regions located especially in the boundary regions where growth from two nuclei has met. These are due to the shortest molecules and crystallization of what would have been the last pools of molten polymer. The second pattern is of progressively thinner layers interleaving between the thickest ones[54] (Figs 6, 7). Here one is observing progressive fractionation during crystallization. More complex, but still systematic, patterns have also been reported. In the case of one block copolymer incorporating hexene monomer, four (possibly five) sites were found differing in lamellar thickness, habit and location within the framework of spherulites.[4] These lamellae are placed transversely to the dominants, thereby enclosing narrow columns of melt. There is still much to learn in this regard, but the potential and practicability of the contributions permanganic etching has to make have been clearly demonstrated.

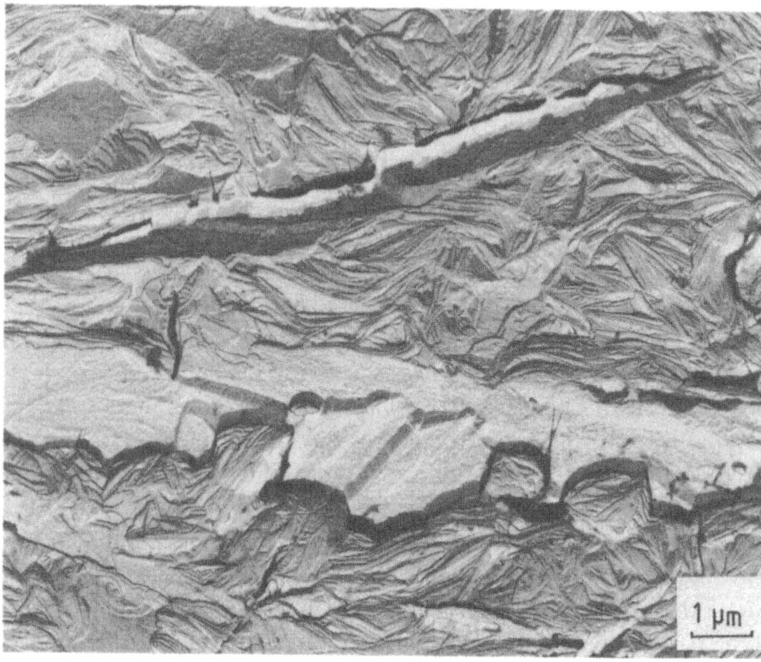

FIG. 8. Linear polyethylene (Rigidex 9) crystallized at 0·26 GPa, 4% as thick, initially-hexagonal lamellae and 96% as thin, always-orthorhombic layers. Note the shear parallel to the chain axis undergone by the thick lamellae. Replica of an etched cut surface.

Morphology of Melt-Crystallized Polymers

Melt-crystallized polymer morphologies are far from being just aggregates of lamellae. Instead, there are systematically ordered patterns within the aggregates. The identification of these is helped if the growth interface can be revealed by quenching during crystallization. Visibility depends upon the quenched portion being recognizably different from the remainder. In isotactic polystyrene, for instance, slow crystallization rates make it possible to quench the matrix surrounding growing objects to the 'amorphous' condition from any growth temperature. But in polyethylene, the very fast crystallization below ~125°C makes it difficult to distinguish quenched matrix from growing object much below this crystallization temperature. However, polyethylene has the advantage of forming non-planar lamellae with

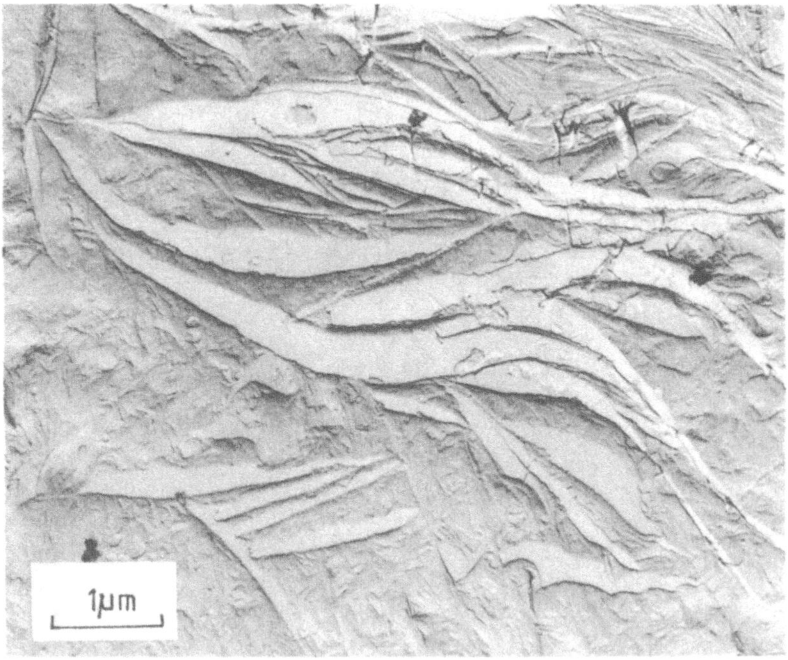

FIG. 9. Initial crystallization at 125°C for 3 min of a linear polyethylene within a quenched matrix. Note the S-shaped dominant lamellae. Replica of an etched cut surface.

oblique habits (i.e. with the **c**-axis inclined to lamellar normals) (Fig. 9). These habits are, in part, sensitive to crystallization temperature and molecular mass, so that they have been used to display isothermal fractionation during melt-growth.[40]

Several polymers crystallized from the melt *in vacuo* or at 1 bar have now been examined using the approaches just described, beginning with linear polyethylene,[13,26–28,38–40] then linear low-density polyethylene,[66] isotactic polystyrene,[33,67] isotactic polypropylene,[44,45] PVF_2,[68] PEEK,[69] poly(4-methylpentene-1)[70] and, most recently, PET.[71] It is generally convenient to start by examining specimens grown relatively slowly at low supercoolings because this maximizes the scope for quenching and consequently for extracting quenched polymer to measure its molecular characteristics. Most of the above polymers have, however, now been examined over a wide and

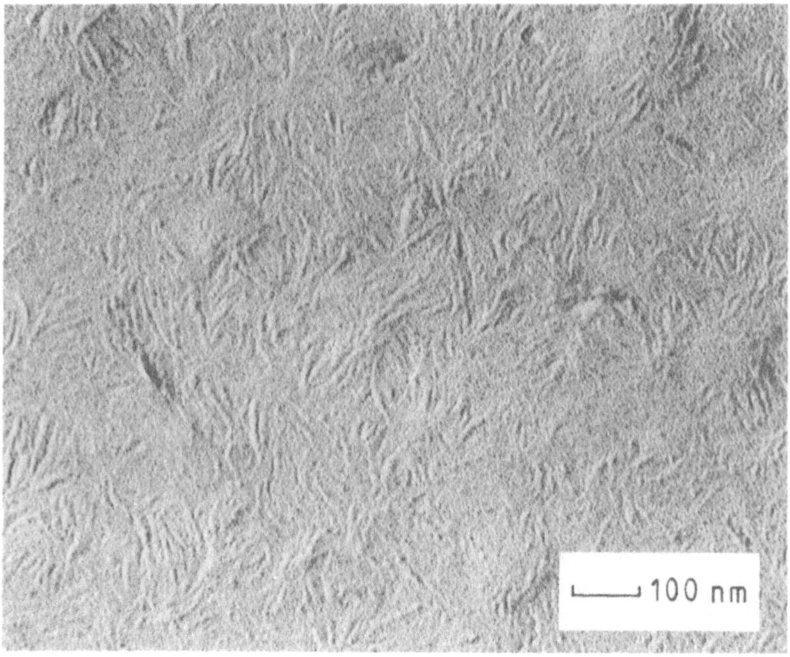

FIG. 10. Splaying sheaf-like development of lamellae in PEEK crystallized from the glass by heating at 163·5°C for 15 s, then quenching. Replica of an etched cut surface.

representative range of crystallization temperatures including, for example, in the case of PEEK, growth from the glassy, or strictly the rubbery, state[69] (Fig. 10). What principally emerges is a common pattern of growth whereby individual, so-called dominant, lamellae[38] grow outwards, branch and splay apart to form a framework within which later growth occurs. In the simplest cases this later growth gives a single population of infilling lamellae; these have been called 'subsidiary',[38] but a more extensive vocabulary is needed to describe the more complicated textures with multiple populations which commonly occur. Subsidiary lamellae have been shown to contain molecules of different length[40,43] and/or branch content,[50,66] in certain cases. Such fractionation effects are most marked in polyethylene. At 128°C, as previously cited, the average molecular length of subsidiary lamellae is about half that of dominant lamellae in typical polymers of $\sim 10^5$ mass (Fig. 3). In the linear low-density material the same factor has been measured for branch content (see above) with segregation placing more branched molecules in subsidiary lamellae (Fig. 4). Differences of molecular length have also been measured in polystyrene but the effects there are less pronounced.[33,67]

Order in Spherulites

The details of melt-crystallized polymer morphologies newly available through permanganic etching are relevant to the long-standing problem of spherulitic growth. What has been shown is a common pattern of development based on individual dominant lamellae providing a branched and splayed framework (Figs 3, 4, 7, 9, 10, 15). This holds throughout the crystallization ranges investigated, although the distinction between dominant and later-forming lamellae is less marked at lower crystallization temperatures. There is also evidence for systematic fractionation between different populations of lamellae crystallizing isothermally and in close proximity. A major element in this pattern is the tangential separation λ between adjacent dominant lamellae at the growing edge of a spherulite. This is a function of branching angle and branching interval. It has been shown, for polystyrene,[67] to be a parameter varying systematically with growth conditions, of temperature, molecular mass (of isotactic polymer and of atactic matrix) and of melt composition. Measurements of λ as a function of these quantities have been made and a preliminary paper published.[67] This is the primary pattern but others have also been recognized in more restricted ranges of variables. Transverse lamellae apparently closing off volumes of melt from the spherulite's envelope[4] have already been mentioned (Fig. 3), but their spacing has not yet been quantified. On the other hand a small number of measurements of the diameter of groupings of dominant lamellae has been made for i-polypropylene as a function of growth rate (by varying crystallization temperature). For one homopolymer, this fell by one order of magnitude (3 to 0·3 μm) between crystallization temperatures of 155 and 115°C, i.e. by some three orders of magnitude less than the change in growth rate.[72]

The patterns identified and measured within actual morphologies of many polymers do not readily conform to pre-existing conceptions based on extrapolation from optical observations. The key concept is that of the fibre, supposed to be generally a group of elongated lamellae with, in one theory,[73] a cross-sectional dimension of order the characteristic length $\delta = D/G$ where D is the diffusion coefficient of species rejected at the growth front which advances with speed G. Fibres would be separated by regions of high concentration of segregated species and their elongated habits have been suggested to result from a cellulation phenomenon set in train when the lateral dimensions of regular lamellae greatly exceed the length δ.[73] There are serious quantitative discrepancies between these suggestions and what actually occurs. It has not yet proved possible to identify the proposed 'fibres' in typical homopolymers although crystallization from heavily-doped melts (e.g. 1:1 isotactic:atactic polystyrene) does produce clusters of lamellae which appear as discrete lines optically when viewed edge-on.[5,67] Nor does the principal parameter, λ, or any other parameter so far identified, have the magnitude or the variation with relevant variables required for an identification with δ.

It has recently been shown for PEEK that the dominant/subsidiary pattern is associated with two types of crystal differing in melting point.[69] The suggestion is that these differ not primarily because of associated molecular lengths but essentially because of their physical location. What is observed is that PEEK crystallized either by cooling from the melt or heating from the glass ('annealing') shows two peaks in its melting endotherm.[74] One is ~15 K above the crystallization or annealing temperature; the other is in the range of 330°C and above and shows less variation with the preparative conditions.

The occurrence of two peaks in a melting endotherm from an initially homogeneous sample is a common phenomenon and in the case of polyethylene is due to reorganization with rising temperature.[4] Thus the lower-temperature peak would be due to residual material, virtually unchanged from the original condition, whilst the upper peak would be given by polymer which had reorganized (primarily by local melting and crystallization) to more stable crystals. The relative proportion of the two peaks varies with heating rate: slower rates allow more reorganization and enhance the size of the upper peak. Conversely, sufficiently fast heating melts all the sample in its original condition and shows only the one (lower) peak. However, this is not the situation for PEEK.

The two melting peaks of PEEK are comparatively insensitive to changes in heating rate, and do not give way to a single peak of one or other component at extreme rates of heating.[69] Moreover, it has been shown, by melting directly from the treatment temperature without intermediate cooling, after various times of treatment, that the higher-melting peak forms first and the lower-melting peak begins to develop later.[69,75] Local melting followed by recrystallization is, therefore, not the essential cause of the two peaks in the melting endotherm. They represent rather two different components within the sample. Moreover, the results of annealing from the glass to a constant temperature for various times show that the lower peak arises because of reorganization of the most unstable of a very broad spectrum of crystallites created in primary crystallization.[69] Annealing for longer times or at higher temperatures reorganizes more of the material into the condition represented by the lower peak. The size and melting temperature of this component both increase accordingly, in the same manner familiar from linear low-density polyethylenes.[4] Permanganic etching brings to this problem the ability to locate the material giving the two melting peaks. It has been shown that the low-melting polymer lies between the dominant lamellae while the dominants themselves are high-melting.[69] Internal evidence suggests that molecular fractionation is not the essential reason for this difference, which relates primarily to location within the crystalline texture.

Crystallization within the restricted confines of the dominants could affect the melting point of the product in various ways. Prealignment of the melt, for example, would lower its entropy, raise its chemical potential and so the melting point of a crystal in equilibrium with it.[76] A lowered melting point would come from a less-ordered crystal or a thinner one. X-ray evidence has, so far, shown no evidence for the former possibility. However, crystals are known from earlier work to be reduced in thickness when they are constrained to fit to small volumes.[77,78] This seems the most likely explanation of both the very low melting points of subsidiary crystals and their broad range, probably reflecting the range of micro-environments in which they form. Because they are so unstable, they are very susceptible to reorganization at the annealing or crystallization temperature. In this way the lower-melting peak develops. Preliminary work has also shown that a similar explanation is appropriate to phenomena in PET.[71]

Origins of Spherulites

The earliest stages of melt growth are a rewarding area of study. They show, for example, features related to nucleation and to the question of whether melt-grown lamellae are chain folded. The lamellae which form have habits which have become central to recent discussions of polymeric crystallization.[79] One can also test directly to see whether lamellae do undergo a habit change leading to fibrous development once their width exceeds the characteristic length δ.[73,80]

The possibility of observing the beginnings of melt growth depends on the practicability of quenching so that the object of interest can be distinguished from the quenching matrix. For the same reasons discussed previously this effectively limits polyethylene to growth at low supercoolings: this has so far precluded observations of the start of banded spherulites (Fig. 11), which form at too high a supercooling. Polystyrene is more amenable. Figure 12 shows early objects in a homopolymer of molecular mass $\bar{M}_m = 1{\cdot}1 \times 10^6$, $\bar{M}_n = 3{\cdot}8 \times 10^5$. For this polystyrene it was not possible to find single lamellae. Instead groups of lamellae could often be seen to form along a central thread and could be presumed to do so at other times. Nucleation along flow lines to give 'shish-kebabs' is a familiar phenomenon but in this

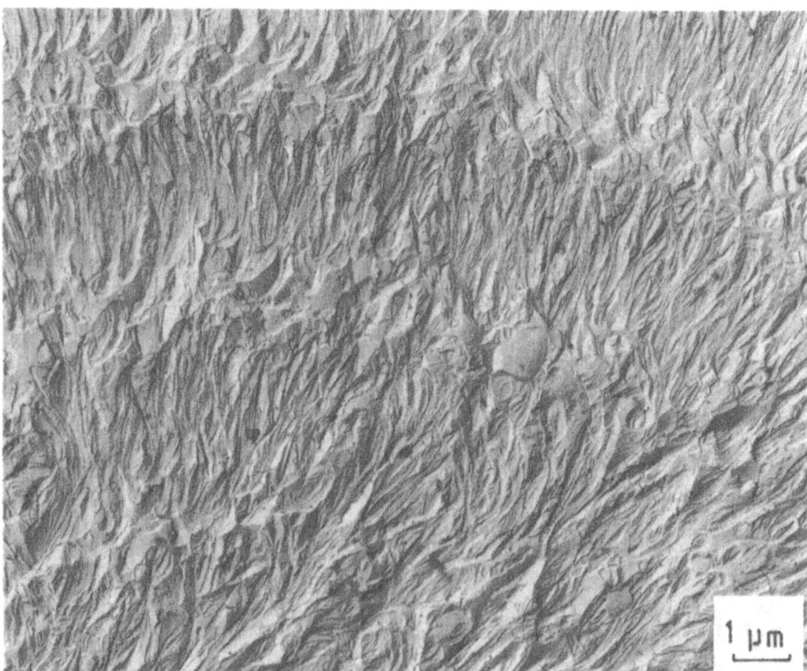

FIG. 11. Portion of a banded spherulite of linear polyethylene (Sclair 2907). This sample has been annealed at 0·5 GPa and 230°C. Replica of an etched cut surface.

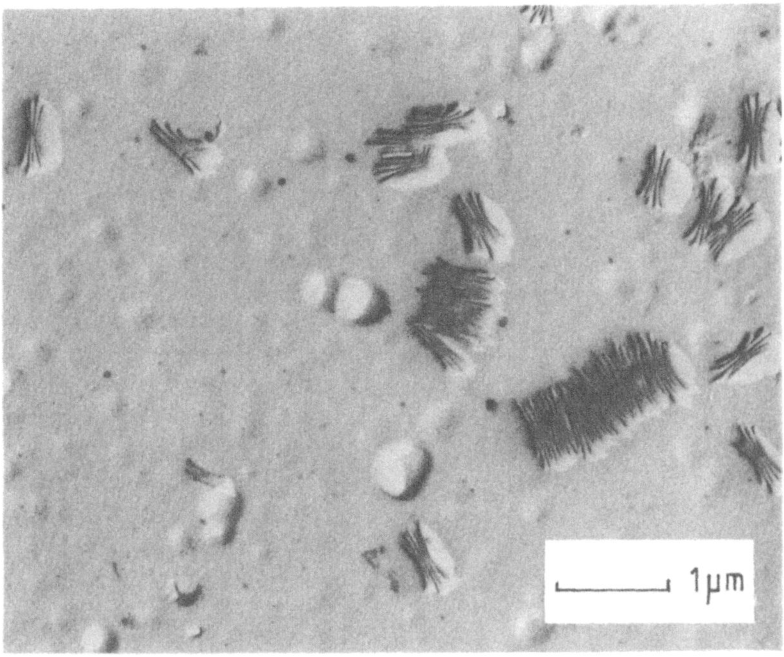

FIG. 12. The beginnings of growth in isotactic polystyrene crystallized from the melt at 220°C for 81 min, then quenched. Replica of an etched cut surface.

instance is from a nominally unperturbed melt. There is the possibility that one is detecting an inherent entanglement associated with very long molecules. In this figure, note that lamellae in small groups are already splaying apart in the way now familiar and widespread in melt-crystallized growth.

When individual lamellae lie in the etched surface, for high crystallization temperatures they show lines perpendicular to the growth faces which give a six-sectored appearance (Fig. 13). Similar patterns found in solution-crystallized polymers are associated with lattice distortions caused by chainfolding.[81,82] It is probable that the same arguments based on symmetry[4] can also be used for these melt-grown lamellae which are effectively separate layers along a thread. Previously[33] this conclusion was resisted when there was the possibility of conicalness being imposed by deformation. Nowadays the question of there being a proportion of regular folding in melt-crystallized polymers is no longer controversial, but it would be

FIG. 13. An individual polystyrene lamella in the same sample as in Fig. 12. This is presumably one member of a shish-kebab (note the central nucleus). The layer is slightly dished, with striations normal to growth faces the same as those found in chainfolded lamellae grown from solution. Replica of an etched cut surface.

pleasant to see equivalent morphological evidence for this conclusion provided both for solution- and melt-grown material.

In polyethylene, nucleation also occurs as shish-kebabs along threads for higher mass polymer ($\bar{M}_m \geqslant 3 \times 10^5$ at 130°C), but single lamellae initiate growth for polymer of 10^5 mass.[83] They have convex growth faces and are elongated along the **b**-axis (Fig. 14). A central screw dislocation is often present, in which case successive turns of the resulting helicoidal terrace can be seen not to be in contact, reinforcing the point noted in studies of multilayer early objects,[33,44] that polymer lamellae behave as if they repel each other.

When individual lamellae are naturally elongated, the problem of constructing a spherulite from them devolves into mechanisms of branching and splaying of lamellae. Details of both these processes have been identified from the first use of permanganic etching to

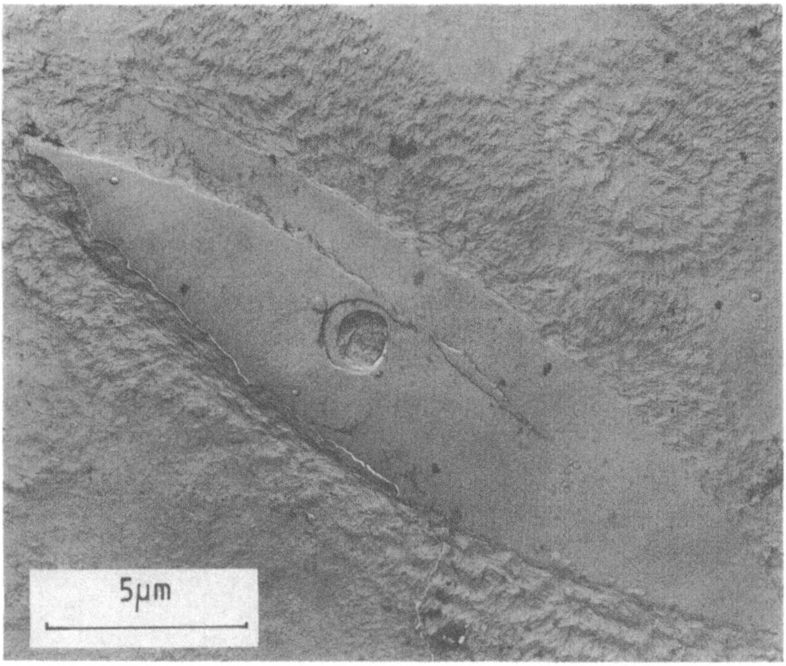

FIG. 14. Linear polyethylene (Sclair 2907) crystal elongated along **b** with convex growth faces formed at 130°C. Successive layers of the helicoidal terrace formed around the etched-out screw dislocation splay apart. Replica of an etched cut surface.

examine spherulites.[13,40] However, when the crystals have high symmetry, as in polystyrene (hexagonal) or poly(4-methylpentene-1) (tetragonal), it has been supposed that there is a change in habit, giving elongated development once crystal dimensions surpass the length δ. Such a change in habit has been claimed to be responsible for changes in optical appearance, especially when coarsely branched objects grown at low supercoolings (axialites) give way to more rounded development (spherulites).[80] Historic evidence for this point of view is provided by four different polystyrene objects, claimed to be at successive stages of developing elongation.[80] On the contrary, our experience is that lamellae grow at constant shape in both polystyrene[68] and poly(4-methylpentene-1).[70] The shape can be regular or twinned so as to give the optical impression of protuberances, although at electron-microscope resolution the twinning origin of these is apparent. Spherulites of both these polymers have to date been found not to show markedly asymmetric crystal habits but to be built from more or less regular lamellae with parallels to multilayer development from solution as described by Khoury and Barnes.[84] In all cases early objects develop by branching, usually at comparatively large angles (25–30° for polystyrene),[5] and splaying apart of dominant lamellae. This leaves its record in the mature object which is often detectable even with the optical microscope. This is especially so for monoclinic polypropylene, whose cross-hatched centre (Fig. 15) gives a four-fold pattern in the appropriate perspective to the developed spherulite. It is also noteworthy that spherulites grown by heating the glass have a similar construction to those crystallized by cooling the melt (Fig. 10). The morphological details of how lamellae are organized in melt-crystallized polymers have thus become available by

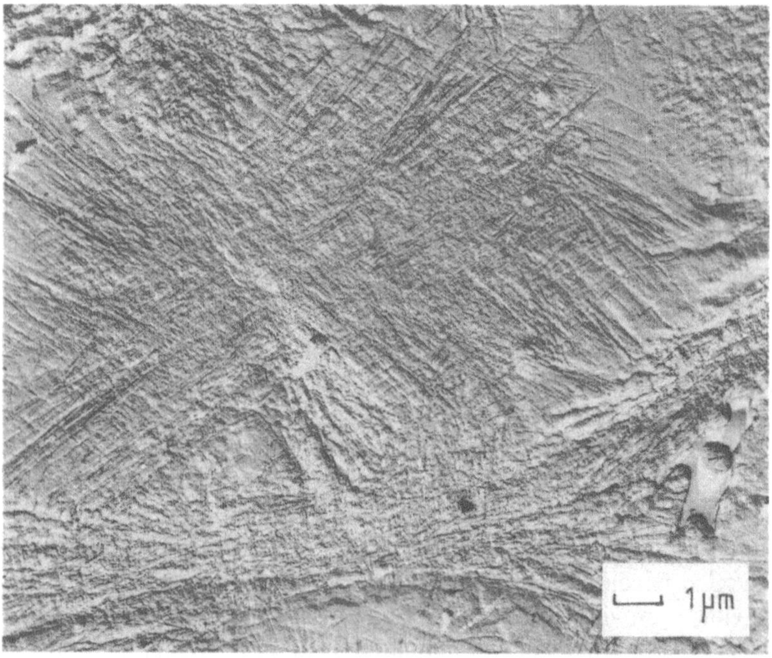

FIG. 15. Two views of early growth at 155°C in isotactic polypropylene (monoclinic structure). One object would appear cross-hatched in plan (above) or sheaf-like when viewed sideways (below). Replica of an etched cut surface.

the application of permanganic etching. In many instances these have been surprising and different from previous supposition. The availability of the facts is likely to lead soon to their being placed in a theoretical context.

Morphology and Problem Solving

The usefulness of polymer morphology depends in large measure not merely on the essential characterization of microstructure but on the record of the sample's history that is contained in the texture.[4] In polymers this record exists essentially because of the specific routes through which a system passes in developing its microstructure, their sensitivity to processing variables and the relative stability of the morphology once long molecules have been organized within it. In particular systems the record can be made easier to read, as illustrated by the use made in this laboratory of anabaric polyethylenes as model systems. Previous discussion of this theme[54] allows present mention to be limited, although interesting extensions to new areas are now in progress.

On the subject of polymer deformation, it is often difficult to distinguish clearly, for example in a drawn polymer, between lamellae which have survived the draw (albeit deformed) and those which have been created during the drawing process. There has been a long argument as to whether sufficient heat could be generated and concentrated for melting to be involved in reformation. This still continues.[85] However, if one deforms thick polyethylene lamellae which can only be formed at high pressure any thick layers present after drawing must have survived the draw. Highly sheared lamellae of this kind surviving at least to draw ratios of 12 but surrounded by a matrix of reformed lamellae have been discussed previously.[54]

This is one example of how it is possible to provide unequivocal answers by suitable choice of system. Another is the question of the location of branches within linear low-density polyethylenes. A comprehensive analysis[66] shows that methyl groups enter the lattice while ethyl and butyl are excluded. This is also controversial. It can be illustrated very clearly by the use of high pressure. In methyl-branched polyethylene, recrystallization at high pressure (0·5 GPa) produces lamellae with thicknesses ranging to ~1 μm (Fig. 16). There is no hindrance to producing thick crystals but the presence of methyl groups within the lattice is indicated by the progressive lowering of melting point at a given thickness for increasing branch content.

FIG. 16. A methyl-branched linear low-density polyethylene (6·25 CH_3/1000 C atoms) crystallized at 0·5 GPa by cooling from the melt at 0·25 K min^{-1}. The complex texture includes very thick lamellae. Replica of an etched cut surface.

However, ethyl and longer-branched polymers do not produce such thick crystals (Fig. 17), but give a maximum thickness of the interbranch distance. (To demonstrate this it is necessary to gain knowledge of fractional crystallization, so that a meaningful figure can be specified.) At the same time there is a single relationship between crystal thickness and melting point appropriate to the polyethylene lattice. The approach can be extended to identify block copolymerization, as mentioned above in relation to systematic fractional crystallization within a spherulite and illustrated previously.[4,5]

A further, new extension of this approach is giving very interesting

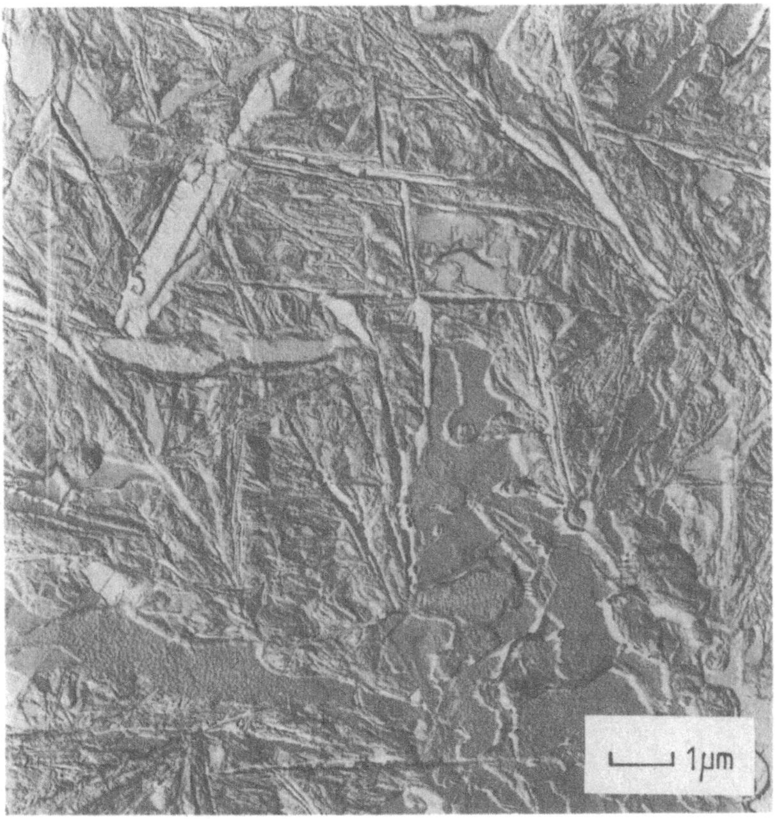

FIG. 17. In an ethyl-branched linear low-density polyethylene (7·5 C_2H_5/1000 C atoms) crystallized alongside the sample of Fig. 16, there are no very thick lamellae. This is a consequence of exclusion of ethyl branches from the crystal lattice, in contrast to methyl branches which are included. Replica of an etched cut surface.

data concerning the effect of electron irradiation on linear polyethylene. This is a subject which demonstrates considerable morphological sensitivity. The gel content of samples for a given dose depends upon crystallization conditions and related matters such as the manner of aggregation of the same lamellae. The conclusions of many studies have been that there is a strong tendency to form cross-links at the fold surfaces of lamellae.[86]

The objective of a new study[89] was two-fold. Firstly it was attempted

1µm

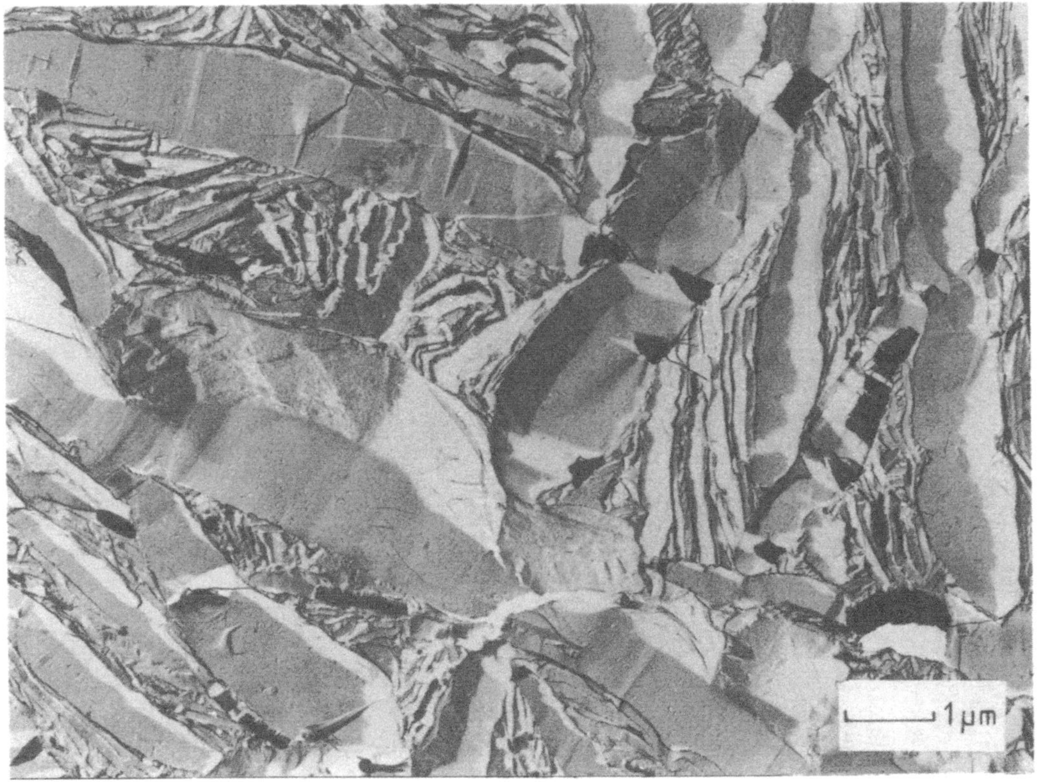

FIG. 18. A comparison of two samples of the same linear polyethylene (Rigidex 140/60), both of which have been irradiated to 1 Mrad, then crystallized from the melt at 0·5 GPa. The sample which had thicker lamellae before irradiation (crystallized at 127·5°C; above) forms less thick anabaric lamellae than the sample initially crystallized at 119·5°C (below). Replicas of etched cut surfaces.

to separate gel and sol molecules in an irradiated sample by suitable annealing and recrystallization procedures. This has been accomplished in a similar way to parallel work by Hikmet and Keller.[87] Secondly it was attempted to use high pressure to form crystals whose thickness is limited by cross-links. Preliminary results illustrated in Fig. 18 show the large differences which ensue, in this case at a very early stage of irradiation. Here are two polyethylene samples differing only in that one was crystallized at 119·5°C, the other at 127·5°C. Both were irradiated to 1 Mrad, which is below the gel point, then recrystallized at 5 kbar. The ranking of thickness has been reversed: the sample crystallized at 127·5°C initially had the thicker lamellae, now its layers are not so thick as those in its companion (which itself is similar to the unirradiated controls).

One expects that, on recrystallization at high pressure, cross-links will be excluded from lamellae as are short chain branches.[54] In this case the greater final lamellar thicknesses shown by the sample which initially had the thinner lamellae suggests that it has the greater separation between cross-links. The reason for this remains to be firmly established. One possibility turns on there being more unreacted radicals after higher temperature crystallization.[86] If these did not react until the sample was melted (at high pressure) and then did so randomly, the qualitative effect observed would result.

The State of the Art

Explanation has been given of how permanganic etching came to be introduced at higher concentrations and temperatures than one would now recommend. Even mixtures solely with sulphuric acid which have been used very successfully in this laboratory and elsewhere do not match reagents incorporating orthophosphoric acid. Since this modification was announced etching has advanced firmly into territory it could previosuly only survey coarsely. The number of such applications is growing rapidly. At first, it was possible to undertake systematic studies of polypropylene. Early work has been published[44,45] with significantly more to follow. Even the so-called smectic modification has yielded to careful analysis at the limits of resolution.[46]

A second topic is that of deformed polymers. It is now possible to reveal very fine detail in drawn polyethylene whereas previously it was very difficult to advance knowledge of atmospherically-crystallized polymer notwithstanding the substantial gains from the high pressure products previously mentioned. This is no longer true and Fig. 19 illustrates an early stage in the deformation of banded spherulites. (Fig. 11). With permanganic etching one can now advance the subject beyond the point where it was necessarily left by Hay and Keller 20 years ago.[88] One does, however, meet a major difficulty in achieving systematic diffraction information. Almost all the applications of

FIG. 19. The early stages of deformation of a banded spherulite of linear polyethylene: draw ratio 1·2. Replica of an etched cut surface.

permanganic etching have relied upon crystal habits. In polyethylene and monoclinic polypropylene these are of low symmetry and can often be used to indicate crystallographic orientation. If not, one always has the possibilities of diffraction from lamellae removed from the surface, confirmed as being practical.[1] Norton and Keller have made particular use of lamellae stripped from the etched surface in their work on polypropylene spherulites.[45] It is possible to pull layers from deformed polyethylene in the same way but there is then the difficulty that the layers removed are not necessarily the ones one would most like to use for diffraction in a systematic study. Moreover, because of the complex changes of phase and orientation already documented for solution-grown lamellae,[4] diffraction information in relation to the observed morphology is essential to adequately characterize the deformed state. The most useful way of obtaining this is from specimens which have been thinned sufficiently to allow transmission electron diffraction, relying once again on rapid working at low intensities. Permanganic etching has thus opened up enormous areas of structural information but has avoided rather than overcome the fundamental difficulty of radiation damage.

ACKNOWLEDGEMENTS

I am indebted to Dr A. M. Freedman, Dr A. M. Hodge, R. H. Olley, Dr I. A. M. Al Raheil, Dr A. S. Vaughan and Dr A. Vokal for the provision of photographs from their work in this laboratory. Professor A. Keller and Dr M. J. Hill kindly supplied Fig. 2.

REFERENCES

1. Olley, R. H., Hodge, A. M. and Bassett, D. C., *J. Polym. Sci., Polym. Phys. Ed.*, 1979, **17**, 627.
2. Olley, R. H. and Bassett, D. C., *Polymer*, 1982, **23**, 1707.
3. Bunn, C. W. and Alcock, T. C., *Trans. Faraday Soc.*, 1945, **41**, 317.
4. Bassett, D. C., *Principles of Polymer Morphology*, 1981, Cambridge University Press, Cambridge.
5. Bassett, D. C., *CRC Crit. Rev. Solid State Mat. Sci.*, 1984, **12**, 97.
6. Anon. *Disc. Faraday Soc.*, 1979, **68**.
7. Ward, I. M., in I. M. Ward (Ed.), *Developments in Oriented Polymers—1*, 1982, Applied Science, London.

8. Isoda, S., Tsuji, M., Ohara, M., Kawaguchi, A. and Katayama, K. *Polymer,* 1982, **23,** 1568.
9. Revol, J.-F. and Manley, R. St J., *J. Mat. Sci.Lett.*, 1986, **5,** 249.
10. Chanzy, H. D., Smith, P. and Revol, J.-F., *J. Polym. Sci. Polym. Lett. Edn.*, 1986, **24,** 557.
11. Yeung, P. H. J. and Young, R. J., *Polymer,* 1986, **27,** 202.
12. Agar, A. W., Frank, F. C. and Keller, A., *Phil. Mag.*, 1959, **4,** 32.
13. Bassett, D. C. and Hodge, A. M., *Proc. Roy. Soc. A,* 1978, **359,** 121.
14. Bassett, D. C., Khalifa, B. A. and Olley, R. H., *Polymer,* 1976, **17,** 284.
15. Keller, A., *Phil. Mag.*, 1957, **2,** 1171.
16. Frank, F. C., *Proc. Roy. Soc. A.*, 1964, **282,** 9.
17. Geil, P. H., *Polymer Single Crystals,* 1963, Wiley, New York.
18. Vaughan, A. S. and Bassett, D. C., submitted to *Polymer.*
19. Grubb, D. T., *J. Mat. Sci.*, 1974, **9,** 1715.
20. Grubb, D. T., Keller, A. and Groves, G. W., *J. Mat. Sci.*, 1972, **7,** 731.
21. Grubb, D. T. and Keller, A., *J. Mat. Sci.*, 1972, **7,** 822.
22. de Korosy, F. and Zeigerson, E., *J. Appl. Polym. Sci.*, 1967, **11**.
23. Kanig, G., *Kolloid Z.*, 1973, **251,** 782.
24. Kanig, G., *Kunststoffe,* 1974, **64,** 470.
25. Hodge, A. M. and Bassett, D. C., *J. Mat. Sci.*, 1977, **12,** 2065.
26. Gohil, R. M. and Phillips, P. J., *Polymer,* 1986, **27,** 1687.
27. Bassett, D. C. and Hodge, A. M., *Polymer,* 1978, **19,** 469.
28. Bassett, D. C., Olley, R. H. and al Raheil, I. A. M., submitted to *Polymer.*
29. Bashir, Z., Odell, J. A. and Keller, A., *J. Mat. Sci.*, 1984, **19,** 3713.
30. Bashir, Z., Hill, M. J. and Keller, A., *J. Mat. Sci. Lett.*, 1986, **5,** 876.
31. Palmer, R. P. and Cobbold, A. J., *Makromol. Chem.*, 1964, **74,** 174.
32. Attenburrow, G. E. and Bassett, D. C., *J. Mat. Sci.*, 1977, **12,** 192.
33. Bassett, D. C. and Vaughan, A. S., *Polymer,* 1985, **26,** 717.
34. Rybnikar, F., *J. Appl. Polym. Sci.*, 1985, **30,** 1949.
35. Rybnikar, F. in *Morphology of Polymers,* B. Sedlacek, Ed., 1986, de Gruyter, Berlin, p. 309.
36. Olley, R. H., Bassett, D. C. and Blundell, D. J., *Polymer,* 1986, **27,** 344.
37. Olley, R. H., *Sci. Progr. Oxf.*, 1986, **70,** 17.
38. Bassett, D. C. and Hodge, A. M., *Proc. Roy. Soc. A,* 1981, **377,** 25.
39. Bassett, D. C., Hodge, A. M. and Olley, R. H., *Proc. Roy. Soc. A,* 1981, **377,** 39.
40. Bassett, D. C. and Hodge, A. M., *Proc. Roy. Soc. A,* 1981, **377,** 61.
41. Anon. *Workshop Proceedings: Morphology of Polyethylene and Cross-linked Polyethylene,* 1981, Electric Power Research Institute, Palo Alto, California.
42. Naylor, K. L. and Phillips, P. J., *J. Polym. Sci., Polym. Phys. Ed.*, 1983, **21,** 2011.

43. Freedman, A. M., Bassett, D. C., Vaughan, A. S. and Olley, R. H., *Polymer,* 1986, **27,** 1163.
44. Bassett, D. C. and Olley, R. H., *Polymer,* 1984, **25,** 935.
45. Norton, D. R. and Keller, A., *Polymer,* 1985, **26,** 704.
46. Grubb, D. T. and Yoon, D. Y., *Polymer Comm.*, 1986, **27,** 84.
47. Freedman, A. M. and Bassett, D. C., paper presented at the Conference on Physical Aspects of Polymer Science, Reading, 1987.
48. Reneker, D. H. and Bolz, L. H., *J. Macromol. Sci.—Chem.*, 1976, **A10,** 599.
49. Bassett, D. C., in *Morphology of Polymers,* B. Sedlacek, Ed., 1986, de Gruyter, Berlin, p. 47.
50. Freedman, A. M., Bassett, D. C. and Olley, R. H., submitted to *J. Macromol. Sci.* (*Phys*).
51. Bassett, D. C., Block, S. and Piermarini, G. J., *J. Appl. Phys.*, 1974, **45,** 4146.
52. Bassett, D. C. and Turner, B., *Nature* (*Phys. Sci.*), 1972, **240,** 146.
53. Bassett, D. C. and Turner, B., *Phil. Mag.*, 1974, **29,** 925.
54. Bassett, D. C., in *Developments in Crystalline Polymers—1,* D. C. Bassett Ed., 1982, Applied Science, London.
55. Bassett, D. C., Khalifa, B. A. and Olley, R. H., *J. Polym. Sci.*, 1977, **15,** 995.
56. Bassett, D. C., *Polymer,* 1976, **17,** 460.
57. Rees, D. V. and Bassett, D. C., *J. Polym. Sci. Part A2,* **9,** 385.
58. Attenburrow, G. E. and Bassett, D. C., *J. Mat. Sci.*, 1979, **14,** 2679.
59. Yasuniwa, M. and Takemura, T., *Polymer,* 1974, **15,** 661.
60. Khalifa, B. A. and Bassett, D. C., *Polymer,* 1976, **17,** 291.
61. Bassett, D. C. and Carder, D. R., *Phil. Mag.*, 1973, **28,** 513.
62. Dlugosz, J., Fraser, G. V., Grubb, D. T., Keller, A., Odell, J. A. and Goggin, P. L., *Polymer,* 1976, **17,** 275.
63. Bassett, D. C. and Keller, A., *Phil. Mag.*, 1962, **7,** 1553.
64. Kawai, T. and Keller, A., *Polymer Lett.*, 1964, **2,** 333.
65. Blackadder, D. A., Keniry, J. S. and Richardson, M. J., *Polymer,* 1972, **13,** 584.
66. Bassett, D. C., Hodge, A. M. and Olley, R. H., to be published.
67. Vaughan, A. S., Bassett, D. C. and Olley, R. H., in *Morphology of Polymers,* B. Sedlacek, Ed., 1986, de Gruyter, Berlin, p. 387.
68. Vaughan, A. S. and Bassett, D. C., to be published.
69. Bassett, D. C., Olley, R. H. and al Raheil, I. A. M., submitted to *Polymer.*
70. Patel, D., Olley, R. H. and Bassett, D. C., paper presented at the Conference on Physical Aspects of Polymer Science, Reading, 1987.
71. Cotton, C., Bassett, D. C. and Olley, R. H., paper presented at the Conference on Physical Aspects of Polymer Science, Reading, 1987.
72. Olley, R. H. and Bassett, D. C., unpublished work.
73. Keith, H. D. and Padden, F. J., *J. Appl. Phys.*, 1963, **34,** 2409.
74. Blundell, D. J. and Osborn, B. N., *Polymer,* 1983, **24,** 953.
75. Cheng, S. Z., Cao, M. Y. and Wunderlich, B., *Macromolecules,* 1986, **19,** 1868.

76. PETERLIN, A., in *Flow-induced Crystallization in Polymer Systems,* R. L. Miller, Ed., 1979, Gordon and Breach, New York, p. 1.
77. BASSETT, D. C. and KHALIFA, B. A. *Polymer,* 1973, **14,** 390.
78. BASSETT, D. C. and KHALIFA, B. A., *Polymer,* 1976, **17,** 275.
79. SADLER, D. M., *Polymer,* 1987.
80. KEITH, H. D. *J. Polym. Sci.*, 1964, **A2,** 4339.
81. BASSETT, D. C., DAMMONT, F. R. and SALOVEY, R., *Polymer,* 1964, **5,** 579.
82. BASSETT, D. C., *Phil. Mag.*, 1964, **10,** 595.
83. AL RAHEIL, I. A. M., BASSETT, D. C. and OLLEY, R. H., paper presented at the Conference on Physical Aspects of Polymer Science, Reading, 1987.
84. KHOURY, F. and BARNES, J. D., *J. Res. NBS,* 1972, **76A,** 225.
85. PHILLIPS, P. J. and PHILPOT, R. J., *Polym. Comm.*, 1986, **27,** 307.
86. KELLER, A., in *Developments in Crystalline Polymers—1*, D. C. Bassett, Ed., 1982, Applied Science, London.
87. HIKMET, R. and KELLER, A., *Radiat. Phys. Chem.*, 1987, **29,** 275.
88. HAY, I. L. and KELLER, A., *J. Mat. Sci.,* 1967, **2,** 538.
89. VOKAL, A., BASSETT, D. C. and OLLEY, R. H., paper presented at the Conference on Physical Aspects of Polymer Science, Reading, 1987.

Chapter 3

ORIENTATION IN LIQUID CRYSTAL POLYMERS

G. R. MITCHELL
J. J. Thomson Physical Laboratory, University of Reading, UK
and
A. H. WINDLE
Department of Metallurgy and Materials Science, University of Cambridge, UK

SUMMARY

This chapter is concerned with the definition and evaluation of molecular anisotropy in liquid crystal polymers. Molecular orientation is a central theme of the liquid crystal state; however, its definition requires considerable care. The basic mathematical relationships for the description of molecular anisotropy are reviewed together with an assessment of the various 'orienting' units ranging from the molecular envelope to particular covalent bonds. The different methods of measuring orientation parameters for liquid crystal polymer samples are compared. Particular emphasis is placed upon X-ray and electron scattering procedures which are able to provide structural as well as orientational information. The experimental methods are highlighted through the use of several examples.

INTRODUCTION

The liquid crystal state is intermediate to crystal and liquid phases and displays molecular organizations in which orientational ordering is the dominant theme while the long-range positional ordering characteristic of a crystal is absent.[1] It is this orientation ordering which is successfully exploited in liquid crystal polymers, to give, on the one hand, high-modulus fibres and mouldings[2] and, on the other, electro-optical effects for information storage.[3,4] This chapter is concerned with the definition and evaluation of molecular orientation in liquid crystal polymers.

In many ways, the greatest problem in the study of anisotropy in non-crystalline polymer systems in general is its definition. For example, what is meant by 'molecular orientation' in a liquid crystal polymer in which just the side-groups are mesogenic? It may imply the anisotropy in the trajectory of the molecular backbone or it could mean the alignment of the side-chains alone. In this case both measures are needed for a full description of the anisotropic molecular organization, but each measure requires careful definition.

There is a wide range of appropriate experimental techniques available to probe orientation, each of which has a particular role in fully evaluating the details of the anisotropic molecular organization. We shall consider each of these methods, but our emphasis will be on the exploitation of X-ray scattering procedures, since these are able to provide both orientational parameters and direct structural information.[5–7]

The evaluation of molecular orientation in liquid crystal polymers is of course only one step in the process of understanding the origin and development of macroscopic anisotropy. The methods by which anisotropy is introduced or controlled will affect the types of measurements which it is possible to make. As previously indicated, the range of liquid crystal polymer materials is wide. Polymer chains formed from so-called rigid mesogenic units (for example **I**, Fig. 1) form liquid crystal phases at elevated temperatures. They respond very slowly to applied electric/magnetic fields, if at all.[8–10] They do show considerable promise as engineering materials in which the macroscopic molecular alignment is introduced mechanically.[11] The introduction of flexible coupling chains into such liquid crystal polymers (**II**, Fig. 1) reduces the temperatures at which the liquid crystal phases are observed, and considerably facilitates alignment using electric/magnetic fields,[12] although in general fibres may still be prepared. Side-chain liquid crystal polymers contain the rigid mesogenic group as a side-group of a flexible polymer backbone (**III**, Fig. 1). A variety of effects may be observed in these materials using electric/magnetic[13–15] and stress fields.[16,17]

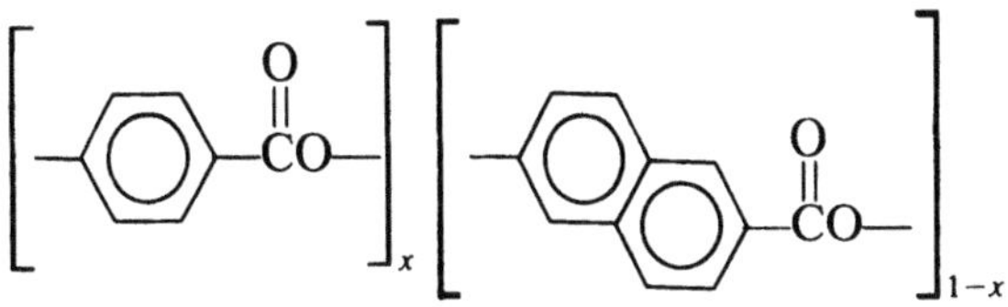

I

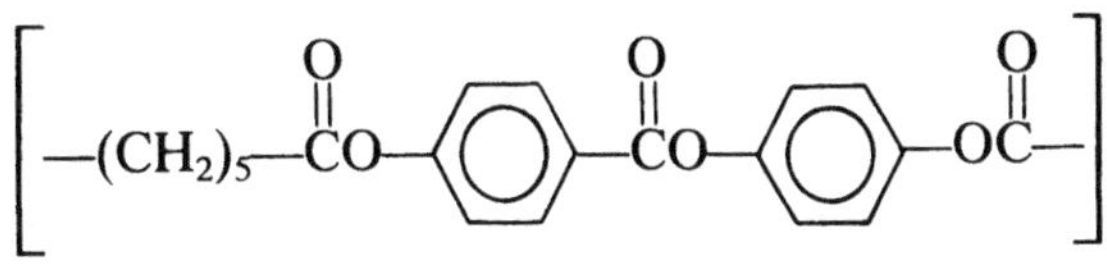

II

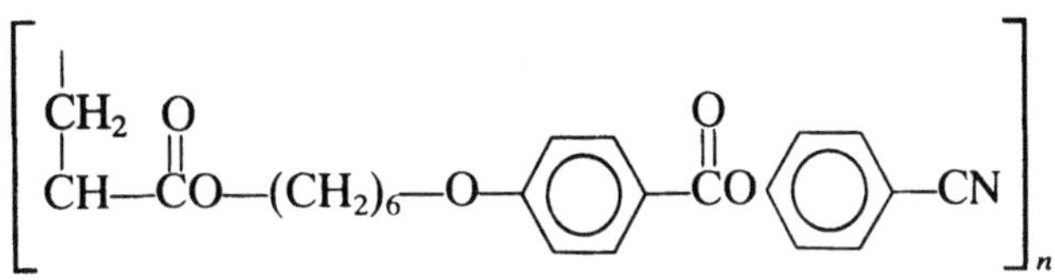

III

FIG. 1. Examples of three types of liquid crystal polymer molecule: **I**, chain made up from rigid aromatic units joined by relatively stiff ester groups; **II**, chain containing rigid units separated by flexible aliphatic sequences; **III**, chain with a flexible backbone but rigid side-groups.

We consider first the various approaches to the definition of molecular anisotropy, before reviewing the spectrum of experimental techniques available. The utilization of wide-angle X-ray scattering procedures for the quantitative measurement of molecular orientation is then treated in detail. This is followed by a discussion of methods which offer information as to the spatial variations of molecular orientation.

DESCRIBING ORIENTATION

Polymers which have been subjected to an external field, whether stress, flow, electric or magnetic, may show a preferred alignment of structural units either about, or normal to, the applied field direction. The quality of alignment achieved is generally not perfect in relation to the field axis. When the probability distribution of unit orientations is symmetrical about any axis, then it can be referred to that axis without the necessity of an external reference frame. Such an axis is known as the director. We shall restrict ourselves initially to situations where this grouping occurs symmetrically about only one axis, in other words a uniaxial system.

How are we to describe this distribution of orientations? Let us consider the angular distribution of the structural units shown in Fig. 2, in which the angle the axis† of the structural unit makes with the director is defined as α. It is possible to plot the angular distribution of these units in polar coordinates (r, α), where the length parameter r is proportional to the number of units at the angle α. Of course, for the example of Fig. 2 the plot would be discontinuous as the number of units is small. However, for a much larger model the surface is continuous and represents the probability of angular distribution $D(\alpha)$. Since the distribution concerned has cylindrical symmetry (we only describe the orientation with a single angle, α), the surface appears as a two-dimensional section; an example is plotted in Fig. 3. If there was no preferred orientation of the structural units, the surface would be spherical.

The shape of such a surface may be described in a mathematical way which has several important advantages. The shape is analysed into harmonic components functions much as one might analyse a wave into a Fourier series. The spherical harmonics, as these components are called, have the necessary property of orthogonality; however, their form is more complicated than the $\cos(nx)$ type of component of a Fourier series for a plane wave. The spherical harmonics are functions of the polar coordinate angle, referred to the director axis. The first four components, abbreviated P_0, $P_2(\cos\alpha)$, $P_4(\cos\alpha)$ and $P_6(\cos\alpha)$ or simply P_0, P_2, P_4, P_6, are defined below and drawn in Fig. 4.

$$P_0 = 1$$

$$P_2 = (3\cos^2\alpha - 1)/2$$

$$P_4 = (35\cos^4\alpha - 30\cos^2\alpha + 3)/8$$

$$P_6 = (231\cos^6\alpha - 315\cos^4\alpha + 105\cos^2\alpha - 5)/16$$

† The assignment of an axis to the structural unit is clear in this example. However, for a 'real' unit the situation is much more complicated, as discussed by Frank.[18] We shall consider this issue more fully later.

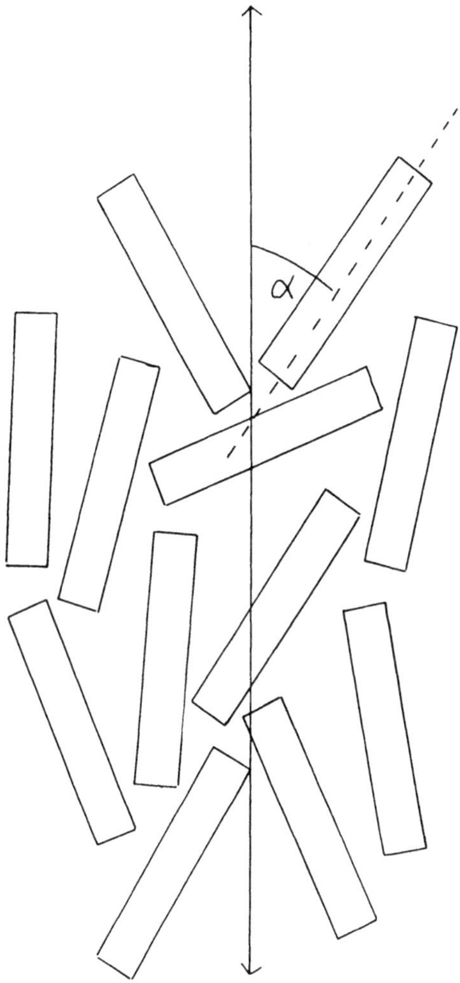

FIG. 2. Anisotropic units with a preferred orientation about a symmetry axis which is known as the director.

These functions are in fact the even members of a Legendre polynomial series. Any surface which encloses a volume of rotation about the axis and has a central symmetry plane normal to this axis may be described by a summation of the component functions each with an appropriate amplitude, written $\langle P_{2n}(\cos\alpha)\rangle$. Loss of the central symmetry plane would then require the odd components of the

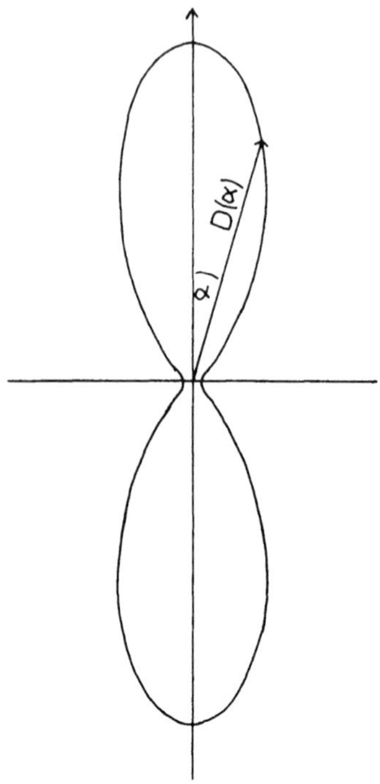

FIG. 3. Section through a polar probability plot of a uniaxial orientation distribution.

series. If the surface is described as $D(\alpha)$, its synthesis can be written:

$$D(\alpha) = \sum_{n=0}^{\infty} (4n+1)\langle P_{2n}(\cos\alpha)\rangle P_{2n}(\cos\alpha) \tag{1}$$

Normally the amplitudes of the harmonics decrease with increasing order, and when the preferred orientation in the system is low, the higher-order terms will be small. However, this need not be the case for liquid crystal polymers where the degree of preferred orientation is often pronounced, for example $\langle P_2(\cos\alpha)\rangle$ may reach ~0·8 and more than a dozen terms may be required adequately to describe $D(\alpha)$.[6]

The inversion of eqn (1) is:

$$\langle P_{2n}(\cos\alpha)\rangle = \int_0^{\pi/2} D(\alpha)P_{2n}(\cos\alpha)\sin\alpha\,\mathrm{d}\alpha \tag{2}$$

and enables the harmonic amplitudes or coefficients to be determined

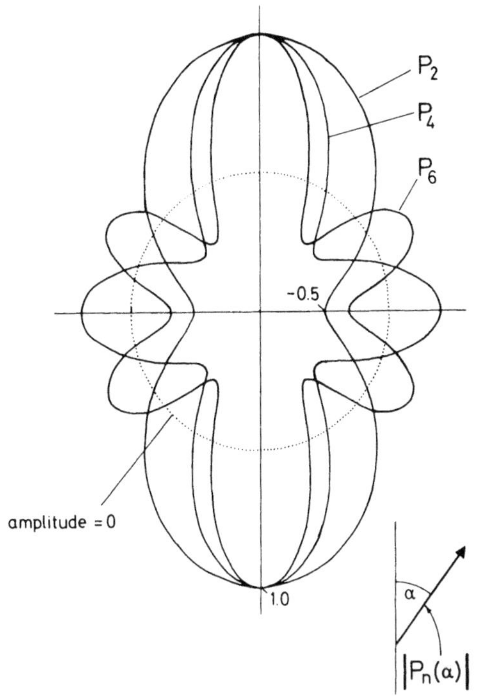

FIG. 4. Sections through the even-ordered spherical harmonic functions, P_0, P_2, P_4 and P_6, which are in effect Fourier-type components of orientation distributions such as that shown in Fig. 3.[86]

from $D(\alpha)$. There are several ways in which $D(\alpha)$ may be normalized. We have selected the definition that:

$$\int_0^{\pi/2} D(\alpha) \sin \alpha \, d\alpha = \pi/2 \tag{3}$$

since that allows us to use normalized Legendre polynomials, i.e. $P_{2n}(\cos 0) = 1$. If the condition expressed in eqn (3) is not correct, then eqn (2) should be normalized accordingly.

The most valuable aspect of expressing the quality of orientation as the amplitudes of a series of spherical harmonic functions can be seen by considering Fig. 5(a). There are rod elements OA which have a preferred orientation about the axis OB described by the harmonic coefficients $\langle P_{2n}^{AB}(\cos \psi) \rangle$. Additionally, the axis OB is itself distributed about the overall symmetry axis OC as described by $\langle P_{2n}^{BC}(\cos \phi) \rangle$. The distribution of the units OA about OC represents

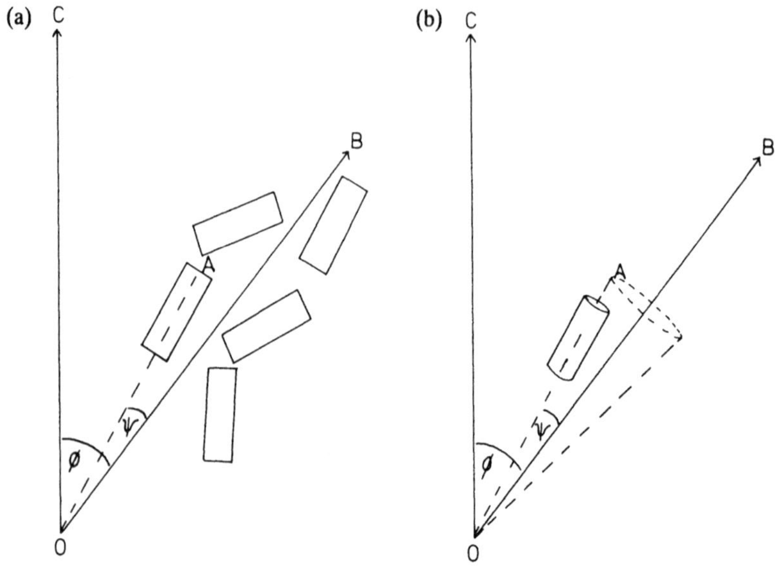

FIG. 5. Diagrams to illustrate the convolution of orientation distributions which may occur in oriented polymers. In (a) there is a distribution of units about one axis, B, which in turn is distributed about a second axis C. In (b) the units are oriented as a cone at a fixed angle to axis B which is again axially distributed about C.

the convolution of the two orientation distributions. A property of the spherical harmonics is that the coefficients of the total distribution are simply the products of the coefficients of the two contributing distributions. Hence:

$$\langle P_{2n}^{\mathrm{AC}}(\cos\alpha)\rangle = \langle P_{2n}^{\mathrm{AB}}(\cos\psi)\rangle \,.\, \langle P_{2n}^{\mathrm{BC}}(\cos\phi)\rangle \tag{4}$$

This property is equivalent to that of the convolution theorem, which is central to diffraction theory.

In Fig. 5(b), we consider another possible situation. Here the rod element OA is at a fixed angle ψ to OB, although distributed uniformly around the cone defined by that angle. In this case the coefficients to describe OA about OB are simply the amplitudes of the harmonic function at ψ; hence

$$\langle P_{2n}^{\mathrm{AB}}(\cos\psi)\rangle = P_{2n}(\cos\psi) \tag{5}$$

and for the total distribution function:

$$\langle P_{2n}^{\mathrm{AC}}(\cos\alpha)\rangle = \langle P_{2n}^{\mathrm{BC}}(\cos\phi)\rangle \,.\, P_{2n}(\psi) \tag{6}$$

As will be seen later, this aspect of orientation description is important in some experimental techniques, where for example OA represents a transition moment at an angle ψ to the molecular chain axis OB which has a preferred orientation with respect to the director OC.

The mathematics of spherical harmonics is an accepted tool in many scientific disciplines and is treated in several classic texts.[19,20] In 1939 the method was applied to orientation in materials by Hermans and Platzek,[21] who used just $\langle P_2 \rangle$, which is often referred to as the Hermans orientation function. Spherical harmonics were applied to liquid crystals in the theoretical work of Maier and Saupe,[22] who again emphasized only $\langle P_2 \rangle$. However, they called it the order parameter and designated it S, setting a nomenclature which is now standard in liquid crystal studies.

THE ORIENTING UNITS

Rod-like Molecules

For many purposes small mesogenic molecules may be approximated to short axially symmetric rods. The axes of such rods are the simple basis for defining their orientation with respect to some director. Molecules are of course not simple rods and some techniques will provide values for the orientation of particular interatomic vectors with the molecule. Such a vector will not necessarily be parallel to the longest axis of the molecule. For example, consider the selectively deuterated molecule (**IV**).[23] It is possible using NMR techniques to

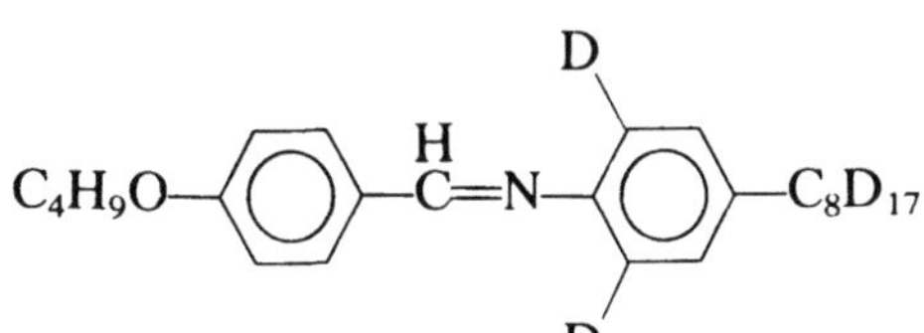

IV

determine the orientation of the C–D bonds on the phenyl rings, and yet there will still be uncertainty as to where the molecular inertial (long) axis should be drawn. The angle between the C–D bond and the chosen molecular axis corresponds to the angle ψ in eqn. (6) and it is needed if measurements of the angular distribution of C–D bonds are to yield the orientational distribution of the long axis. It follows from the shape of the $P_2(\cos\alpha)$ function that any uncertainty in determining the angle ψ will have a minimum influence on the

$\langle P_2(\cos\alpha)\rangle$ values for the molecular axis when ψ is close to 0 or 90°. Of course any flexibility within the molecules will lead to further uncertainty in ψ and independent structural information may be necessary in some cases. However, by chosing three orthogonal bond vectors or transition moments for the same rigid structural unit, both the angles those directions make with the molecular axis, and the orientation of the long axis itself, may be determined at the same time.

Polymer Chains

If a polymer molecule is totally rigid then there would be little doubt as to the choice of chain axis to which to refer the orientation. However, no mesogenic polymer molecules are completely rigid, and in many cases they are surprisingly flexible, and even in a liquid crystalline phase may not be completely straight. The issues involved in describing the orientations of a polymer chain are best understood by considering first a fairly flexible one. In a conventional polymer which has been mechanically extended along one axis, a molecule might look like Fig. 6. The molecule has been 'stretched out' in a vertical direction and the end-to-end vectors of all the similar molecules will show preferred orientation about the extension axis. Preferred orientation involving whole chains can be measured by small-angle neutron scattering from deuterated/undeuterated mixtures. Such experiments give information about the anisotropy and orientation of the molecular envelope and not about the details of the chain itself.[24] However, if we are concerned with the orientation of the more localized regions of the chain, such as that emphasized as 'A' in Fig. 6, then it is apparent that the orientation of the segments about the external extension axis will be less than that of the end-to-end vectors. In fact the orientation distribution of such local segments will be the convolution of the orientation of the segments about the end-to-end vector (i.e. with respect to ψ), with the orientation of those end-to-end vectors about the external axis (i.e. with respect to ϕ). From eqn (6) we may express this more simply as the product of their respective spherical harmonic coefficients. If the length of the segmental unit is increased, as for example 'B' in Fig. 6, then the orientation distribution of those segments about the end-to-end vector of the chain is sharpened and hence the overall distribution of the chain segments with respect to the external axis is enhanced. In other words, by changing the scale of our structural unit we alter the apparent level of orientation. To emphasize this important point

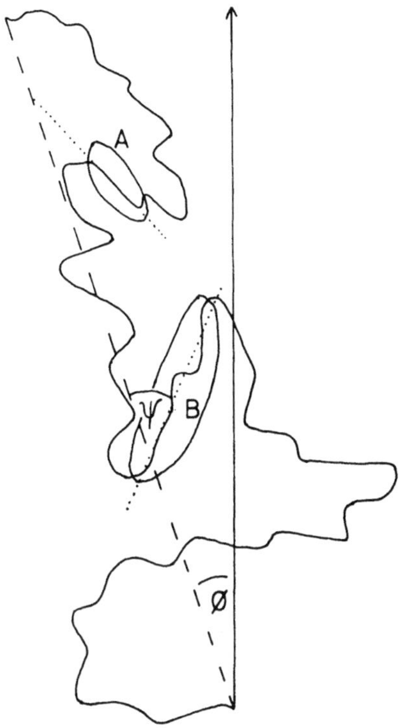

FIG. 6. A schematic diagram of a molecule of a flexible-chain polymer stretched in the vertical direction. The annotated regions of the chain are referred to in the text.

further, the orientation of particular covalent bonds along the chain referred to the external axis will be less than for a chain segment. Furthermore, there may be additional freedom for a side-group such that the level of orientation of bonds within that group may be negligible even though the polymer backbone may be highly aligned. Thus any orientation measurement for a polymer system must be qualified by a clear indication of the scale with which that measurement corresponds.

Figure 6 depicts a flexible molecule which has been mechanically stretched, illustrating the case where the molecule is coupled to the stress field through its ends, and the quality of orientation decays as the information is passed down the chain to the local segments and individual chemical groups or bonds. In addition there can be an element of orienting influence through the molecular environment of a particular anisotropic segment, which can lead to orientation without directly involving the chain ends.[25–27] Indeed, the stress field may couple directly to the chain segments, as has been observed in liquid crystal elastomers.[17,28] In essence, similar phenomena occur for liquid crystal polymer molecules containing electroactive groups. Segments widely separated along the chain may be oriented independently by suitable electric or magnetic fields.[13–15]

In considering orientation in polymers, it is important to define quite precisely what part of the molecule is under consideration. 'Molecular orientation' and 'chain orientation' are meaningless phrases in the general context. The choice of orienting unit is often determined naturally by the type of technique being used for measurement; infrared (IR) spectroscopy will give the orientation of particular bond types along the chain, wide-angle X-ray scattering that of a local region of chain and small-angle neutron scattering that of the molecular envelope. Liquid crystal polymers where mesogenic groups are either connected end to end via flexible spacers (**II**, Fig. 1) or attached through flexible links to a polymer backbone (**III**, Fig. 1) probably present the least difficulty in this respect, as the mesogenic unit will usually be the orienting unit of most interest. In polymers such as aromatic polyesters (**I**, Fig. 1), with so-called semi-rigid chains, the selection of orienting unit is not so clear-cut. However, the definition of local chain segments in terms of the region which contributes to coherent X-ray scattering is particularly useful and will be described in detail later.

The Question of Scale

The discussion in the previous section was based on the existence of some single external axis, typically that of an applied field, to which the preferred orientation of the various units was referred. The mutual orientational ordering characteristic of a liquid crystal will, if present uniformly throughout a sample, define an external director in the absence of any applied field, rather as a large single crystal of a mineral will show external symmetry. Consider, however, a liquid crystal polymer such as that giving the Schlieren texture of Fig. 7 in which there is preferred orientation on a local scale as indicated by the presence of birefringence, but a variation in director orientation across the field of view as shown by the obvious texture. Thus, within a volume much smaller than the texture there will be a characteristic value of $\langle P_2(\cos\alpha)\rangle$, if indeed it is measurable. If the volume of

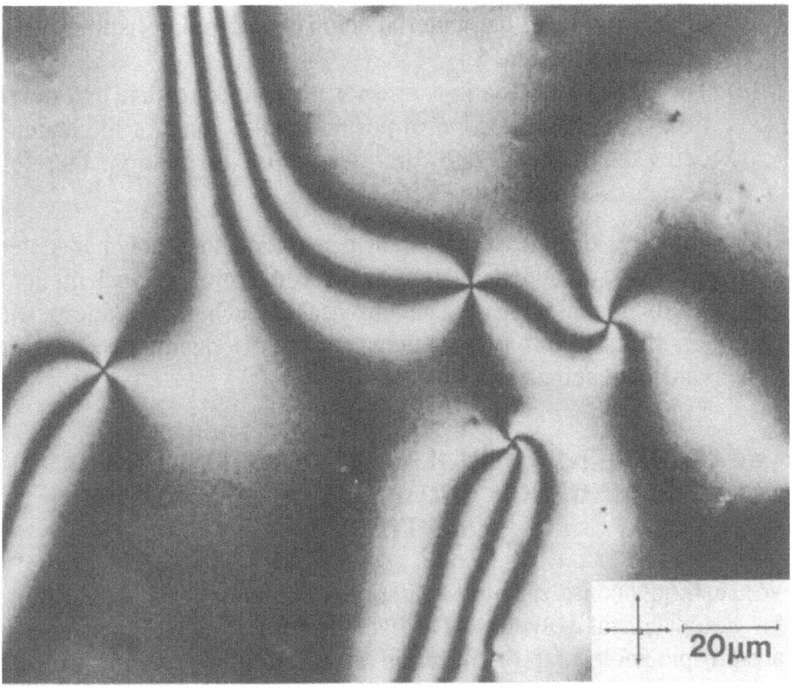

FIG. 7. A texture of a main-chain liquid crystal polymer, **X**, as seen in transmitted light between crossed polars. By analogy with small molecule liquid crystal parlance, it is known as a Schlieren texture. The fine granular nature of the background is caused by crystallization on cooling; otherwise the microstructure is unchanged from that in the liquid crystal state. (Photograph courtesy of T. J. Lemmon, Department of Materials Science and Metallurgy, University of Cambridge.)

interest is enlarged the director may change in orientation but, much more significantly, the degree in orientational order will decrease, until for sufficiently large volumes the value of $\langle P_2(\cos\alpha)\rangle$ will fall to zero. Orientational parameters in general are therefore extensive properties, and hardly characteristic parameters of the material.

Thus in liquid crystal materials there are three distinct conditions which are relevant to the measurement of orientation. They have been identified by Zannoni[29] as the following.

(a) The external director is determined by a field which may influence the magnitude of orientational order. This latter qualification is only of relevance to stress or flow fields.

(b) The orientation of the director is arbitrary, but it is independent of the volume of the material being considered, as is the level of orientation $\langle P_2(\cos \alpha) \rangle$.
(c) The director fluctuations are not negligible, so that its orientation varies as a function of position and volume of the material being considered, and the degree of orientation therefore decreases as the sample size increases.

In summary, the measurement of orientation in a polymer system must always be qualified by statements concerning the orientational unit and its spatial extent. The absence of such qualifications will, except in the simplest of situations, make any quantitative orientational values meaningless.

MEASUREMENT OF ORIENTATION BY SPECTROSCOPY, BIREFRINGENCE AND SMALL ANGLE NEUTRON SCATTERING

We shall now consider how orientational parameters may be measured for liquid crystal polymers. The fact that these materials possess an anisotropic molecular organization means that almost any property will be directional. These would include mechanical properties, thermal expansion, conductivity, etc.; for example, Frenzel and Rehage[30] have used volume changes to deduce an order parameter versus temperature plot using a model based upon the theory of Maier and Saupe. However, until such time as these properties have a reliable and general basis linked to molecular anisotropy, we must rely on more specific properties and phenomena directly related to the underlying molecular organization. In this section we will consider methods other than X-ray and electron scattering.

Absorption Spectroscopy

In principle, one of the simplest methods for the determination of the degree of preferred axial orientation is based on the anisotropic absorption properties for electromagnetic radiation of various wavelengths. The parameter measured in such an experiment is the dichroic ratio R, which is given by:

$$R = A_{\parallel}/A_{\perp} \tag{7}$$

where $A_{\parallel}$ and $A_{\perp}$ are the absorption coefficients for plane polarized radiation with the electric field parallel and normal to the director axis, respectively. Infrared (IR) and ultraviolet (UV) spectra of liquid crystal polymers are rich in absorption bands, which may however be rather broad. Under certain circumstances it is possible to assign these absorption peaks uniquely and hence measure an orientation parameter for a particular bond and excitation. If the angle between the vibration transition moment of the absorbing unit and the local chain axis, ψ, is 0 or 90°, the orientation function $\langle P_2(\cos\alpha)\rangle$ is given by:

$$\langle P_2(\cos\alpha)\rangle = (R-1)/(R+2) \quad (\psi = 0°) \tag{8}$$

and

$$\langle P_2(\cos\alpha)\rangle = 2(1-R)/(2+R) \quad (\psi = 90°) \tag{9}$$

and in general when $\psi \neq 0°$, then from eqn (6)

$$\langle P_2(\cos\alpha)\rangle = \{1/P_2(\cos\psi)\}(R-1)/(R+2) \tag{10}$$

Thus we need both to identify the unit responsible for the absorption band employed and to make assumptions about the local chain conformation to determine the value of ψ. It is sometimes possible, by making a number of dichroic measurements using an orthogonal set of vibration moments, to avoid the latter requirement.

Following earlier work on low molar mass liquid crystal systems,[31] the IR region of the electromagnetic spectrum has been the most popular. For example, Noel *et al.*[32] have used it to measure the orientation parameter for mechanically aligned samples of the random copolymer **V** as a function of temperature. They employed measure-

V

ments of two IR bands (1605 cm^{-1} and 1508 cm^{-1}) which were well defined and associated with vibrations of the aromatic rings. They set ψ to zero and their results, shown in Fig. 8, illustrate a rapid fall in the orientation parameter $\langle P_2(\cos\alpha)\rangle$ with increasing temperature up to the clearing point where liquid crystalline order is lost completely. The values were in broad agreement with a measurement made using wide-angle X-ray scattering.

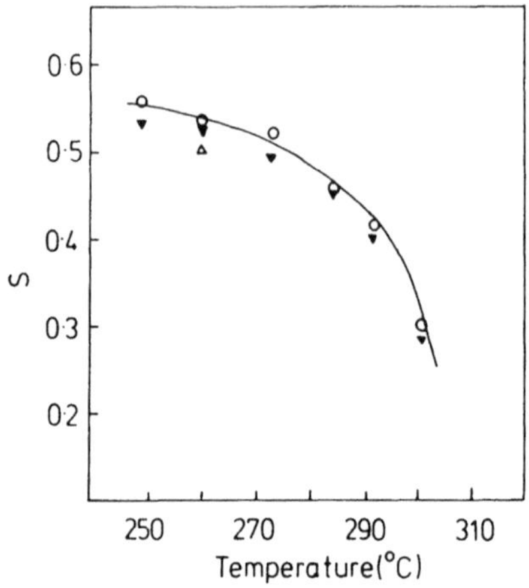

FIG. 8. Temperature dependence of the degree of preferred orientation by IR spectroscopy in a liquid crystal random copolyester (**V**). The orientation is explained as the order parameter, $S = \langle P_2(\cos\alpha)\rangle$.[32] ○, Oriented samples in the nematic state; ▼, oriented samples quenched to the solid from the nematic state; △, X-ray measurement for comparison made at 260°C.

The same principle, but applied to UV spectra, has been exploited for measuring orientation in liquid crystal polymers by groups in Germany and the USSR.[33,34]

Absorption of visible light is not normally strong for polymer molecules but the general effect is exploited through the addition of anisotropic dye molecules which, when dissolved in the polymer mesophase as probes, tend to adopt the same degree of orientational perfection as the host phase. The relationship between the orientation of the two components is of particular concern to those with an interest in molecular models of the liquid crystal phase.[35] However, in many cases the dyes are added for technological application purposes rather than for scientific investigation. This dye approach has been used by Finklemann and Rehage[36] to measure the orientation parameter as a function of temperature for a siloxane-based side-chain liquid crystal polymer system. The orientation of the probe molecule appears to mirror accurately that of the mesogenic component of the polymer phase as measured by another technique. It is of course possible to incorporate chemically the dye molecule into the polymer system. Measurements on a polyacrylate side-chain liquid crystal polymer system have been made by Benthack *et al.*[37] They found order parameters for the dye unit attached to the polymer backbone to be significantly lower than observed for the dye as a solute. It was reasoned that the bulky dye unit when coupled directly to the polymer backbone was more disruptive than when unattached. This type of dual measurement may be of considerable value in building up a complete structure for both the mesogenic units and the polymer backbone.

Absorption measurements in the visible/UV region have been employed widely in the study of liquid crystal polymers displaying cholesteric phases.[36,38] In that case 'absorption' occurs through the selective scattering of light due to the regularly twisted structure.

Probe techniques are also used with small molecules which have characteristics which enable their orientation to be measured by electron spin resonance (ESR).[39,40] By using a computer simulation of the angle-dependent spectra, Wassmer and colleagues[40] were able to obtain both orientational parameters and correlation times for the spin probe. Using magnetically and electrically aligned polyacrylate-based side-chain liquid crystal polymers, they were able to show considerable differences for spin probes dissolved in polymers with different coupling chain length.

These techniques of electromagnetic absorption spectrometry readily give the orientation parameter $\langle P_2(\cos\alpha)\rangle$ with respect to external axes. Combined with the appropriate form of microscopy the methods can, in principle, give orientational information on the scale of the microstructure.[41] However, the absorption property is a second-rank tensor, which means that it is only able to give the orientational information contained in the first spherical harmonic component, $\langle P_2(\cos\alpha)\rangle$. The higher-order components and hence the full orientation function are inaccessible. The techniques do have the advantage of speed of measurement and provide a real-time evaluation procedure.

Optical Birefringence

Birefringence is the difference between two principal refractive indices of the polymer. Where it is optically uniaxial, the principal axes are normal to and perpendicular to the director. Of course, some smectic phases are optically biaxial as are some interesting biaxial nematic polymers.[42,43] It is important to appreciate that birefringence is an optical property of a representative volume of the polymer with dimensions at least those of the wavelength of light. It is not a molecular property but is directly proportional to $\langle P_2(\cos\alpha)\rangle$, at least in the uniaxial case. In one sense it is a ready measure of orientation, but the constant of proportionality is not easily determined. If the birefringence of the material when perfectly orientated, Δn_0 (sometimes known as the intrinsic birefringence), is available, then

$$\langle P_2(\cos\alpha)\rangle = \Delta n/\Delta n_0 \tag{11}$$

The calculation of the intrinsic birefringence for a given molecular organization is almost impossible at present. It involves both molecular and macroscopic polarizability and the form of the Lorenz–Lorentz or internal fields relation relevant to an anisotropic molecular organization.[44] The orientation parameter is given by:

$$\langle P_2(\cos\alpha)\rangle = \frac{\frac{1}{3}(\alpha_e + 2\alpha_o)}{(\alpha_{\parallel} - \alpha_{\perp})}\,\frac{(n_e^2 - n_o^2)}{\frac{1}{3}(n_e^2 + 2n_o^2)^{-1}} \tag{12}$$

where the subscripts 'e' and 'o' refer to the principal axes of the optical continuum, and $\alpha_{\parallel}$ and $\alpha_{\perp}$ are the molecular polarizabilities referred to the principal molecular axes for this property. However, access to the value for polarizability is not at all straightforward, and Finklemann and Rehage[36] have used the technique of Haller *et al.*[45] to estimate the value of the first term on the right-hand side of eqn (12). Basically it depends on the independent knowledge that $\langle P_2(\cos\alpha)\rangle$ of a liquid crystal phase depends upon the temperature according to $-\log(1 - T/T_{n-i})$, where T_{n-i} is the nematic to isotropic transitions temperature. It is then found that

$$\langle P_2(\cos\alpha)\rangle = \left\{\frac{(n_e^2 - n_o^2)}{\frac{1}{3}(n_e^2 + 2n_o^2)^{-1}}\right\}^A \tag{13}$$

where the power A can be determined from log–log plots. The method may be easily extended to cholesteric phases.[36] Other approaches to this problem have been established and the reader is directed to them.[46,47]

The simplicity of measurements of birefringence hide their physical complexity. It may be beneficial to obtain the intrinsic value (eqn (11)) using measurements coupled to an experimentally more complicated method of orientation measurement, and then to use birefringence as a secondary technique. As with absorption methods, refractive index measurements can only give $\langle P_2(\cos\alpha)\rangle$ but not the higher-order components of the orientation distribution.

Nuclear Magnetic Resonance

Nuclear magnetic resonance (NMR) is a technique of considerable versatility in polymer science. It is used universally as a probe of chemical configurations, it provides information on the dynamics and relaxation times of a polymer system and it offers a route to the determination of orientation parameters, the exact route depending on the particular nuclei employed. In principle quadrupolar, dipolar, shielding tensor and indirect spin coupling interactions can all be employed;[48] however, in practice only the first two have any universal appeal. Dipolar coupling using proton NMR offers the simplest approach in terms of material preparation and will be considered first.

The proton NMR signal observed for spin–spin dipolar coupling splits into two components for any anisotropic distribution of that dipolar interaction. In the fast motion limit the splitting is given by:

$$\Delta\nu = \left(\frac{3}{2\pi}\right)(\gamma^2 h)\langle P_2(\cos\alpha)\rangle / r_{ij}^3 \tag{14}$$

where γ = gyromagnetic ratio and h = Planck constant. Thus we measure $\Delta\nu$, and from a knowledge of the values of r_{ij}^3, the distance between protons i and j, determine values of $\langle P_2(\cos\alpha)\rangle$. In the solid phase the NMR peaks are considerably broadened due to the restricted molecular motion, and anisotropy is most easily handled in the rigid lattice limit.[49] However, the high mobility of the liquid crystal phase means that eqn (14) can be used. Real molecular systems will contain many protons and therefore present complicated spectra. In some systems, notably co-polyesters, individual aromatic ring spin systems may be considered isolated[50,51] and hence interpretation is simplified. Simplification may also be achieved by selective deuteration, deuterons having negligible dipolar coupling. The orientation parameters measured of course apply to the proton–proton vector. A knowledge of intrachain geometry is required to determine the angular relationship between this vector and the chain axis, and hence the molecular orientation parameter using eqn (6). It is common practice to employ the dominant interaction between *ortho* and *meta* protons on each aromatic ring, for which the orientation axis and dipolar vector may be considered coincident. Intrachain geometrical information is also required to give r_{ij}^3, so any molecular flexibility or conformational freedom adds to the problem.

Proton NMR has been extensively utilized in the study of orientational ordering in main-chain flexible liquid crystal polymer systems.[52–61] Martins *et al.*[59] have employed proton NMR to measure the order parameter as a function of temperature for a polyester of the form **VI**. This system aligns readily in the magnetic field of the NMR

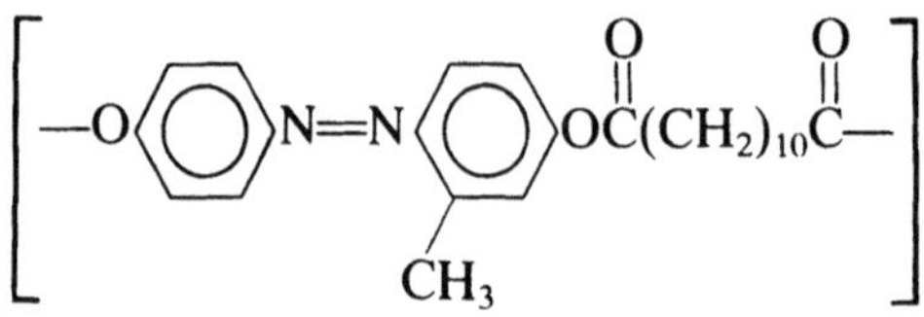

VI

spectrometer. The similarity of the rigid component in the polymer to the classic low molar mass liquid crystal *P*-azoxyanisole (PAA) considerably facilitated the analysis. By simulating the NMR spectra for the set of 'independent' spin subsystems, they were able not only to extract information as to the orientation parameter of the rigid core, but also details of the spacer conformation. They concluded that the 'flexible' spacer takes up extended chain conformations with orientational parameters comparable with the mesogenic core. We reproduce in Fig. 9 their data for the orientation parameter $\langle P_2(\cos\alpha)\rangle$ of the polyester as a function of temperature, compared with that for PAA. It is clear that the orientational order in the polymer system is significantly higher than for the low molar mass analogue; moreover, the temperature dependence on approaching the

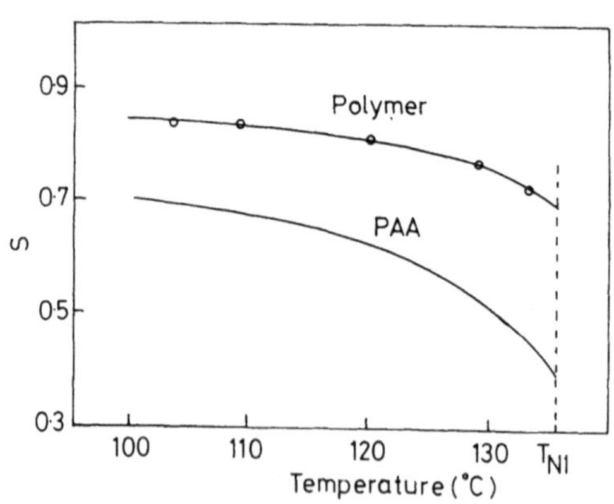

FIG. 9. The preferred orientation, $S = \langle P_2(\cos\alpha)\rangle$ measured as a function of temperature by proton NMR. The data are for polymer **VI** and are compared with results from the small molecule liquid crystal PAA (adapted from ref. 59).

nematic–isotropic transition is reduced. This indicates some measure of chain-induced stiffening which is not altogether surprising.[62]

One of the problems in the experiment just described was the complexity of the NMR signal, where many dipolar interactions overlapped in terms of frequency. Bruckner *et al.*[61] reduced these effects by working with a semi-deuterated polymer system, **VII**. The

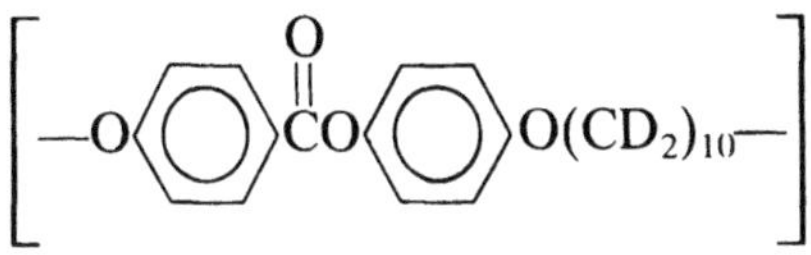

VII

proton NMR spectra are dominated by the neighbouring protons on the aromatic rings. Bruckner *et al.* assumed that this inter-proton vector would be parallel to the chain extension axis and used eqn (14) directly. Their results, which are reproduced in Fig. 10, show an almost linear relationship between $\langle P_2(\cos\alpha)\rangle$ and temperature, which supports the chain 'stiffening' concept. We shall return to this example when considering deuterium NMR, but it is appropriate to emphasize that from the same study the authors measured the ratio of phases in a nematic/isotropic biphasic regime in addition to orienta-

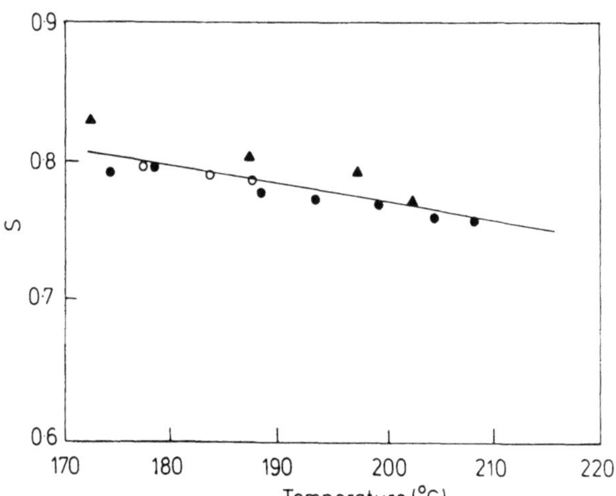

FIG. 10. Data for the orientation parameters $S = \langle P_2(\cos\alpha)\rangle$ as a function of temperature for molecule (**VII**) obtained from deuterium NMR (● and ○) compared with values from the same polymer obtained with proton NMR (▲).[61] Only flexible sequences in the molecule were deuterated.

tion measurements: such is the adaptability of NMR techniques. Martin *et al.*[63] used proton NMR to study the effect of surface preparation on the speed of alignment of thermotropic co-polyester with flexible coupling chains. In these experiments the magnetic field of the spectrometer itself (4·7 T) was used to induce macroscopic molecular alignment. The use of micro-grooved sample tubes not only enhanced the rate of alignment but also the magnitude of the maximum orientation, the latter increasing by a factor of two.

Proton NMR has also been used in the study of orientation in side-chain liquid crystal polymers.[64] A value of $\langle P_2(\cos\alpha)\rangle$ of 0·45 was measured for the nematic phase of one acrylate-based system.

Deuterium NMR is a quadrupole interaction, sensitive to anisotropy arising through the coupling between the nuclear quadrupole moment of the deuteron and the local electric field gradient. It has proved to be a particularly powerful tool in the study of both low molar mass liquid crystals[48,65] and polymeric mesophases.[61,66–69] The only real disadvantage, apart from the capital expense of the NMR spectrometer, is that the technique requires the preparation of special deuterated samples. However, selective deuteration enables non-equivalent sites to be labelled, and the order parameter of individual C–D bonds determined. For rapid motion around the long molecular axis (i.e. correlation times $<10^{-7}$ s) the splitting due to anisotropy is given by:

$$\Delta\nu = \frac{3e^2qQ}{2h}\langle P_2(\cos\alpha)\rangle \qquad (15)$$

where $(3e^2qQ/2h)$, the quadrupole coupling constant, equals 168 kHz. Thus if the angle the C–D bond makes with the molecular axis is known, then the value of $\langle P_2(\cos\alpha)\rangle$ for the molecular orientation may be evaluated using eqn (6). Alternatively if the orientation parameter is known from other experiments, perhaps from proton NMR, then the splitting may be used to provide mobility information relating to the C–D bond. Invariably, ψ, the angle between the C–D vector and the molecular axis, is far removed from 0° or 90° and hence errors in comparison with proton NMR may be greater. Interpretation of spectra is more cumbersome for samples in which slow motions preclude the use of eqn (15).[70] It has, however, formed the basis of a study of the molecular motion and orientation in a side-chain liquid crystal polymer based upon an acrylate backbone, in which one of the aromatic rings of the mesogenic core had been selectively deuterated.[71]

Reference was made previously to a proton NMR study which gave the data of Fig. 9. Samulski *et al.*[67] studied the same polyester but with the flexible coupling chain deuterated as well. They showed that the coupling chain had enhanced orientational ordering compared with that displayed by a pendant chain of a low molar mass analogue. However, the fact that several different quadrupolar splittings were observed showed that the methylene units were not identical, with the implication that the coupling chain conformation was not restricted to the all-*trans* case. A more recent study by Bruckner *et al.*[61] was also mentioned earlier. Proton NMR was used to provide a measure of the mesogenic core, while considerable conformational and orientational data concerning the deuterated coupling chain were obtained using deuterium NMR. Whilst there are five CD_2 groups which are non-identical in this system, the experimental spectra show only one pair of peaks, which suggests a highly extended chain conformation, with all CD_2 groups in similar conformations. These general conclusions are substantiated using predicted chain conformations from steric energy calculations. Orientational parameters derived from the deuterium NMR spectra agree well with those obtained using proton NMR (Fig. 10), which provides further evidence of the high level of conformational ordering displayed by the coupling chains in the nematic phase. These results are supported by NMR experiments from Mueller *et al.*[68,69]

It is appropriate to emphasize that in each of the NMR experiments described, the authors were also able to extract information relating to two-phase structures, correlation times and conformational isomer populations.

Small-Angle Neutron Scattering

Small-angle neutron scattering may be used to measure the size of the molecular envelope, which is usually expressed as the radius of gyration of the molecule. This is achieved by dissolving partially deuterated polymers in undeuterated material. Experiments of this type have been reported by Kirste and Ohm[72] and by Keller *et al.*[73] Both studied methacrylate-based side-chain liquid crystal polymers in which only the backbone was deuterated. Kirste and Ohm[72] used the

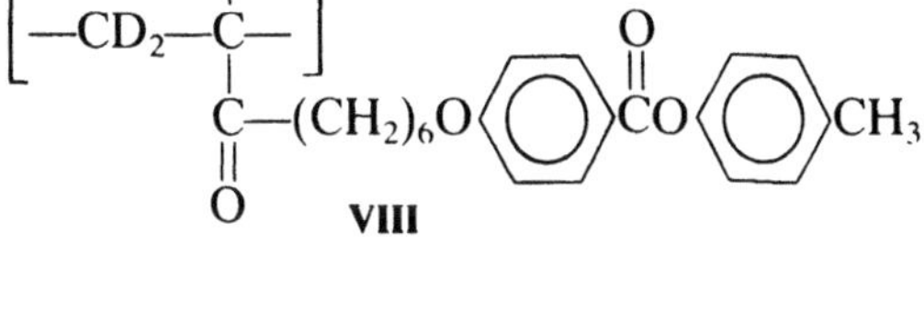

VIII

molecule **VIII**, whilst those studied by Keller *et al.* differed only in the terminal group of the mesogenic unit. Kirste and Ohm[72] found that the radius of gyration of the polymer backbone, which is the parameter obtained when the side-groups are undeuterated, was the same in the nematic phase as it was in the isotropic phase for a sample with no macroscopic orientation. A similar result was observed for a similar polymer exhibiting a smectic phase. However, for a sample in the glass which had previously been aligned in a 13 T magnetic field, anisotropic small-angle neutron scattering was observed. The radius of gyration in the direction of the macroscopic alignment (magnetic field direction) was about 25% less than that normal to the alignment direction. Since the induced macroscopic alignment would correspond to the orientation of the mesogenic units, this result indicates a preference for the polymer backbone trajectory to be normal to the mesogenic side-group despite the six-unit alkyl coupling chain. Similar results were observed by Keller *et al.*[73] although the anisotropy in the radius of gyration was slightly smaller. These authors however managed to align magnetically the smectic phase of their polymer. There was considerably more anisotropy in the values of the radii of gyration, which differed by a factor of four. The magnitude of the radius of gyration parallel to the director or normal to the smectic layer planes was of the order of the length of one mesogenic side-chain, indicating that the backbone is strongly confined in the smectic layers.

Small-angle neutron scattering is clearly a powerful tool in the study of molecular anisotropy. It does however require the preparation of special deuterated samples. It would be particularly illuminating to measure the mesogenic core alignment using IR spectroscopy or X-ray scattering, and correlate that with the backbone alignment observed with neutron scattering.

Diamagnetic Susceptibility

The magnetic susceptibility χ is defined as

$$\chi = \mu_0 M/B \tag{16}$$

where M is the magnetization, B the magnetic induction, and μ_0 the permeability in vacuum. For an anisotropic material the susceptibility is a second-rank tensor, and the magnetic anisotropy is directly related to the order parameter via:

$$\Delta\chi = \Delta\chi_0 \langle P_2(\cos\alpha)\rangle \tag{17}$$

where $\Delta\chi_0$ is the diamagnetic anisotropy of the orienting unit, i.e. that for perfect orientational order. Thus, diamagnetic anisotropy is equivalent to the dielectric property of birefringence, and as we discussed earlier is therefore limited to providing information about $\langle P_2(\cos\alpha)\rangle$. However, the development of a molecular model for diamagnetic anisotropy is more straightforward than for birefringence.[46] This arises from the fact that the diamagnetic moments of the molecules are very small and their interaction so small that it may be ignored. Therefore, the magnetic field 'operating' on an orienting unit is equal to the applied field, i.e. there is no internal field problem in diamagnetism. Usually $\Delta\chi$ is positive and $\Delta\chi_0$ is independent of temperature. Measurements of $\Delta\chi$ made using a Faraday balance have been reported in the literature (e.g. refs 74, 75). The most comprehensive example of the use of this method is the study of Sigaud *et al.*[76] They measured $\Delta\chi$ as a function of time, temperature and field strength for a main-chain liquid crystal polymer (**IX**) which was aligned in a field of 2 T. Sigaud *et al.*[76] used a

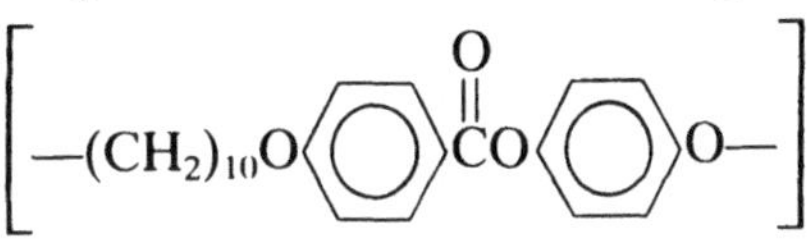

IX

molecular additivity method[77] to determine $\Delta\chi_0$. They found that the alkyl chain contribution to the magnetic anisotropy was significant, although dependent upon the particular chain conformation chosen, and obtained orientational parameters of 0·37 for the monomer, 0·49 for the dimer and 0·6 for the polymer at the nematic–isotropic transition. These results can be interpreted as indicating a high conformational order of the coupling chain. Diamagnetic susceptibility measurements have also been employed by Dupre and colleagues[78,79] for measurements of the order parameter in lyotropic polymer liquid crystal systems.

Dielectric Spectroscopy

Dielectric relaxation spectroscopy is widely used to study molecular dynamics of conventional and liquid crystal polymers.[80,81] Since the mesogenic groups of a side-chain liquid crystal polymer contain strong permanent dipoles, the technique may be utilized to study the reorientation and, as Attard *et al.*[82–84] have shown, the level of orientational order of those mesogenic groups. These authors have studied siloxane-based side-chain liquid crystal polymers with cyano-terminated phenyl benzoate mesogenic groups. The technqiues involve the computer-fitting of various Fuoss–Kirkwood lineshapes to the experimental spectra. From a detailed analysis, values for $\langle P_1(\cos\alpha)\rangle$ are obtained. Although this is a complex procedure, it provides an attractive route to the measurement of the quality of homeotropic (i.e. normal to the sample surface) orientation in thin films.

MEASUREMENT OF ORIENTATION BY X-RAY SCATTERING

This section is concerned with the measurement of orientational parameters using X-ray scattering techniques. Such methods have been established for many years for polymers,[85,86] and for low molar mass liquid crystals,[87,88] often in a rather qualitative manner. However, in the last few years a more quantitative approach to the measurement of orientation parameters using X-ray scattering has been developed, and the application of this technique to liquid crystal polymers will be described in detail.

X-Ray Scattering

The X-ray scattering from liquid crystal polymers, as with other disordered materials, is diffuse, a typical pattern containing two or three broad peaks (Fig. 11). It is the presence of these broad peaks, or rather the absence of sharp Bragg reflections, which demonstrates the lack of three-dimensional translational order. Figure 11 was obtained from the nematic phase of a side-chain liquid crystal polymer;[89] a smectic phase would give rise to additional scattering from the layer structure. The diffuse scattering peaks of Fig. 11 arise from variations in the electron density on the scale of 1–100 Å. The challenge is to extract information relating to an anisotropic distribution of those electron density differences and relate that to the molecular axis. Figure 12 shows one quadrant of the scattering intensity $I(s, \alpha)$ together with equatorial and meridional sections, recorded for the same side-chain liquid crystal polymer as was used for Fig. 11 but for a sample which has been macroscopically aligned.[89] The basic geometry used to record this scattering pattern is shown in Fig. 13(a) and (b), but the reader is directed elsewhere for full details of experimental procedures (see refs 5, 86, 89). The principal result of introducing

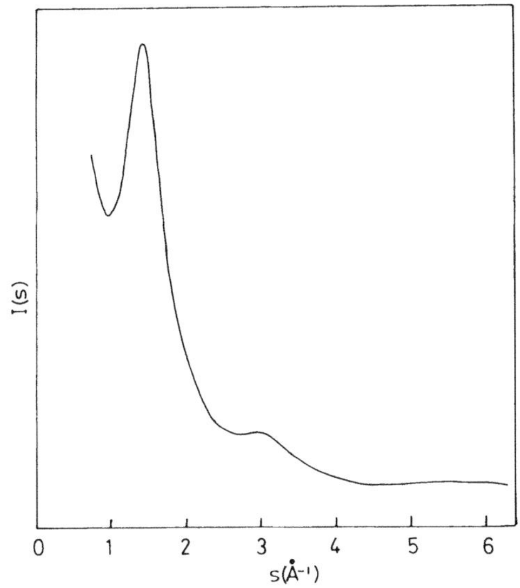

FIG. 11. X-ray scattering intensity from an unaligned side-chain liquid crystal polymer (**III**) plotted against the scattering vector $s = 4\pi \sin \theta / \lambda$.[89]

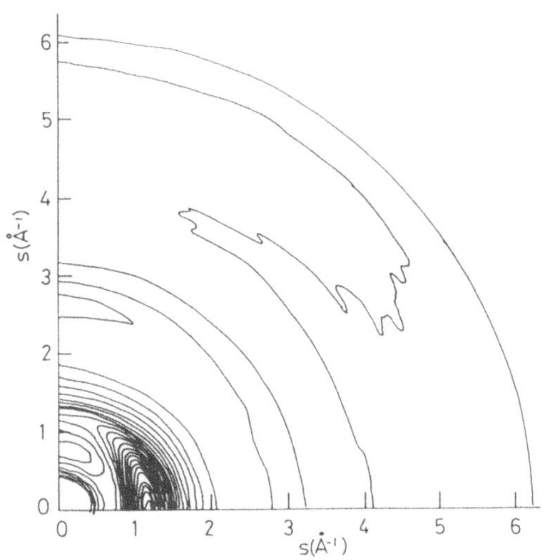

FIG. 12. Plots of X-ray scattered intensity for the same sample as Fig. 11 but after alignment in a magnetic field (field direction vertical). The strong interchain peak is readily apparent on the equator.

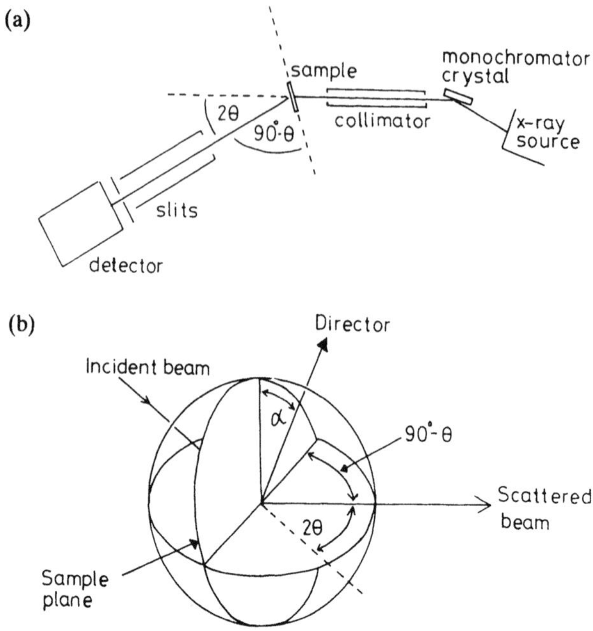

FIG. 13. Diagrams illustrating the geometry of the X-ray scattering experiments.

alignment has been to produce an anisotropic scattering pattern, each of the diffuse peaks in s intensifying about either the meridional ($\alpha = 0°$, parallel to alignment axis) or equatorial ($\alpha = 90°$, normal to alignment axis) section. The peak which intensifies on to the equatorial section relates to a real-space correlation of 4–6 Å, while those peaks most intense on the meridional section correspond to real-space correlations of ~2 Å and 1·3 Å. These distances immediately suggest that the equatorial peak is associated with *inter*molecular correlations, while those on the meridional section relate to *intra*molecular correlations.

In order to illustrate the imprint of an orientation distribution on the scattering pattern, we shall consider first a crystalline array of rigid rods, with perfect transverse and axial register. Figure 14(a) shows a schematic of such an organization and the resultant X-ray scattering pattern. The diffraction spots on the equatorial section correspond to the spacing between the rods. Loss of axial register does not affect those equatorial maxima, but the remaining sharp diffraction spots are smeared out into continuous layer lines (Fig. 14b). Where there is only

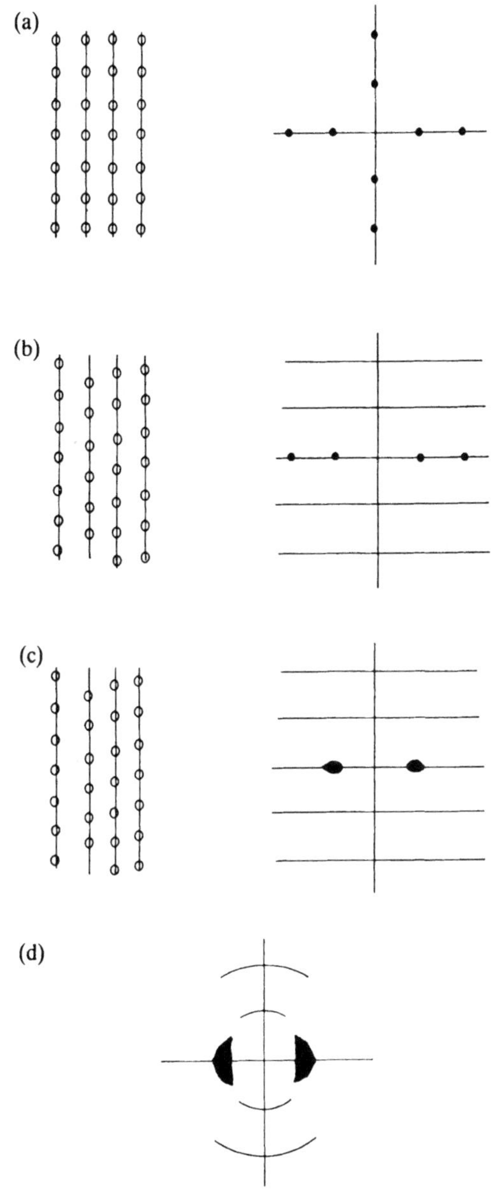

FIG. 14. Schematic X-ray patterns for assemblies of perfectly aligned rod molecules with reducing levels of positional order. (a) Full crystalline order; (b) crystalline lateral packing but no longitudinal register; (c) no long-range positional order; (d) rods with less-than-perfect preferred orientation.

disordered liquid-like packing of the rods in the transverse plane (Fig. 14c) the equatorial maxima are blurred into a diffuse scattering pattern. At this stage the model consists of long rigid rods with no translational symmetry but with perfect orientational ordering. We can reduce this orientational ordering in two principal ways. One approach is for the quality of local orientational order to be reduced as shown in Fig. 15(a). The alternative is to limit the extent of the perfect orientational order and hence produce a 'domain'-like model (Fig. 15b). It is relatively easy to predict the resultant pattern of the latter case, for it will correspond to a superposition of the scattering pattern from each domain, and as a result the peaks in the scattering pattern will appear arced (Fig. 14d). The other option (Fig. 15a) is a more complex situation in which the spatial correlations are linked, and interact with the orientational correlations. It is therefore not surprising that orientation measurements using wide-angle X-ray scattering have relied on the former option, that of a perfectly aligned set of infinitely long rods. We shall now consider this approach in a more formal manner.

The X-ray scattering for a sample with a macroscopic but less-than-perfect orientation, $I(s, \alpha)$, may be written as the convolu-

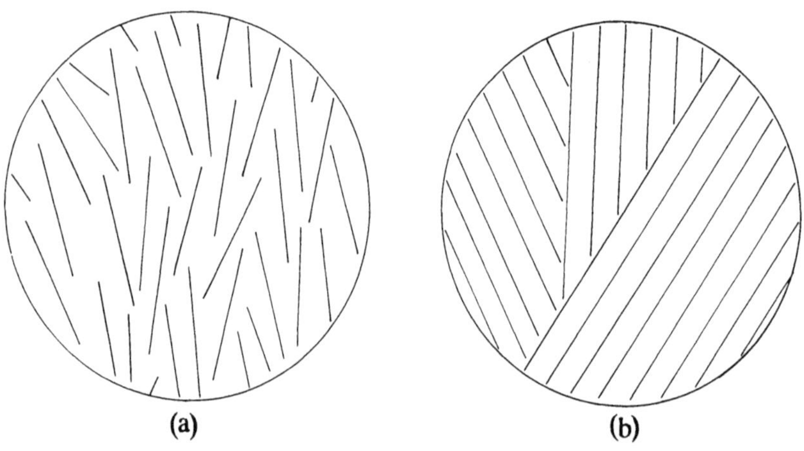

(a) (b)

FIG. 15. Diagram illustrating different organizations of partially oriented rods. Although the overall degree of preferred orientation is similar for both diagrams, in (a) the rods are misoriented independently of their relative positions, whilst in (b) they are perfectly aligned within domains, the macroscopic degree of orientation being dependent on the misalignment between domains.

tion of the scattering for a perfectly aligned structural unit $I_m(s, \alpha)$ with the orientation distribution function $D(\alpha)$:

$$I(s, \alpha) = I_m(s, \alpha)_* D(\alpha) \tag{18}$$

For cases in which each of these functions has uniaxial symmetry with an inversion centre, the convolution may be written as:[90–93]

$$I_{2n}(s) = I^m_{2n}(s) D_{2n} \left\{ \frac{2\pi}{4n+1} \right\} \tag{19}$$

where $I_{2n}(s)$, $I^m_{2n}(s)$ and D_{2n} are the coefficients of a spherical harmonic expansion (see eqns (1) and (2)) of the functions $I(s, \alpha)$, $I^m(s, \alpha)$ and $D(\alpha)$. For example, $I_{2n}(s)$ is given by:

$$I_{2n}(s) = (4n+1) \int_0^{\pi/2} I(s, \alpha) P_{2n}(\cos \alpha) \sin \alpha \, d\alpha \tag{20}$$

with analogous definitions for $I^m_{2n}(s)$ and $D(\alpha)$. The total scattering pattern may be resynthesized from:

$$I(s, \alpha) = \sum_{n=0}^{\infty} I_{2n}(s) P_{2n}(\cos \alpha) \tag{21}$$

or

$$I(s, \alpha) = \sum_{n=0}^{\infty} \frac{2\pi}{4n+1} I^m_{2n}(s) D_{2n} P_{2n}(\cos \alpha) \tag{22}$$

Thus in order to obtain values for the orientation parameters D_{2n}, we use the expression:[6,91]

$$D_{2n} = \left(\frac{4n+1}{2\pi} \right) I_{2n}(s) / I^m_{2n}(s) \tag{23}$$

The use of eqn (23) requires a knowledge of the spherical harmonics, $I^m_{2n}(s)$, for a perfectly aligned structural unit. Before proceeding to discuss values for $I^m_{2n}(s)$ it is appropriate to emphasize the limitations of eqn (23). It is an obvious requirement that $I(s, \alpha)$, $I_m(s, \alpha)$ and $D(\alpha)$ have the appropriate axial and mirror symmetry that enables them to be expressed in terms of a series of even-order spherical harmonics. Equally the expression of the convolution of eqn (18) in the form of eqn (19) or as a deconvolution eqn (23) requires the convolution to involve the scattering from independent structural units. In other words, eqn (19) ensues from the separation of the spatial and orientational correlations.

It is usually more convenient to rewrite eqn (23) in terms of the normalized amplitudes of the spherical harmonics:

$$\langle P_{2n}(\cos\alpha)\rangle_d = \langle P_{2n}(\cos\alpha)\rangle / \langle P_{2n}(\cos\alpha)\rangle_{\mathrm{m}} \tag{24}$$

where for example $\langle P_{2n}(\cos\alpha)\rangle$ is given by:

$$\langle P_{2n}(\cos\alpha)\rangle = \frac{\int_0^{\pi/2} I(s,\alpha) P_{2n}(\cos\alpha)\sin\alpha\,\mathrm{d}\alpha}{\int_0^{\pi/2} I(s,\alpha)\sin\alpha\,\mathrm{d}\alpha} \tag{25}$$

This is in essence the normalized version of eqn (2).

We have introduced one model appropriate to the use of these equations, namely a set of regularly spaced and infinitely long rods (Fig. 14d). The scattering from such a set of rods will be confined to the equatorial plane ($\alpha = 90°$). This corresponds to the scattering (in terms of α) taking the form of a delta-function and hence $I_{2n}^{\mathrm{m}}(s)$ components are given by:[91]

$$\langle P_{2n}(\cos\alpha)\rangle_{\mathrm{m}} = \frac{(2n)!}{(-1)^n 2^{2n}(n!)^2} \tag{26}$$

and hence the first few terms are:

$$\begin{aligned} P_2^{\mathrm{m}} &= -\tfrac{1}{2} \\ P_4^{\mathrm{m}} &= \tfrac{3}{8} \\ P_6^{\mathrm{m}} &= -\tfrac{5}{16} \\ P_8^{\mathrm{m}} &= \tfrac{35}{128} \end{aligned} \tag{27}$$

Thus in practical terms, the measurement of orientation using this approach reduces to the following stages.

(a) Measure the azimuthal dependence of intensity $I(\alpha)$ at a scattering vector corresponding to the maximum of this inter-chain peak.
(b) Use eqn (25) to evaluate $\langle P_{2n}(\cos\alpha)\rangle$.
(c) Use the values given for P_{2n}^{m} in eqn (27) and eqn (24) to give the orientation parameters $\langle P_{2n}(\cos\alpha)\rangle_{\mathrm{d}}$.

Mitchell and Windle[6] have used this approach to evaluate the orientation parameters for a random thermotropic co-polyester. Figure 16 shows the recorded X-ray scattering intensity function $I(s,\alpha)$, and

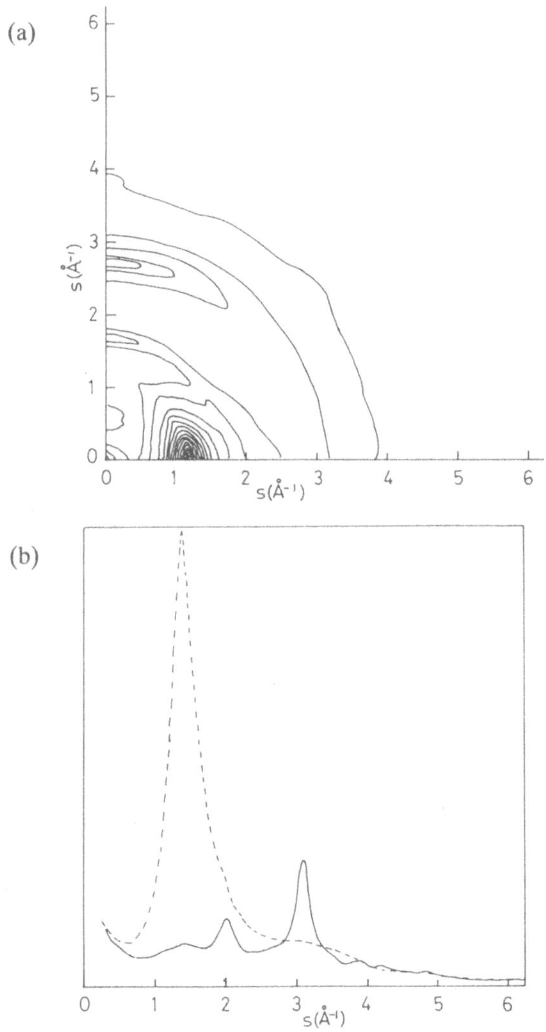

FIG. 16. (a) Experimental intensity function for the oriented random copolymer **X**. The orientation was introduced by melt extrusion. (b) Equatorial (- - - -) and meridional (——) sections of (a).

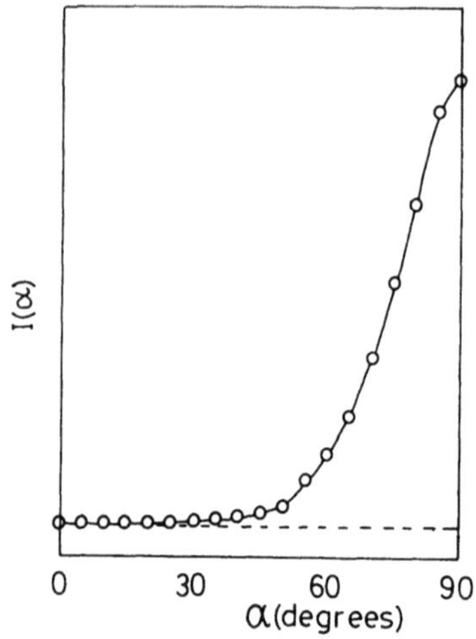

FIG. 17. Azimuthal distribution of intensity measured from the data of Fig. 16 at $s = 1{\cdot}4\,\text{Å}^{-1}$, the scattering position of the main interchain peak. The broken line represents the background level used in the analysis.[6]

Fig. 17 the azimuthal distribution of intensity around the ring at $s = 1{\cdot}4\,\text{Å}^{-1}$, which corresponds to the interchain peak. The X-ray scattered intensity recorded at that scattering vector will include some scattering due to air and incoherent scattering. It is common practice to assume that the intensity level at $\alpha = 0°$ may be taken as a background figure and subtracted from the azimuthal profile, and this is shown in Fig. 17. This process is rather arbitrary, although it yields results in good agreement with other techniques. One of the principal advantages of the use of X-ray scattering analysis is that it yields values for $\langle P_{2n}(\cos\alpha)\rangle_d$ for $n > 1$ in contrast to many other techniques. These higher moments are much more sensitive to the particular characteristics of a model, and therefore offer more detailed information. The complete set accessible through X-ray scattering procedures may be used to create the complete orientation distribution function $D(\alpha)$ (eqn (1)). The complete orientation distribution function derived from the data of Fig. 17 is shown in Fig. 18. It should be noted that these measurements were made at room temperature, where the polymer structure is more ordered,[41] and thus the orientation measurement is a spatial average over regions with different degrees of positional order.

Analyses similar to the above but in general reporting only $\langle P_2(\cos\alpha)\rangle_d$ have been followed by a number of researchers for a

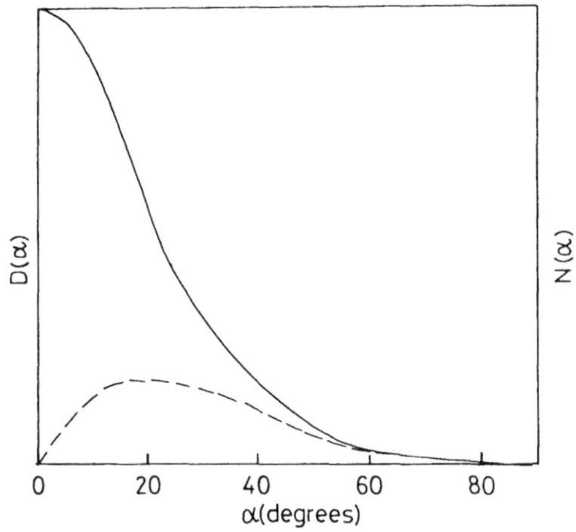

FIG. 18. The measured orientation distribution function, plotted as a section $D(\alpha)$ (——) and as $N(\alpha)$ (- - - -), the proportions of chain segments at α: $N(\alpha) = D(\alpha) \sin \alpha$. The curves are derived from the experimental data of Figs. 16 and 17.[6]

variety of liquid crystal polymers.[94–98] Liebert *et al.*[94] have measured the order parameter $\langle P_2(\cos \alpha)\rangle_d$ for a co-polyester with flexible coupling chains of five alkyl units, after alignment in a magnetic field of 0·3 T for 24 h. A range of co-polyesters with flexible links were considered by Iannelli *et al.*[95] They reported $\langle P_2(\cos \alpha)\rangle_d$ as a function of temperature for samples aligned in a magnetic field after prealignment as fibres. Reported values of $\langle P_2(\cos \alpha)\rangle_d$ were in the range 0·7–0·45; those for polymers with an even number of alkyl units being significantly higher. Blumstein *et al.*[96] compared the molecular orientation for samples of a polyester with a spacer of nine alkyl units aligned in a 12 T magnetic field ($\langle P_2(\cos \alpha)\rangle_d \simeq 0{\cdot}6$) and by fibre drawing ($\langle P_2(\cos \alpha)\rangle_d \simeq 0{\cdot}88$). However, even higher values of $\langle P_2(\cos \alpha)\rangle_d$ were recorded for polyesters with eight alkyl units. Takeuchi *et al.*[98] have measured $\langle P_2(\cos \alpha)\rangle_d$ for a random co-polyester (the same as that employed in ref. 6) as a function of applied shear rate. Despite a linear increase in the orientation function from 0·6 to 0·8 over a shear rate increase from 10^2 to $10^4\,s^{-1}$, the linear thermal expansion coefficient showed a discontinuous change, the latter being of importance in the potential use of the material as an optical fibre coating.

Each of the studies reviewed above made several assumptions, although they were not always implicitly stated, in order to evaluate the molecular orientation parameters using X-ray scattering. These are best understood by looking again at the basic model shown schematically in Fig. 14(d), of a set of rigid, infinitely long collinear rods. 'Real' molecules are of course not infinite in length, particularly not for mesogenic groups in side-chain liquid crystal polymers. Leadbetter and Norris[87] and De Vries[99] have evaluated the effect on the final orientation parameter of a finite-length molecule, in their case ~20 Å. It was found that although there is smearing of the diffraction pattern out of the equatorial plane, the effect was negligible for orientation parameters when $\langle P_2(\cos\alpha)\rangle < 0{\cdot}8$. We have considered some liquid crystal polymer systems which have orientation parameters in excess of this level, so significant errors may be introduced if these 'simple' X-ray scattering procedures are employed. There also may be some further problems associated with the diffuse nature of the scattering within the equatorial plane. We have already noted that the subtraction of a background was rather arbitrary, there being no reason in a highly disordered structure why there should be *no* intramolecular scattering at the same scattering vector as the main interchain peak. If there is some such intrachain scattering, the background determination will be correspondingly complicated, independently of α.

This effect and that due to the scale of the spatial correlations were explored by Mitchell,[100] using a model comprising a set of parallel chains with aromatic moieties, the lateral packing being random. The scattering from such a model, embedded in a spherical enclosure of varying size, was calculated using procedures detailed elsewhere.[5] From such scattering the orientational average $\langle P_2(\cos\alpha)\rangle_m$ was derived and the results are reproduced in Fig. 19. The limiting value for an infinite model would be $-0{\cdot}5$ as given earlier.[27] The apparent correlation between $\langle P_2(\cos\alpha)\rangle_m$ and the radius of the model results from two factors. The first is directly due to the limited correlation volume, but in fact the dependence is very slight. The factor of greater importance is the enhancement of the intrachain scattering due to the increasing intrachain correlation, which has been evaluated recently by Mitchell using models set in a cylindrical enclosure. An example was addressed in which the X-ray scattering from a random co-polyester of hydroxybenzoic and hydroxynaphthoic acids (70:30 mole ratio) was recorded and the azimuthal distribution of the interchain peak obtained gave $\langle P_{2n}(\cos\alpha)\rangle = -0{\cdot}25$. Taking a domain size appropriate to the breadth of the scattering peaks a final orientation parameter of 0·54 is determined. This is in contrast to the conventional approach,

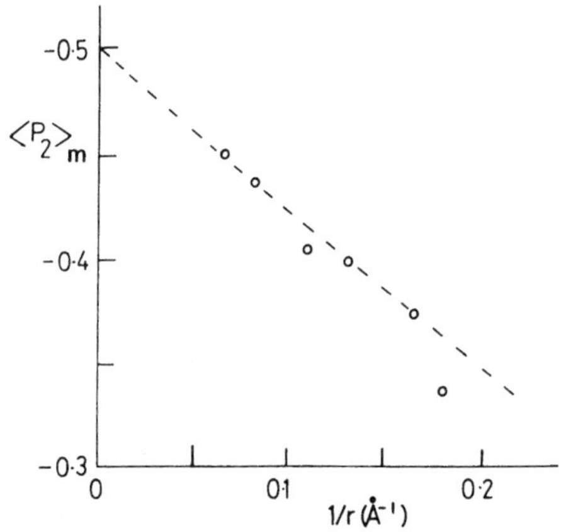

FIG. 19. Results of calculations of orientation functions from the scattering determined from models with perfect alignment, short-range positional order, but of varying size as defined by r.

which after background subtraction would give $\langle P_2(\cos\alpha)\rangle = -0{\cdot}31$ and hence $\langle P_2(\cos\alpha)\rangle_d = 0.62$.

The final assumption made in the reported studies is that of uniaxial symmetry. In nematic phases of low molar mass liquid crystal polymers, small deviations from uniaxial packing (i.e. from rod packing to packing of lath-like chains) have been reported. However, a number of rigid chain copolyesters have been reported to exhibit biaxial optical properties and a possible interpretation is that they possess a biaxial nematic phase.[41–43] It is important to stress here that the procedures described above require local axial symmetry in addition to a global uniaxial orientation. The possibility of a sheet-like structure, for example, would clearly distort any calculation of $\langle P_{2n}(\cos\alpha)\rangle_d$ for the interchain peak. Thus, whilst the anisotropy of the intense interchain peak provides a convenient method of evaluating all terms of the orientation distribution function, there are clear problems of interpretation for certain samples. These arise simply because we have chosen to interpret the most complicated part of the scattering pattern. The interchain peak arises from spatial correlations between neighbouring chains, in other words a many-chain model. To date no satisfactory quantitative model has been reported which details the local molecular interactions in a liquid crystal phase, whether polymeric or not. The possibility of using the anisotropy of the intrachain scattering, which can be considered as a single-molecule problem, has not featured largely in the literature. It does however have considerable potential and justifies a more detailed consideration in the following section.

Intrachain Scattering

The scattering pattern of the macroscopically aligned side-chain liquid crystal polymer shown in Fig. 12 has two components. The first centred on the peak at $s \simeq 1{\cdot}4\,\text{Å}^{-1}$ is interchain in nature and was discussed fully above. The second at $s > 2\,\text{Å}^{-1}$ arises from spatial correlations within individual molecular segments. It has been shown on a number of occasions[5,89,101–103] that this division is a valid approach, and that we may consider scattering at scattering vectors greater than $2\,\text{Å}^{-1}$ as arising solely from individual chain segments. Thus we may utilize eqn (19) to determine $\langle P_{2n}(\cos\alpha)\rangle_{\rm d}$ although we require values of $I^{\rm m}_{2n}(s)$. The intrachain scattering is continuous in nature, and thus in principle eqn (19) may be applied for values of $s > 2\,\text{Å}^{-1}$ rather than in a restrictive manner as in the case of the interchain peak.

For liquid crystal polymers comprised of completely rigid units with no conformational freedom, the calculation of $I^{\rm m}_{2n}(s)$ using data from model compounds etc would present no problems. However, no liquid crystal polymer falls into that category, even the so-called rigid-chain polymer systems having considerable flexibility. Thus to proceed we need a method of evaluating the chain structure.

The answer to this requirement lies in eqn (19). Considering only scattering vectors greater than $2\,\text{Å}^{-1}$, the observed intrachain scattering $I_{2n}(s)$ is that for a perfectly aligned molecular segment weighted with the orientation parameters D_{2n}. Since these are not functions of s, the observed harmonics are those for a perfectly aligned system simply rescaled by the orientation parameters. Thus, in all ways other than absolute magnitudes, the observed harmonics as a function of the scattering vector s do not depend on the level of orientation. A material (assuming the molecular conformation unchanged) with a higher level of molecular orientation simply displays high-order spherical harmonics of greater magnitude. Therefore, we may compare the observed harmonics $I_{2n}(s)$ with those calculated from viable molecular models. This comparison may only be made in terms of peak positions, peak breadths and relative magnitudes. When there is an appropriate match, this not only gives the molecular conformation, but by insertion of the calculated $I^{\rm m}_{2n}(s)$ functions into eqn (19), the orientation parameters are derived. Thus we obtain, at the same time, structural and orientational parameters. This is of particular significance as we know both the level of orientation and what the orienting unit is. In other words, any ambiguity is removed from the description of molecular anisotropy. These procedures are not at all limited by the approximations discussed in the interchain case. Moreover, since we are now dealing with a single-molecule problem, we only require the macroscopic or global orientation distribution to be uniaxial; the local environment is of no concern, although (as we will discuss later) some information on the local orientation may be extracted.

The procedures which employ the intrachain scattering to evaluate the molecular orientation parameters have been applied successfully to main-chain and side-chain liquid crystal polymers[5,6,17,28,89,100] and to non-crystalline polymers in general.[7,12,27,86] They have recently been reviewed by Windle.[101] We shall explore their use through the example of a semi-rigid thermotropic co-polyester **X**.[5,103] The X-ray

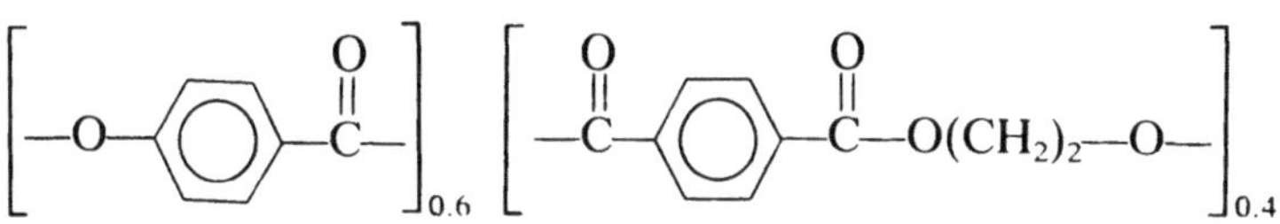

X

measurements were performed at room temperature, on a sample in which a macroscopic alignment had been introduced by melt extrusion. The static molecular organization is believed to be essentially that of the liquid crystal phase exhibited at temperatures above 250°C based on optical microscopy[104] and on X-ray diffraction patterns recorded from samples in the liquid crystal phase and aligned using a magnetic field.[10] The X-ray scattering was measured as described above and normalized using well-established procedures,[5,105] to give the s-weighted reduced intensity function $si(s, \alpha)$ shown in Fig. 20. $si(s, \alpha) = kI_{\text{exp}}(s, \alpha) - I_{\text{comp}}(s) - \sum_i f_i^2(s)$ where I_{exp} is the fully corrected experimental data, $I_{\text{comp}}(s)$ is the calculated incoherent scattering, $\sum_i f_i^2(s)$ is the calculated independent coherent scattering and k is the normalization constant.[5]

The first five even spherical harmonic terms describing the data of Fig. 20, derived using eqn (20), are shown in Fig. 21. The portion of the curve relating to the intense interchain peak at $s \simeq 1{\cdot}4\ \text{Å}^{-1}$ is not drawn for some of the higher harmonic components since it is not

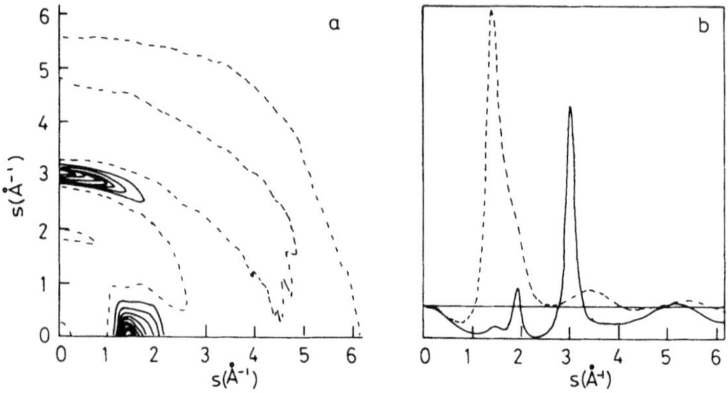

FIG. 20. (a) Experimental, s-weighted, reduced-intensity function for random co-polyester **X**. The broken contours are negative and the extrusion axis is vertical. (b) Meridional (——) and equatorial (- - - -) sections of the intensity function in (a).[103]

relevant for the present calculations. The principal features of the intrachain scattering are those relating to the peaks at $s \simeq 3$ Å^{-1} and at 5·5 Å^{-1}, although it should be noted that only the peak at $s \simeq 3$ Å^{-1} has a component in all harmonics. This may be related to the nature of this particular diffraction maximum and the reader is directed elsewhere for full details (ref. 106). Although in principle the complete intrachain scattering may be employed in the orientation measurement, intense peaks offer a reduction in the associated errors, and in this case the peak at $s \simeq 3$ Å^{-1} was utilized.

A model for the intrachain scattering need only take account of structural information relevant to an isolated polymer chain. For the relatively rigid molecule considered here, the number of conformational parameters is limited. However, the scattering $I^{\mathrm{m}}_{2n}(s)$ must represent an 'average' chain and therefore include any variation in chemical and conformational sequences. Bond lengths and angles may be obtained from model compounds, whilst the rotations of the phenyl groups do not greatly affect the scattering intensity distribution. The model spherical harmonic components are calculated using:[5]

$$\mathrm{NI}^{\mathrm{m}}_{2n}(s) = (-1)^n(4n+1)\sum_{j}^{N}\sum_{k}^{N} f_j(s)f_k(s)j_{2n}(r_{jk})P_{2n}(\cos\alpha_{jk}) \qquad (28)$$

where r_{jk} is the distance between the jth and kth atoms with scattering factors $f_j(s)$ and $f_k(s)$, and $\cos\alpha_{jk}$ is the angle between the orientation

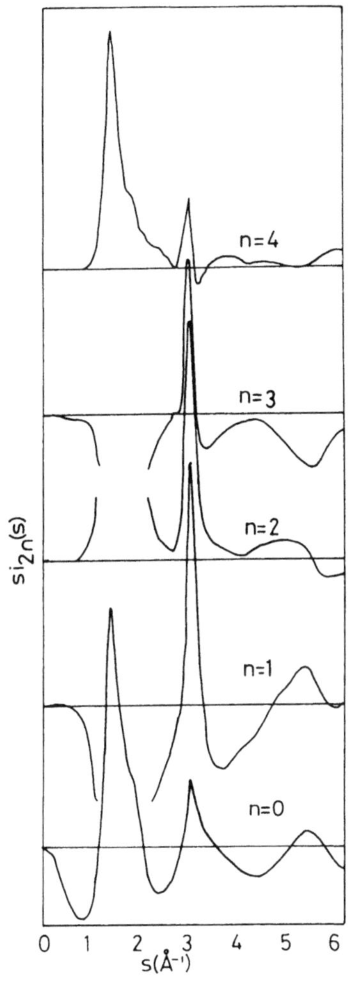

FIG. 21. The first five even spherical harmonic functions calculated from the scattering data of Fig. 20 as a function of s.[103]

axis and the interatomic vector r_{jk}. It is impossible to reduce the computations required in the calculation of eqn (28) and these are described in ref. 5. The s-weighted version of these calculated harmonics is displayed in Fig. 22. The curves shown are for a molecular chain with a random distribution of chemical segments with a level of intrachain positional order of 40–50 Å. The scattering was averaged over 150 subchains. The calculated spherical harmonics are

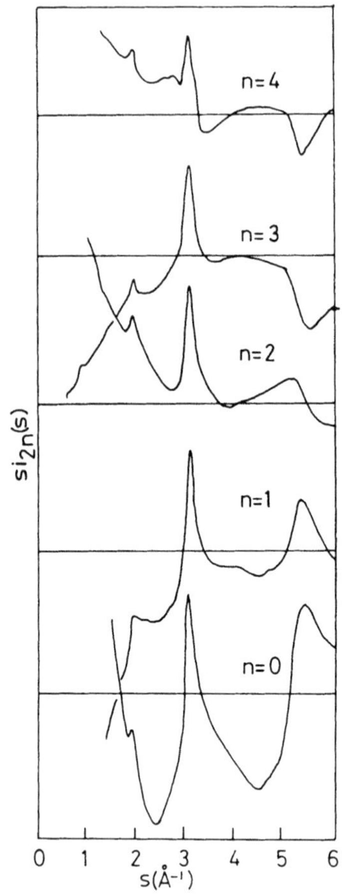

FIG. 22. The first five harmonic functions calculated from a refined model of the random copolymer molecular chain.[103]

in broad agreement in terms of peak positions although the peak widths and relative heights do not match those displayed by the experimental functions (Fig. 21). To improve the match, models were calculated for differing levels of intrachain order. The breadth of the peak at $s \simeq 3$ Å^{-1} in the calculated $si_2^m(s)$ as a function of intrachain correlation length is shown in Fig. 23. Note that the relationship is not linear. We have chosen to match the breadth of the peaks in the $si_2(s)$ curves, since their breadths from eqn (19) are independent of orientation. This would not be the case for the breadths of the meridional peaks in the complete scattering pattern. The breadth of

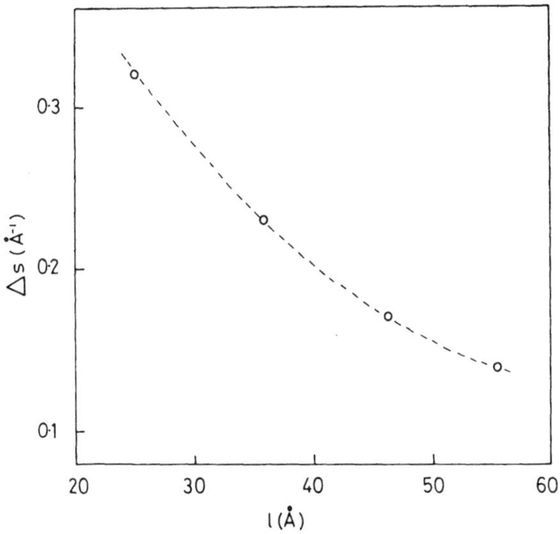

FIG. 23. Plot of the breadth at half height, As, for the main meridional peak at $s \simeq 3$ Å^{-1} in $si_2(s)$ calculated from models of the random co-polyester molecule with different levels of conformational disorder giving different correlation lengths, l.[103]

the peak in the experimentally derived functions is ~0·2 Å^{-1} and would increase with reducing orientation. The model terms $\langle P_{2n}(\cos\alpha)\rangle_m$ for various correlation lengths are shown in Fig. 24. These are obtained from eqn (28) by:

$$\langle P_{2n}(\cos\alpha)\rangle_m = \frac{I_{2n}^m(s)}{\{I_0^m(s) + \sum_i f_i^2(s) + I_{comp}(s)\}} \tag{29}$$

Using the correlation length obtained from matching the breadths of the peaks in terms of s, values of $\langle P_{2n}(\cos\alpha)\rangle_m$ may be read off from the curves in Fig. 24. Following this route for the particular sample considered, a value was obtained of $\langle P_2(\cos\alpha)\rangle_d = 0{\cdot}64$ with $\langle P_4(\cos\alpha)\rangle_d \simeq 0{\cdot}33$ and $\langle P_6(\cos\alpha)\rangle_d \simeq 0{\cdot}2$, the higher components being equally accessible.

This example serves to illustrate the lengthy nature of the analysis of the intrachain scattering to give orientational parameters. However, it must be accepted that the final product, that of orientational parameters for a defined and evaluated structure, is of considerable value. It provides a sound route to the measurement of orientation using X-ray scattering, in contrast to the methods employing the interchain

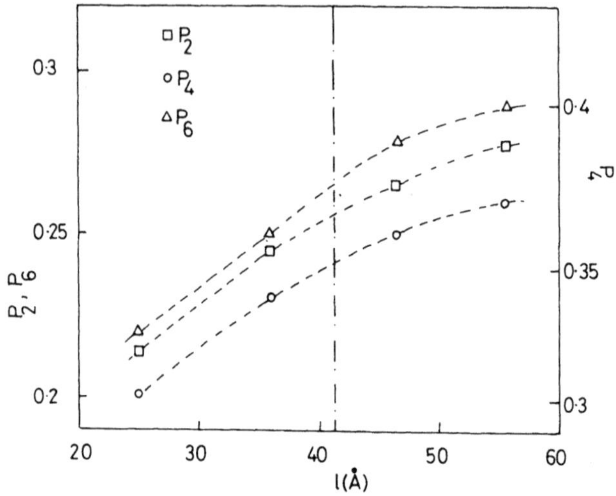

FIG. 24. Plot of orientation functions $\langle P_{2n}(\cos\alpha)\rangle_m$ derived from model molecules as a function of the correlation lengths of the models. The vertical broken line represents the correlation length determined from the experimental scattering.[103]

scattering. However, since the two parts of the scattering pattern, namely the interchain and the intrachain scattering, arise from different correlations there is the possibility of extracting even more information and we shall consider that next.

Local Orientation

We have considered two routes to the extraction of molecular orientation parameters from X-ray scattering patterns of liquid crystal polymers. The method employing the intrachain scattering provides a measure of the global orientation of the sample $\langle P_{2n}(\cos\alpha)\rangle_{\text{global}}$. In contrast, the interchain scattering approach measures the anisotropy in intermolecular correlations. We could ascribe to a small volume a local orientation $\langle P_{2n}(\cos\alpha)\rangle_{\text{local}}$. The total or global orientation is then given by:

$$\langle P_{2n}(\cos\alpha)\rangle_{\text{global}} = \langle P_{2n}(\cos\alpha)\rangle_{\text{local}}\langle P_{2n}(\cos\alpha)\rangle_{\text{domain}} \quad (30)$$

where $\langle P_{2n}(\cos\alpha)\rangle_{\text{domain}}$ describes the orientation of the directors of these small volumes. The concept is shown schematically in Fig. 25. In obtaining the orientation parameters from the interchain scattering we have assumed that the local orientation is perfect, i.e. in essence that

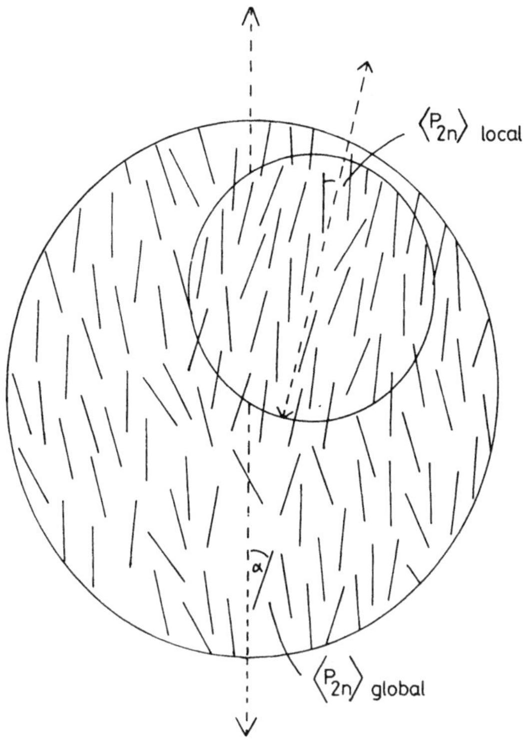

FIG. 25. Sketch illustrating the difference between global and local orientation.

the method provides a measure of $\langle P_{2n}(\cos\alpha)\rangle_{\text{domain}}$. Since the intrachain scattering provides the global orientation, the local orientation parameters may be derived using:

$$\langle P_{2n}(\cos\alpha)\rangle_{\text{local}} = \frac{\langle P_{2n}(\cos\alpha)\rangle_{\text{global}}}{\langle P_{2n}(\cos\alpha)\rangle_{\text{domain}}} \tag{31}$$

In a study of the random co-polyester of hydroxybenzoic and hydroxynapthoic acids Mitchell[100] derived a value for $\langle P_2(\cos\alpha)\rangle_{\text{global}}$ of 0·53. Taking account of the improvements to the interchain approach discussed, the value of $\langle P_2(\cos\alpha)\rangle_{\text{domain}}$ obtained was 0·54, giving a local orientation of 0·98. Although the errors build up in any such calculation, it is clear that the local orientation is substantially higher than the global value. This would be in disagreement with the theories of nematic ordering based upon a mean-field approach,[22] but in support of alternative treatments which provide a high level of local ordering.[107] A similar analysis on another co-polyester gave a local orientation parameter $\langle P_2(\cos\alpha)\rangle_{local}$ of 0·83 against a global value of 0·64.[6]

In summary, wide-angle X-ray scattering offers an almost unique opportunity to assess properly the molecular segmental orientation in liquid crystal polymers, by providing both orientational parameters and a structural model of the orienting unit. Furthermore, it is possible to obtain some information as to the level of local orientation, in contrast to the macroscopic or global orientation normally quoted.

SPATIAL VARIATION IN ORIENTATION

The techniques we have discussed so far have produced orientation parameters relating to a macroscopic average. In the last section, we considered how the local orientation may be greater than the global orientation level, and earlier (pp. 117–23) we emphasized the importance of scale in any orientation parameter. This part of the chapter is concerned with techniques and procedures which provide information as to the variation of the director orientation and the level of orientation from place to place in the sample. In principle any technique discussed could be made to sample in a selected area. For example, infrared spectroscopy and Raman scattering coupled with microscopy[108,109] are being more widely used. These are of course just more sophisticated versions of the polarizing optical microscope used at and since the advent of liquid crystal science.[110] X-ray scattering patterns may be obtained from micron-sized sample areas using appropriate instruments.[43] However, electron scattering and electron microscopy are a ready source of information sampled from small areas.

Optical Microscopy

Optical microscopy using polarized light is a classic technique, and liquid crystal systems can show a variety of often very beautiful textures.[111] The basic experimental procedure utilizes plane-polarized white light, passed through both the sample and a ‘crossed’ analyser. Thus the parts of the sample with either isotropic optical properties or with their in-plane optical director† parallel or perpendicular to the analyser direction appear dark. The isotropic material will appear dark irrespective of the analyser/polarizer direction (normally the analyser is at 90° to the polarizer and ganged with it on rotation). It is possible to resolve the 0°/90° degeneracy by the use of a phase plate such as a sensitive tint plate, which is inserted between the polars, with its principal optical axis at 45° to those of the analyser and polarizer. The best spatial resolution of the light microscope will be of the order of 1 μm. The fact that the sample appears dark when the local optical director is parallel or perpendicular to the analyser orientation provides a measure of mapping out the spatial variation of the director orientation. Such studies of this type have been reported for thermotropic co-polyesters by Mackley *et al.*[112] and by Viney and Windle.[104] An example is reproduced in Fig. 26, which shows the resolved in-plane director map about two disclinations.

It is important to appreciate that the in-plane optical director may not necessarily be equivalent to the projection of the molecular director of the liquid crystal on to the specimen plane. One of the most strident demonstrations of this non-equivalence is provided by the work of Viney *et al.*[41–43] Thin microtomed sections of the random co-polyesters, **I** and **X**, were examined by X-ray scattering and shown to have a high level of preferred molecular orientation. However, when the same thin slices were viewed in the polarizing microscope, no suggestion of a macroscopic orientation was seen. Indeed, rotation of the sample with respect to the polarizer/analyser orientation gave no discernible change in the total transmitted light. However, on a local scale (<5 μm) the microstructure displayed the expected $(\sin\theta)/2$ relationship in intensity upon rotation. From these observations it was concluded that the co-polyester possessed biaxial optical properties, with the possible implication that the liquid crystal phase was itself a biaxial nematic. Biaxial optical properties occur where there are three different principal refractive indices and no cylindrical

† The principal axes of the optical properties are also the axes of an ellipsoidal representation surface known as the optical indicatrix. Extinction between crossed polars occurs when the plane of the polarized light is parallel to either of the semi-major axes of the elliptical cross-section of the indicatrix which is parallel to the specimen plane. The orientation of the polarizer at which extinction occurs defines the in-plane optical director for any point on the sample.

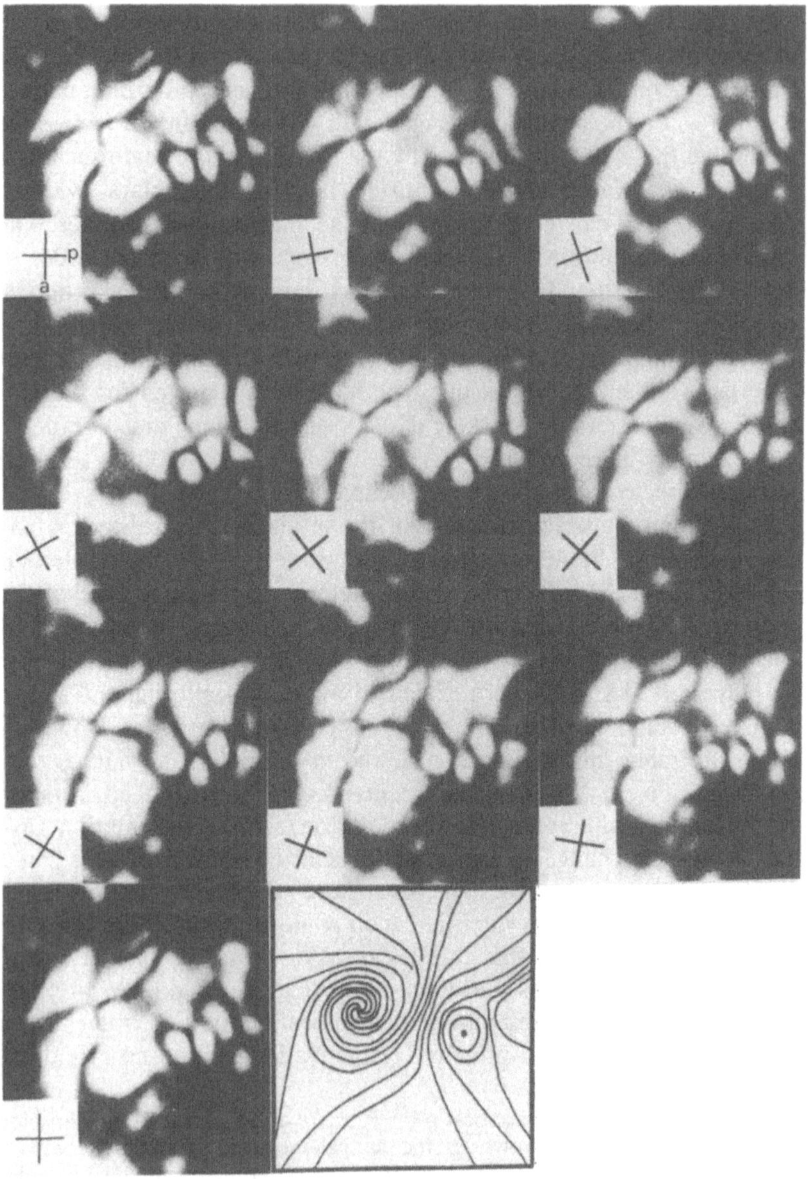

FIG. 26. A sequence of optical micrographs for different orientations of the crossed polars, showing the changes in the four-fold 'brush' patterns around disclinations of unit strength in polymer **X**. The director map gives an indication of the local changes in director orientation; however, an equivalent pattern could be drawn where the streamlines are orthogonal to those drawn at every point.[104]

symmetry about any of these axes. In this general case there is no reason why any of the principal axes of the optical-property ellipsoid should be identified with the polymer chain axis. This is in contrast to uniaxial properties where the axis of rotation will be parallel to the symmetry axis of the polymer structure, typically the chain axis.

As with small molecule liquid crystals, the variation of optical director with orientation can give rise to a wide range of textures in mesogenic polymers. It is important to appreciate that chain entanglements, which are not precluded just because a polymer is liquid crystalline, will greatly increase relaxation times and inhibit the attainment of equilibrium structures. Random co-polyesters with a degree of polymerization (DP) in excess of 100 appear to have a very fine, micron-scale texture which shows little tendency to coarsen even though the polymer is held at elevated temperature in the liquid crystal phase (Fig. 27a). Where the DP is of the order of 50 and/or the chain contains some flexible links of the $(CH_2)_2$ type, the texture rapidly coarsens to give domains within which the optical orientation remains relative uniform. For example, the polymer (**XI**) has the microstructure shown in Fig. 27(b), in which the domain size is of the order of 10 μm. The misorientation is concentrated in the domain boundaries, the structure of which can be analysed by systematic observations in the polarizing microscope.[113] Similar effects may be observed in side-chain liquid crystal polymer systems, with a relatively rapid relaxation of the mesogenic side-groups, although the speed of attainment of a true equilibrium structure is severely restricted by the slow relaxation of the long polymer backbone **XI**.[114]

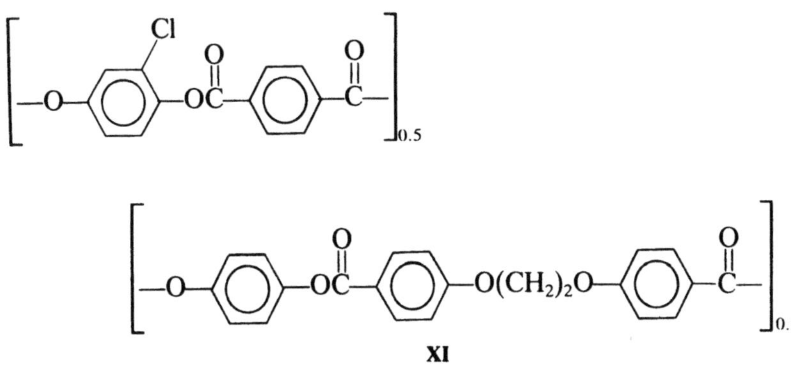

XI

In summary it should be noted that although optical microscopy appears to be a simple and straightforward technique, interpretation of

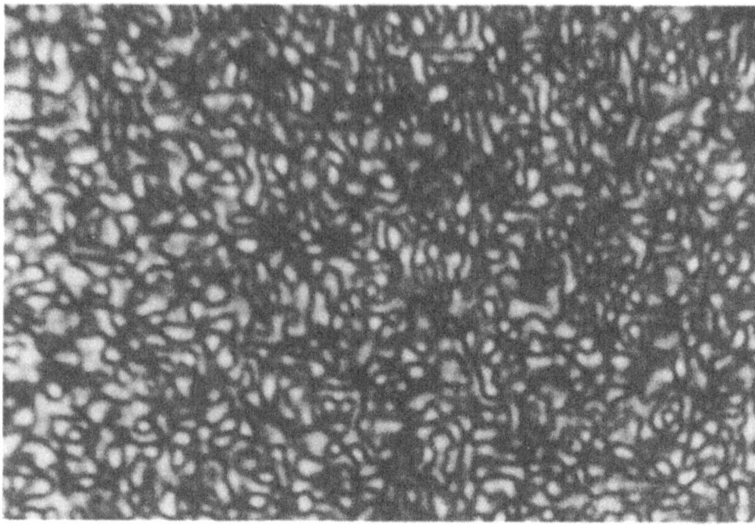

(a)

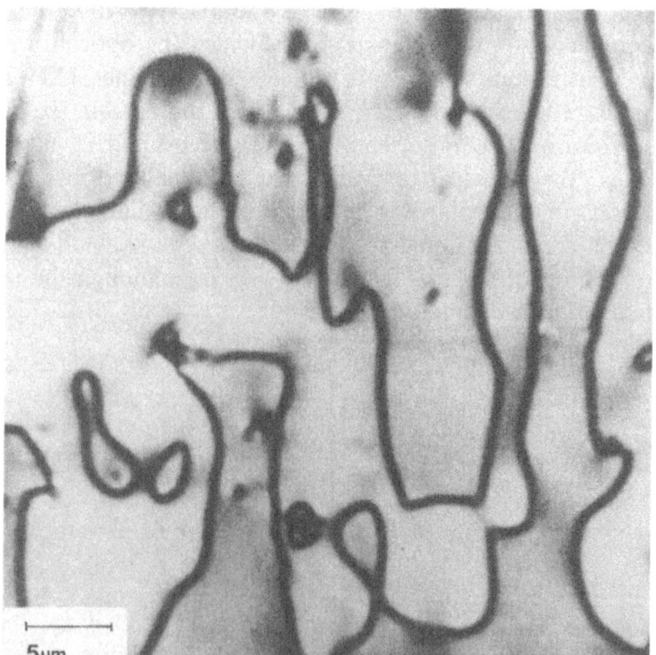

(b)

FIG. 27. Microstructures of liquid crystal copolyesters obtained in polarized light showing the variation of orientation with position in: (a) polymer **X** of a degree of polymerization greater than 100. (Photograph courtesy of T. J. Lemmon, Department of Materials Science and Metallurgy, Pembroke Street, Cambridge.); (b) polymer **XI** with a degree of polymerization of approximately 50.[41]

its results may not be. In any serious study attention must be directed to building a complete picture through the use of additional wave-plates to give circularly polarized light, phase contrast, tilting stages and observation of the diffraction pattern in the back focal plane of the microscope.

Electron Microscopy

Liquid crystals do not lend themselves to electron microscopy, as a liquid phase cannot be easily handled within a vacuum column unless cumbersome 'wet stages' are employed. However, the fact that polymeric liquid crystals can be quenched to the glassy or partially crystalline states without any apparent disturbance of the local orientation characteristic of the liquid mesophase means that the electron microscope becomes a most useful tool. It is now providing new insights into the local molecular orientation in some liquid crystal polymer systems.[115–123]

At one level, the electron microscope may be thought of as a high-resolution microscope, rather like the optical microscope, but routinely operating with substantially sub-micron resolution. Under those circumstances bright field images reveal detail around disclinations[115] or walls between regions of different orientation.[117,121] Alternatively the electron microscope may be seen as a diffraction apparatus with an extremely fine incident beam. Thus, for example, diffraction patterns taken from 150 nm-diameter regions of a random copolyester show levels of orientation (as indicated by the degree of arcing) substantially above the macroscopic average.[118]

However, the real advantage of electron microscopy is the ability to obtain real-space information on the sub-micron scale, coupled with opportunity to probe the molecular organization using diffraction techniques on the same sample.

Donald and Windle[119] have used the technique of 'dark-field' diffraction imaging to provide information relating to the spatial distribution of selected orientations. Thus by imaging an area using only part of an arced diffraction spot, the image highlights those 'domains' with that particular orientation.

A particular application of diffraction contrast has been in the study of the changes in local orientation which occur on annealing[116,121] thin films of liquid crystalline co-polyesters. Diffraction imaging, using the main equatorial reflection, was able to show the variation of director orientation as projected on to the specimen plane. On the other hand,

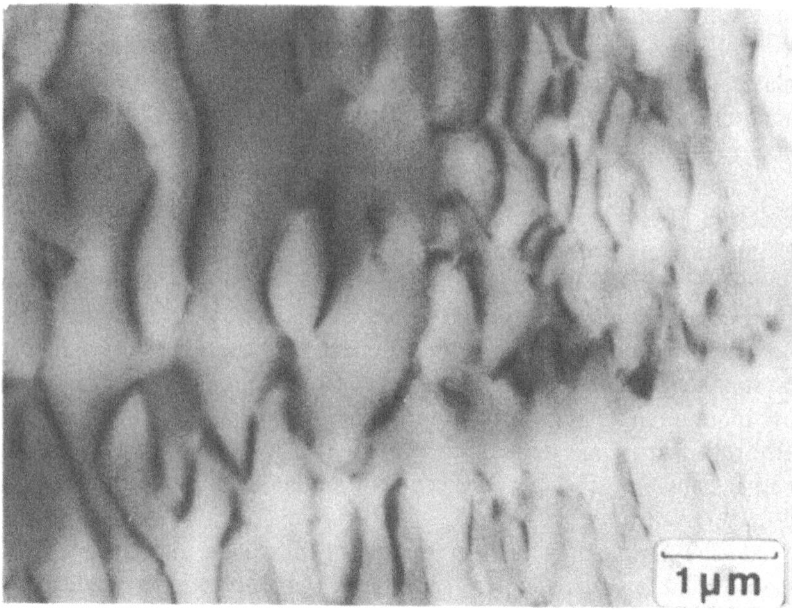

FIG. 28. Diffraction contrast in a transmission electron micrograph showing domain walls as dark on account of the pronounced out-of-plane tilt of the molecules in their vicinity.[116]

diffraction contrast available with the aperture placed over the central, 000, spot clearly showed regions, such as the domain boundaries in Fig. 28, where there was considerable out-of-plane tilt in the director.

There are of course difficulties associated with transmission electron microscopy. Notable amongst these are problems relating to specimen preparation. Ultramicrotomy carries with it the risk of mechanical damage, whereas casting on to substrates can lead to the modification of the structure by specific molecular interactions at the surfaces.

A Combined Optical and Electron Microscopy Study of Banding

One of the most intriguing orientation phenomena seen in rigid-chain liquid crystal polymers is that of banding. The texture appears in samples which have been sheared and oriented in extensional flow, as contrasting bands normal to the direction of extensional flow. The phenomenon was observed in Kevlar fibres by Dobb *et al.*[123] It has since been reported in other lyotropic liquid crystal polymers[124,125] and in thermotropic random copolymers.[126,127] In the latter studies a special effort has been made to combine optical and electron microscopy to determine the nature of the local molecular orientation which gives rise to the banding. Between crossed polars, the banding appears as in Fig. 29, for a sheared sample of polymer **X**. The shear axis is horizontal and the photographs correspond to different rotations of the crossed polars with respect to this axis. Analysis of the contrast indicates that the in-plane optical director meanders in a sinusoidal path about the shear axis. The important issue in this case is finding the extent to which the optical director can be identified with the molecular-chain axis. The matter has been resolved by complementing

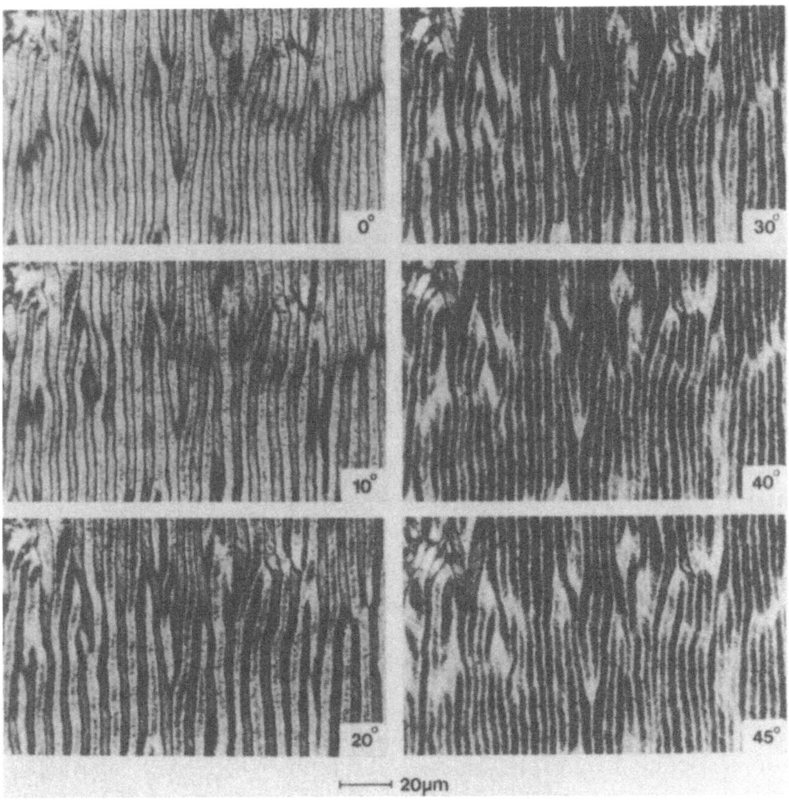

FIG. 29. Banded textures observed between crossed polars in a specimen of sheared and annealed thermotropic co-polyester (**X**). The six micrographs show the same field of view for different orientations of the crossed polars.[132]

the polarized light data with transmission electron microscopy and micro-electron diffraction.[116,128] Banding in samples prepared by shearing on to rocksalt was observed in dark-field diffraction contrast although it was not visible in bright-field. The contrast is thus confirmed as being due to periodic variations in molecular orientation rather than thickness changes, so it is possible, at least qualitatively, to correlate the light and electron images.[130] More detailed analysis shows that the trajectories of the optical and molecular directors are not exactly the same; there is also some evidence that the form of the meandering changes slightly from sample to sample and on annealing.[131]

The mechanism involved in the formation of the bands is not properly understood, and there is also some contention as to whether they form during shear or on stress relaxation afterwards. However, their appearance under suitable conditions of melt or solution flow is an almost universal characteristic of main-chain 'rigid' liquid crystal polymers.

CONCLUSIONS

The measurement of molecular anisotropy is central to developing an understanding of the molecular organization and properties of liquid crystal polymers. It is not a straightforward characteristic to define. There is, however, considerable variety in the techniques which can be employed for this purpose and in the information those techniques deliver. Perhaps the first question we should address with respect to this range of techniques is what exactly do we measure? For infrared and nuclear magnetic resonance spectroscopy, it is clear that we extract the average orientation of a particular molecular vector. Through the use of additional structural data this information may usually be converted to a molecular orientation parameter. This approach is particularly useful for a rigid chain segment, for, by using a number of such molecular vector orientation measurements, the so-called ordering matrix may be established. For flexible molecular segments, the techniques must be coupled with molecular modelling for a detailed interpretation. Deuterium NMR is a particularly powerful tool in this respect as it is able, at least in principle, to obtain orientational information about different molecular sites. ^{13}C NMR offers perhaps even more power, but is restricted at present through the lack of data relating to the anisotropy of the shielding tensors. In contrast, a complete analysis of the X-ray scattering from aligned samples provides details of both the orientation parameters and the orienting unit. Since it requires the latter to be defined before the orientational information may be extracted, the user is confronted with the general issue of orientation definition. Furthermore, X-ray scattering provides orientational information which is on the same scale as the spatial correlations present in the liquid crystal polymer. Thus the orientation description carries conformational data, and the approach allows an estimate of the local orientation, as against global orientation, to be assessed. Measurements such as birefringence and diamagnetic anisotropy do not provide structural information, certainly not at a molecular level. They are really bulk property parameters, which depend on a molecular model for a full interpretation.

The usefulness of any techniques cannot be divorced from the availability of appropriate equipment and the ease of sample preparation. NMR spectroscopy makes the most demands on machines and budgets. The remaining techniques use machinery normally available in the physical science arena, although perhaps the requirements for the quantitative X-ray measurements are slightly unusual. It is interesting to note that the budgetary requirements seem to be inversely related to our general understanding of the physical processes. Thus, for example, birefringence is a very simple method but very difficult to interpret at a molecular level, whilst deuterium NMR and X-ray scattering, although they are more complex techniques requiring considerable machinery, are supported with a detailed and thorough understanding in terms of molecular parameters.

The techniques of small-angle neutron scattering and deuterium NMR require the synthesis of particular materials. Equally, probe methods such as ESR spectroscopy and UV spectroscopy using dye molecules also involve the use of non-standard materials. The methods involving IR and UV/visible light are most appropriate to the study of thin films, and of course electro-optic effects in their device configuration. On the other hand X-ray scattering and NMR employ larger samples, with results more appropriate to the bulk material.

The timescale of the measurement is often significant. The orientation measurements we have reviewed in previous sections may have been made on samples in orienting fields in the liquid crystal phase; alternatively they may have examined samples in which the orientation imposed had been retained by cooling rapidly to room temperature.

The former *in situ* approach may involve the measurement of a static orientation, or some 'real-time' effect. Where short measurement times are of the essence, it is sometimes possible to achieve reliable measurements either by increasing the sensitivity, as in Fourier transform infrared spectroscopy, or by using sources of much more intense radiation such as synchrotron-generated X-rays.

Finally, in terms of techniques, we must remember that for most of the techniques discussed, the evaluation of molecular anisotropy represents only one part of the information that may be extracted. We would therefore employ X-ray scattering for structural studies on the scale 1–100 Å although, as we have demonstrated, such information comes as part of the orientation definition. NMR spectroscopy can provide invaluable information on the molecular dynamics, as can ESR and dielectric relaxation spectroscopy.

While concerning ourselves with methods of measurement, we must also take care to consider what we mean by 'molecular orientation'. What is the 'orienting unit'? What is the scale? Electron microscopy (pp. 165–6) clearly shows that locally (over ~100 nm) the molecular orientation of some liquid crystal polymers is significantly greater than the global average. This view is supported by detailed X-ray scattering (pp. 140–52). These experimental observations must not only form part of any realistic model of the polymeric liquid crystal state, but also should be borne in mind when specifying orientation parameters predicted from models or derived by experiment. Each of the techniques we have considered probes one aspect of molecular anisotropy. By combining their strengths we are in a position to tackle the question of molecular orientation, which is a central issue in any understanding of liquid crystal polymers.

REFERENCES

1. Chandrasekhar, S., *Liquid Crystals,* 1977, Cambridge University Press, Cambridge.
2. Acierno, D., La Mantia, F. P., Polizzotti, G., Ciferri, A. and Valenti, B., *Macromolecules,* 1982, **15,** 1455.
3. Coles, H. J. and Simon, R., in *Polymer Liquid Crystals,* A. Blumstein, Ed., 1983, Plenum, New York, p. 351.
4. Talroze, R. V., Shibaev, V. P., Sinitzyn, V. V. and Plate, N. A., in *Polymer Liquid Crystals,* A. Blumstein, Ed., 1983, Plenum, New York, p. 331.

5. MITCHELL, G. R. and WINDLE, A. H., *Colloid Polym. Sci.*, 1982, **260,** 754.
6. MITCHELL, G. R. and WINDLE, A. H., *Polymer,* 1983, **24,** 1513.
7. MITCHELL, G. R. and WINDLE, A. H., *Polymer,* 1983, **24,** 285.
8. KRIGBAUM, W. R. in *Polymer Liquid Crystals,* Ciferri, A., Krigbaum, W. R. and Meyer, R. B., Eds, 1982, Academic Press, New York, p. 275.
9. VINEY, C., *Faraday Disc. Chem. Soc.*, 1985, **79,** 228.
10. FORD, J. R. and MITCHELL, G. R., *Mol. Cryst. Liq. Cryst.*, in preparation.
11. DOBB, M. G. and MCINTYRE, J. E., *Advances in Polymer Science.*, 1984, **60/61,** Springer-Verlag, Berlin, p. 61.
12. GILLI, J.-M. and SIXOU, P., *Mol. Cryst. Liq. Cryst.*, 1984, **113,** 179.
13. FINKLEMANN, H., *Phil. Trans. R. Soc. London,* 1983, **A309,** 105.
14. COLES, H. J., *Faraday Disc. Chem. Soc.*, 1985, **79,** 201.
15. TAL'ROZE, R. V., SHIBAEV, V. P. and PLATE, N. A., *Polym. Sci. USSR,* 1983, **25,** 2863.
16. KOCK, H. J., FINKLEMANN, H., GLEIM, W. and REHAGE, G., in *Polymer Liquid Crystals,* Blumstein, A., Ed., 1983, Plenum, New York, p. 275.
17. DAVIS, F. J. and MITCHELL, G. R. *Polym. Commun.*, 1987, **28,** 8.
18. FRANK, F. C., *Phil. Trans. R. Soc. London,* 1983, **A309,** 71.
19. HOBSON, E. W., *Theory of Spherical and Ellipsoidal Harmonics,* 1931, Cambridge University Press, London.
20. MORSE, P. M. and FRESHBACH, H., *Methods of Theoretical Physics,* 1955, McGraw-Hill, New York.
21. HERMANS, P. H. and PLATZEK, P., *Kolloid Z.*, 1939, **88,** 65.
22. MAIER, W. and SAUPE, A., *Z. Naturforsch.*, 1958, **13a,** 564.
23. DELOCHE, B., and CHARVOLIN, J., *J. Phys. (Paris),* 1976, **37,** 1497.
24. HADZIIANNAOU, G., WANG, L. H., STEIN, R. S. and PORTER, R. S., *Macromolecules,* 1982, **15,** 880.
25. BROWN, D. J. and WINDLE, A. H., *J. Mat. Sci.*, 1984, **19,** 1997.
26. ERMAN, B. and MONNERIE, L., *Macromolecules,* 1985, **18,** 1985.
27. MITCHELL, G. R., *Brit. Polymer J.*, 1985, **17,** 111.
28. MITCHELL, G. R., DAVIS, F. J., and ASHMAN, A. S., *Polymer,* 1987, **28,** 639.
29. ZANNONI, C., in *Molecular Physics of Liquid Crystals,* Luckhurst, G. and Gray, G. W. Eds, 1979, Academic Press, London.
30. FRENZEL, J. and REHAGE, G., *Makromol. Chem.*, 1983, **184,** 1685.
31. MAIER, W. and ENGLERT, G., *Z. Physik. Chem.*, 1959, **19,** 168.
32. NOEL, C., LAUPRETRE, F., FRIEDRICH, C., FAYOLLE, B. and BOSIO, L., *Polymer,* 1984, **25,** 808.
33. RINGSDORF, H., SCHMIDT, H. W., BAUER, G. and KIEFER, R., *ACS Polym. Prepr. (Polym. Chem.),* 1983, **24,** 306.
34. TALROZE, R. V., SHIBAEV, V. P., SINITZYN, V. V., PLATE, N. A. and LOMONOSOV, M. V., *ACS Polym. Prepr. (Polym. Chem.),* 1983, **24,** 309.
35. EMSLEY, J. W., HASHIM, R., LUCKHURST, G. R. and SHILSTONE, G. N., *Liquid Crystals,* 1986, **1,** 437.

36. Finklemann, H. and Rehage, G., in *Advances in Polymer Science,* M. Gordon, Ed., 1984, **60/61,** Springer-Verlag, Berlin, p. 99.
37. Benthack, H., Finklemann, H. and Rehage, G., *J. Phys.* (*Paris*), 1983, **44,** 672.
38. Asada, T. in *Polymer Liquid Crystals,* Ciferri, A., Krigbaum, W. R. and Meyer, R. B., Eds, 1982, Academic Press, New York, p. 9.
39. Wassmer, K. H., Ohmes, E., Kothe, G., Portugall, M. and Ringsdorf, H., *Makromol. Chem. Rapid Commun.*, 1982, **3,** 281.
40. Mueller, K., Wassmer, K. H., Lenz, R. W. and Kothe, G., *J. Polym. Sci., Polym. Lett. Ed.*, 1983, **21,** 785.
41. Windle, A. H., Viney, C., Golombok, R., Donald, A. M. and Mitchell, G. R., *Faraday Disc. Chem. Soc.*, 1985, **79,** 55.
42. Viney, C., Mitchell, G. R. and Windle, A. H. *Polym. Commun.*, 1983, **24,** 145.
43. Viney, C., Mitchell, G. R. and Windle, A. H., *Mol. Cryst. Liq. Cryst.*, 1985, **129,** 75.
44. Kelker, H. and Hatz, R., *Handbook of Liquid Crystals,* 1980, Verlag Chemie, Weinheim, p. 260.
45. Haller, J., Huggins, H. A., Lilenthal, H. R. and McGuire, T. R., *J. Phys. Chem.*, 1973, **77,** 950.
46. De Jeu, W. H. *Physical Properties of Liquid Crystalline Materials,* 1980, Gordon and Breach, New York.
47. Tough, R. J. A. and Bradshaw, M. J., *J. Phys.* (*Paris*), 1983, **44,** 447.
48. Emsley, J. W., in *Nuclear Magnetic Resonance of Liquid Crystals,* Emsley, J. W., Ed., 1983, Reidel, Dordrecht, p. 279.
49. McBreity, V. J., *Polymer,* 1974, **15,** 503.
50. Molchanov, Yu. V., Privalov, A. F., Bilibin, A. Yu. and Skorokhodov, S. S., *Mol. Crys. Liq. Crys. Lett.*, 1985, **1,** 171.
51. Mitchell, G. R., and Ishii, F., *Polym. Commun.* 1985, **26,** 34.
52. Blumstein, A., Gauthier, M. M., Thomas, O. and Blumstein, R. B., *Faraday Disc. Chem. Soc.*, 1985, **79,** 33.
53. Yoon, D. Y., Bruckner, S., Volksen, W., Scott, J. C. and Griffin, A. C., *Faraday Disc. Chem. Soc.*, 1985, **79,** 33.
54. Martins, A. F., Volino, F. and Blumstein, R. B., in *Recent Advances in Liquid Crystalline Polymers,* Chapoy, L. L. Ed., 1985, Elsevier Applied Science Publishers, London, p. 279.
55. Volino, F., Alloneau, J. M., Giroud-Godquin, A. M., Blumstein, R. B., Stickles, E. M. and Blumstein, A., *Mol. Cryst. Liq. Cryst. Lett.*, 1984, **102,** 21.
56. Volino, F. and Blumstein, R. B., *Mol. Cryst. Liq. Cryst.*, 1984, **113,** 147.
57. Volino, F., Gautheir, M. M., Giroud-Godquin, A. M. and Blumstein, R. B., *Macromolecules,* 1985, **18,** 2620.
58. Blumstein, A., *Polym. J.*, 1985, **17,** 277.
59. Martins, A. F., Ferreira, J. B., Volino, F., Blumstein, A. and Blumstein, R. B., *Macromolecules,* 1983, **16,** 279.
60. Blumstein, R. B., Stickles, E. M., Gauthier, M. M., Blumstein, A. and Volino, F., *Macromolecules,* 1984, **17,** 177.

61. Bruckner, S., Campbell-Scott, J., Yoon, D. Y. and Griffin, A. C., *Macromolecules,* 1985, **18,** 2709.
62. de Gennes, P. G., in *Polymer Liquid Crystals,* Ciferri, A., Krigbaum, W. R. and Meyer, R. B., Eds, 1982, Academic Press, New York, p. 115.
63. Martin, P. G., Moore, J. S. and Stupp, S. I., *Macromolecules,* 1986, **19,** 2461.
64. Piskunov, M. V., Kostromin, S. G., Stroganov, L. B., Shibaev, V. P. and Plate, N. A., *Makromol. Chem. Rapid Commun.,* 1982, **3,** 443.
65. Charvolin, J. and Deloche, B., in *The Molecular Physics of Liquid Crystals,* Luckhurst, G. and Gray, G. W., Eds, 1979, Academic Press, London, p. 343.
66. Toriumi, H., Furuya, H. and Abe, A., *Polym. J.,* 1985, **17,** 895.
67. Samulski, E. T., Gauthier, M. M., Blumstein, R. B. and Blumstein, A., *Macromolecules,* 1984, **17,** 479.
68. Mueller, K., Hisgen, B., Ringsdorf, H., Lenz, R. W. and Kothe, G., *Mol. Cryst. Liq. Cryst.,* 1984, **113,** 167.
69. Mueller, K., Meier, P. and Kothe, G., *Progress in NMR Spectroscopy,* 1985, **17,** 211.
70. Spiess, H. W., in *Developments in Oriented Polymers,* Vol. 1, Ward, I. M., Ed., 1982, Applied Science, London.
71. Geib, H., Hisgen, B., Pschorn, U., Ringsdorf, H. and Spiess, H. W., *J. Amer. Chem. Soc.,* 1982, **104,** 917.
72. Kirste, R. G. and Ohm, H. G., *Makromol. Chem. Rapid Commun.,* 1985, **6,** 179.
73. Keller, P., Carvalho, B., Cotton, J. P., Lambert, M., Moussa, F. and Pepy, G., *J. Phys. (Paris) Lett.,* 1985, **46,** L-1065.
74. Noel, C., Monnerie, L., Archard, M. F., Hardouin, F., Sigaud, G. and Gasparoux, H., *Polymer,* 1981, **22,** 578.
75. Liebert, L., Strzelecki, L., van Luyen, D. and Levelut, A. M. *Eur. Polym. J.,* 1981, **17,** 71.
76. Sigaud, G., Yoon, D. Y. and Griffin, A. C., *Macromolecules,* 1983, **16,** 875.
77. Flygare, W. H., *Chem. Rev.,* 1974, **74,** 653.
78. Dupre, D. B., in *Polymer Liquid Crystals,* Ciferri, A., Krigbaum, W. R. and Meyer, R. B., Eds, 1982, Academic Press, New York, p. 166.
79. Dupre, D. B., Duke, R. W. and Samulski, E. T., *J. Chem. Phys.,* 1977, **66,** 2748.
80. Kresse, H., Kostromin, S. and Shibaev, V. P., *Makromol. Chem. Rapid Commun.,* 1982, **3,** 509.
81. Zentel, R., Strobl, G. R. and Ringsdorf, H., *Macromolecules,* 1985, **80,** 960.
82. Attard, G. S., *Mol. Phys.,* 1986, **58,** 1087.
83. Attard, G. S. and Williams, G., *Liquid Crystals,* 1986, **1,** 253.
84. Araki, K. and Attard, G. S., *Liquid Crystals,* 1986, **1,** 301.
85. Wilchinsky, Z. W., *Adv. X-ray Anal.* 1963, **6,** 231.
86. Windle, A. H., in *Developments in Oriented Polymers* Vol 1, Ward, I. M., Ed., 1982, Applied Science, London.

87. LEADBETTER, A. J. and NORRIS, E. K., *Mol. Phys.*, 1979, **38,** 669.
88. BHATTACHARJEE, B., PAUL, S. and PAUL, R., *Mol. Phys.*, 1981, **44,** 139.
89. MITCHELL, G. R. and ASHMAN, A. S., *Mol. Cryst. Liq. Cryst.*, in preparation.
90. MITCHELL, G. R. and LOVELL, R., *Acta Cryst.*, 1981, **A37,** 189.
91. LOVELL, R. and MITCHELL, G. R., *Acta Cryst.*, 1981, **A37,** 135.
92. DEAS, H. D., *Acta Cryst.*, 1952, **5,** 54.
93. RULAND, W., *Colloid Polym. Sci.*, 1977, **255,** 833.
94. LIEBERT, L., STRZELECKI, L., VAN LUYEN, D. and LEVELUT, A. M., *Eur. Polym. J.*, 1981, **17,** 71.
95. IANNELLI, P., ROVIELLO, A. and SIRIGU, A., *IUPAC Symposium on Non-crystalline Order in Polymers, Naples, Italy,* 1985, p. 69.
96. BLUMSTEIN, A., VILASAGAR, S., PONRATHNAM, S., CLOUGH, S. B., BLUMSTEIN, R. and MARET, G., *J. Polym. Sci., Polym. Phys. Ed.*, 1982, **20,** 877.
97. MARET, G. and BLUMSTEIN, A., *Mol. Cryst. Liq. Cryst.*, 1982, **88,** 295.
98. TAKEUCHI, Y., YAMAMOTO, F. and YAMAKAWA, S., *Polym. J.*, 1984, **16,** 579.
99. DE VRIES, A. J., *J. Chem. Phys.* 1972, **56,** 4489.
100. MITCHELL, G. R., *IUPAC Symposium on Non-crystalline Order in Polymers, Naples, Italy,* 1985, p. 73.
101. WINDLE, A. H., *Pure Appl. Chem.*, 1985, **57,** 1627.
102. MITCHELL, G. R., LOVELL, R. and WINDLE, A. H., *Polymer,* 1982, **23,** 1273.
103. MITCHELL, G. R. and WINDLE, A. H., *Polymer,* 1982, **23,** 1269.
104. VINEY, C. and WINDLE, A. H., *J. Mat. Sci.*, 1982, **17,** 2661.
105. MITCHELL, G. R., in *Order in the Amorphous 'State' of Polymers,* Keinath, S. E., Miller, R. L. and Rieke, J. K., Eds, 1986, Plenum, New York.
106. MITCHELL, G. R., and WINDLE, A. H., *Colloid Polym. Sci.*, 1985, **263,** 230.
107. FABER, T. E., *Phil. Trans. R. Soc. London,* 1983, **A309,** 115.
108. COLES, H. J. and TIPPING, J., *Mol. Cryst. Liq. Cryst. Lett.*, 1985, **2,** 23.
109. BRUCKER REPORT, Vol. 1, 1986, p. 2.
110. LEHMANN, O., *Z. Krist.*, 1890, **118,** 646.
111. DEMUS, D. and RICHTER, L., *Textures of Liquid Crystals,* 1978, Verlag Chemie, Weinheim.
112. MACKLEY, M. R., PINAUD, F. and SIEKMANN, G., *Polymer,* 1981, **22,** 437.
113. DONALD, A. M., VINEY, C. and WINDLE, A. H., *Phil. Mag.* (*B*), 1985, **52,** 925.
114. BARNES, N. R. and MITCHELL, G. R., *Polymer,* in preparation.
115. THOMAS, E. L. and WOOD, B. A., *Faraday Disc. Chem. Soc.*, 1985, **79,** 229.
116. DONALD, A. M. and WINDLE, A. H., *J. Mat. Sci.*, 1984, **19,** 2085.
117. DONALD, A. M., *J. Mat. Sci. Lett.*, 1984, **3,** 44.
118. DONALD, A. M. and WINDLE, A. H., in *Recent Advances in Liquid*

Crystalline Polymers, Chapoy, L. L. Ed., 1985, Elsevier, London, p. 187.
119. DONALD, A. M., *Phil. Mag.*, 1983, **47A,** L13.
120. SAWYER, L. C., *J. Polym. Sci. Polymer Lett.*, 1984, **22,** 347.
121. DONALD, A. M. and WINDLE, A. H., *Polymer,* 1984, **25,** 1253.
122. SAWYER, L. C. and JAFFEE, M., *J. Mat. Sci.*, 1986, **21,** 1897.
123. DOBB, M. G., JOHNSON, D. J. and SAVILLE, B. P., *J. Polym. Sci., Phys. Ed.*, 1977, **15,** 2201.
124. KISS, G. and PORTER, R. S., *Mol. Cryst. Liq. Cryst.*, 1980, **60,** 267.
125. VINEY, C. and WINDLE, A. H., unpublished work.
126. DONALD, A. M., VINEY, C. and WINDLE, A. H., *Polymer,* 1983, **24,** 155.
127. VINEY, C. and WINDLE, A. H., *Phil. Mag.*, in press.
128. DONALD, A. M. and WINDLE, A. H., *J. Mat. Sci.*, 1983, **18,** 1143.
129. VINEY, C., DONALD, A. M. and WINDLE, A. H., *J. Mat. Sci.*, 1983, **18,** 1137.
130. DONALD, A. M. and WINDLE, A. H., *Colloid Polym. Sci.*, 1983, **261,** 793.
131. VINEY, C., DONALD, A. M. and WINDLE, A. H., *Polymer,* 1985, **26,** 870.
132. VINEY, C. and WINDLE, A. H., *Polymer* 1986, **27,** 1325.

Chapter 4

APPLICATION OF HIGH-RESOLUTION ^{13}C NMR TO CRYSTALLINE POLYMERS

RYOZO KITAMARU
Institute for Chemical Research, Kyoto University, Japan

SUMMARY

The basic principles of high resolution ^{13}C NMR for solids, and some applications for studying the semicrystalline structures of polyethylene and celluloses are reviewed. Characteristic ^{13}C spectra, determined by three principal values of the chemical shift tensor, are obtainable by applying a sufficiently large resonant oscillating magnetic field to ^{1}H nuclei (DD). The determination of the principal axes of the chemical shift tensor of orthorhombic polyethylene is reviewed. The DD technique can be applied to oriented samples because the molecular chain axis is parallel to the tensor's principal axis having the largest principal value. Thus ^{13}C NMR with DD for uniaxial samples parallel to the static magnetic field shows a sharp crystalline line corresponding to the largest principal value and a non-crystalline line having the same chemical shift as observed in solution. The detailed study of the phase structure of a highly oriented sample is discussed.

The average of the three principal values is observable using a combination of dipole–dipole decoupling (DD) with magic-angle spinning (MAS). High resolution spectra can be obtained for solid samples in any form, even if the molecule contains many carbon species. Detailed information on the phase structure of samples can be obtained from the differences in magnetic relaxation for different phases. Two examples of such studies are reviewed.

The first concerns polyethylene with a lamellar morphology. DD/MAS ^{13}C NMR spectra show a sharp crystalline line (at 33·0 ppm *from the chemical shift of tetramethylsilene) and a rather broad non-crystalline contribution near 31 ppm which can be resolved into two lines at 31·0 ppm (a liquid-like amorphous phase) and at 31·3 ppm. The latter represents the crystalline–amorphous interface with a* trans-*rich chain conformation and a restricted chain motion. Although the presence of this interface had been predicted it is interesting that it can be detected by NMR.*

The second topic reviewed is a quantitative analysis of DD/MAS ^{13}C NMR spectra of native celluloses. The resonant lines of C1, C4 and C6 carbon species for these samples are deconvoluted into the crystalline and non-crystalline components using the difference in the spin-lattice relaxation time between both components. The chemical shifts of these lines are determined separately for the both components of samples, and correlated to the torsion angles ϕ and ψ about β-1,4-glycosidic linkage and χ, about the exocyclic C5–C6 bond, respectively. On the basis of these results, the characteristic chain conformations of the crystalline and non-crystalline components of native and regenerative cellulose are discussed.

High-resolution NMR in the solid state of matter has been developed fairly recently.[1] Since this technique can detect the local structure of molecules via chemical shift and magnetic relaxation, it has been possible to obtain detailed information on chain conformation as well as chain dynamics of macromolecules not only in the crystalline state but also in the non-crystalline, glassy or rubbery state. This chapter gives a brief description of the basic principles of solid-state high-resolution ^{13}C NMR as well as its recent application to crystalline polymers.

PRINCIPLES OF HIGH-RESOLUTION ^{13}C NMR IN SOLIDS[2]

Dipolar Decoupling

In order to understand the principle of solid-state high-resolution NMR we consider the Hamiltonian of an ensemble of nuclei possessing spin in a static magnetic field $\mathbf{B}_0$. The Hamiltonian $\mathcal{H}$ to be considered of this spin system can be written as follows.

$$\mathcal{H} = \mathcal{H}_Z + \mathcal{H}_S + \mathcal{H}_J + \mathcal{H}_D \qquad (1)$$

Here, $\mathcal{H}_Z$ represents a direct interaction of spins with the external magnetic field $\mathbf{B}_0$. This term, the so-called Zeeman energy, gives the zeroth-order absorption line in the NMR spectrum. It is described as

$$\mathcal{H}_Z = -\sum_{\mathbf{I}} \gamma\hbar\mathbf{I} \cdot \mathbf{B}_0 \tag{2}$$

Here, **I** represents the spin operator and γ the gyromagnetic ratio, h is Planck's constant and $\hbar = h/2\pi$. In a rectangular coordinate system with its z-axis parallel to $\mathbf{B}_0$, eqn (2) reduces to

$$\mathcal{H}_Z = -\sum_{\mathbf{I}} \gamma\hbar I_Z B_0 \tag{2$'$}$$

where I_Z is the Z-component of **I**. Since only one sort of spin species is usually measured, $\mathcal{H}_Z$ becomes equal to $-N\gamma\hbar I_Z B_0$ for a sample containing N spins. $\mathcal{H}_S$ describes the indirect interaction of spins with $\mathbf{B}_0$ via electrons, and gives different shifts to the zeroth-order absorption line. In the case of high-resolution NMR in solution, the absorption lines are mainly determined by this term. In ^{13}C NMR, absorption lines with different shifts are observable for individual carbons in a molecule because the electron environment of each carbon is different. $\mathcal{H}_D$ and $\mathcal{H}_J$ describe the direct and indirect (via electron) dipolar interaction between spins, respectively. The former gives the linewidth of the zeroth-order line and the latter, the so-called J-coupling or spin-coupling term, brings splitting or satellites to the lines determined by $\mathcal{H}_S$. In addition to these Hamiltonians there are some other terms, e.g. quadrupole Hamiltonians. In this chapter we restrict ourselves to consider ^{13}C NMR for a spin system all of whose nuclei have spins of either zero or $\frac{1}{2}$ and therefore such further terms will be omitted.

In a nuclear magnetic resonance measurement in solution, the direct dipolar interaction $\mathcal{H}_D$ actually disappears because, due to rapid molecular motion, the interspin (internuclear) vectors are rapidly space-averaged within the time-scale of a measurement. Hence, the terms $\mathcal{H}_S$ and $\mathcal{H}_J$ are detectable as sharp lines or splittings in a high-resolution spectrum of ^{1}H or ^{13}C and they can be related to detailed molecular structure or conformation of the substance investigated. In a solid, on the contrary, the directions of the internuclear vectors are stationary even if they are distributed randomly in space. Then, $\mathcal{H}_D$ gives a very wide linewidth to the zeroth-order absorption line and completely masks all lines due to $\mathcal{H}_S$ and $\mathcal{H}_J$.

Therefore, in usual NMR measurements one observes only a very broad absorption spectrum which is determined by $\mathcal{H}_Z$ and $\mathcal{H}_D$. However, even if the spectrum obtained is very broad, some information on the substance investigated can be obtained, and particularly on the crystalline structure or molecular motion via the dipolar interaction $\mathcal{H}_D$. Nevertheless, a spectrum in which the terms due to $\mathcal{H}_S$ and $\mathcal{H}_J$ are detectable cannot be obtained and, therefore, it is termed the broad-line NMR spectrum.

The first step to obtaining a high-resolution spectrum from solid matter is to eliminate the contribution from $\mathcal{H}_D$. For this purpose two techniques have been developed to date. One is known as 'dipolar decoupling' and is generally applicable in the NMR of rare spins within one kind (at present) of other abundant spins. The principle of this technique is very simple. Consider ^{13}C NMR for substances where ^{13}C and ^{1}H are the only spin-possessing nuclei. The natural abundance of ^{13}C is only 1·1%. The dipole–dipole interaction between ^{13}C atoms themselves can be neglected in the first approximation. Then it is sufficient for measuring the magnetization of ^{13}C only to eliminate the dipole–dipole interaction between ^{13}C and ^{1}H. This can be done by applying a resonant oscillating magnetic field to protons. If the amplitude of this oscillating field is sufficiently large, the directions of ^{1}H spins are rapidly changed and the dipolar interactions between ^{13}C and ^{1}H pairs as well as the J-coupling between all spin-spin pairs can be removed due to the averaging on the time-scale of the measurement. The amplitude (field strength) of the ^{1}H resonant magnetic field must be large enough. A field strength greater than at least 1·0 mT is preferable for usual solid-state ^{13}C NMR. This technique is known as proton dipolar decoupling (DD). As is well known, proton decoupling is usually used also for high-resolution ^{13}C NMR in solution in order to remove the splitting of lines due to spin-coupling $\mathcal{H}_J$. However, the field strength used for such solution NMR work is very small in comparison with that needed for DD. We will refer to this proton decoupling with small field strengths as proton scalar decoupling.

The second technique to eliminate the term $\mathcal{H}_D$ is the so-called multiple-pulse NMR method developed by Waugh *et al.*[3] The term $\mathcal{H}_D$ is eliminated by averaging in spin space with the use of a specially devised four-pulse sequence (WA–HU–HA). This technique is generally applicable to NMR of systems with abundant spins, for instance ^{1}H NMR of organic compounds. However, the multiple-pulse technique is not widely used at present due to difficulties in controlling the pulse sequence. Since there are not many examples of application to crystalline polymers, we do not describe this technique further.

Chemical Shift Tensor

When the terms $\mathscr{H}_D$ as well as $\mathscr{H}_J$ have been eliminated by the previously described DD or multiple-pulse techniques, the chemical shift term $\mathscr{H}_S$ can be detected and one expects a high-resolution spectrum. However, not all necessary conditions have yet been fulfilled. This is immediately obvious from the explicit form of $\mathscr{H}_S$:

$$\mathscr{H}_S = \gamma\hbar\mathbf{I} \cdot \boldsymbol{\sigma}\mathbf{B}_0 \tag{3}$$

Here, $\boldsymbol{\sigma}\mathbf{B}_0$ represents the shielding field induced by electrons surrounding the spin against $\mathbf{B}_0$ which depends on a local structure; $\boldsymbol{\sigma}$ is a coefficient, usually <100 ppm, termed the chemical shift, which determines the shielding field against $\mathbf{B}_0$. The chemical shift depends on the detailed environment of a nucleus to which spin $\mathbf{I}$ belongs. Hence, in the case of macromolecules $\boldsymbol{\sigma}$ can be defined for each chain conformation as well as for an interchain alignment. Mathematically, it is a linear operator which operates on $\mathbf{B}_0$ and produces another vector $\boldsymbol{\sigma}\mathbf{B}_0$. Since the vector $\boldsymbol{\sigma}\mathbf{B}_0$ is not necessarily parallel to $\mathbf{B}_0$, $\boldsymbol{\sigma}$ must be considered as a second-rank tensor. Therefore, even when the terms $\mathscr{H}_D$ and $\mathscr{H}_J$ are eliminated in the ^{13}C NMR spectrum by dipolar decoupling, the chemical shift Hamiltonian $\mathscr{H}_S$ cannot be observed as a sharp line due to the anisotropy of $\boldsymbol{\sigma}$. In the case of NMR in solution, the molecular conformation undergoes rapid fluctuation over all allowable conformations with rapid change of direction due to fast molecular motion. Accordingly, $\boldsymbol{\sigma}$ reduces to a scalar quantity which can be ascribed to each carbon species in the primary chemical structure and a high-resolution spectrum becomes obtainable. In contradistinction to liquids, yet another device is needed in order to obtain a high-resolution spectrum of solids because of the anisotropy of $\boldsymbol{\sigma}$. Equation (3) can be explicitly written in matrix notation as follows:

$$\begin{aligned}\mathscr{H}_S &= \gamma\hbar\mathbf{I} \cdot \boldsymbol{\sigma}\mathbf{B}_0 \\ &= \gamma h(I_x I_y I_z)\begin{pmatrix} \sigma'_{11} & \sigma'_{12} & \sigma'_{13} \\ \sigma'_{21} & \sigma'_{22} & \sigma'_{23} \\ \sigma'_{31} & \sigma'_{32} & \sigma'_{33} \end{pmatrix}\begin{pmatrix} 0 \\ 0 \\ B_0 \end{pmatrix} \\ &= \gamma B_0\hbar(\sigma'_{13}I_x + \sigma'_{23}I_y + \sigma'_{33}I_z)\end{aligned} \tag{4}$$

Here, σ'_{ij} designates the ijth element of $\boldsymbol{\sigma}$ in the rectangular coordinate system (laboratory frame). As can be seen from eqn (4), $\mathscr{H}_S$ comprises three terms. The first two terms do not commute with the Zeeman energy $\mathscr{H}_Z$ and are generally named non-secular terms. These non-secular terms contribute only negligibly to the absorption spectrum according to the general theory developed by Kubo and Tomita.[4] Hence, the experimentally detected dipolar-decoupled spectrum is considered to contain only the last term from eqn (4), including σ'_{33} in the first approximation.

As is well known, the 3×3 matrix σ'_{ij} can be diagonalized by an appropriate orthogonal coordinate transformation (rotational transformation), provided it is a symmetric matrix; generally it is considered to be symmetric because of its physical meaning. If the principal-axes frame of $\boldsymbol{\sigma}$, where $\boldsymbol{\sigma}$ is expressed by a diagonal matrix, is transformed to the laboratory frame by a rotational transformation $\mathbf{R}(\alpha, \beta, \gamma)$ which is defined by three Eulerian angles α, β and γ, then the representations of $\boldsymbol{\sigma}$ in both frames are related to each other by the equation

$$\boldsymbol{\sigma}' = \mathbf{R}\boldsymbol{\sigma}_\mathrm{p}\mathbf{R}^{-1} \tag{5}$$

Here, $\boldsymbol{\sigma}_\mathrm{p}$ and $\boldsymbol{\sigma}'$ are the representations of $\boldsymbol{\sigma}$ in the principal-axes frame and the laboratory frame, respectively; $\boldsymbol{\sigma}_\mathrm{p}$ is defined by its three principal values σ_{11}, σ_{22}, σ_{33} as follows:

$$\boldsymbol{\sigma}_\mathrm{p} = \begin{pmatrix} \sigma_{11} & & 0 \\ & \sigma_{22} & \\ 0 & & \sigma_{33} \end{pmatrix} \tag{6}$$

where it is customarily designated that $\sigma_{11} \leqslant \sigma_{22} \leqslant \sigma_{33}$. The rotational transformation $\mathbf{R}$ correlates the scalar components of the same vector in both frames. If vectors $\mathbf{r}(x, y, z)$ and $\mathbf{r}'(x', y', z')$ represent the same vector in the principal-axes frame and the laboratory frame, respectively, we have the relation

$$\mathbf{r}' = \mathbf{R}\mathbf{r} \tag{7}$$

or in matrix notation,

$$\begin{pmatrix} x' \\ y' \\ z' \end{pmatrix} = \begin{pmatrix} R_{11} & R_{12} & R_{13} \\ R_{21} & R_{22} & R_{23} \\ R_{31} & R_{32} & R_{33} \end{pmatrix} \begin{pmatrix} x \\ y \\ z \end{pmatrix} \tag{7a}$$

The rotational transformation **R** is the so-called Unitary and hence if $\tilde{\mathbf{R}}$ denotes the transpose of **R**, the inverse matrix $\mathbf{R}^{-1}(\mathbf{R}\mathbf{R}^{-1}=\mathbf{1})$ is equal to $\mathbf{R}: \mathbf{R}^{-1}=\tilde{\mathbf{R}}$, and furthermore $\det \mathbf{R} = \det \mathbf{R}^{-1} = 1$ from the mathematical requirement of keeping the vector length the same in both frames. **R** comprises three simple rotations: (1) a rotation $\mathbf{R}_z(\alpha)$ around the z-axis by an angle α in the principal-axes frame; (2) subsequent rotation $\mathbf{R}_y(\beta)$ around the y-axis by an angle β in the new frame; and (3) a final rotation $\mathbf{R}_z(\gamma)$ around the z-axis by an angle γ in the last frame, as shown in Fig. 1. Then it is represented as the consecutive product of three rotational transformations and explicitly

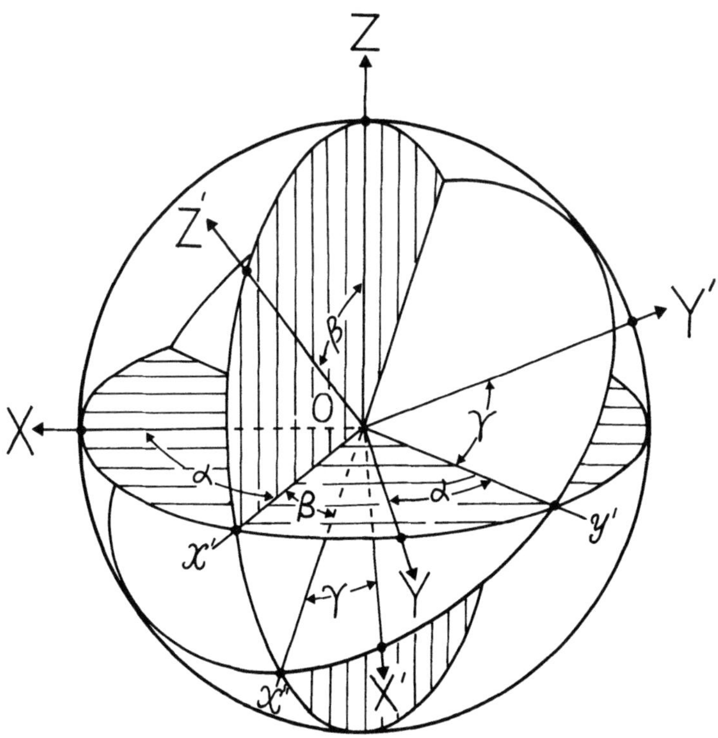

FIG. 1. Schematic representation of $R(\alpha, \beta, \gamma)$ which transforms the principal-axes frame (X, Y, Z) to the laboratory frame (X', Y', Z'). $R(\alpha, \beta, \gamma) = R(\gamma)R(\beta)R(\alpha)$; $R(\alpha) = (X, Y, Z) \rightarrow (x', y', z)$; $R(\beta) = (x', y', z) \rightarrow (x'', y', z')$; $R(\gamma) = (x'', y', z') \rightarrow (X', Y', Z')$.

expressed as a function of α, β and γ as follows:

$$\mathbf{R}(\alpha,\beta,\gamma)=\mathbf{R}_z(\gamma)\mathbf{R}_y(\beta)\mathbf{R}_z(\alpha)$$

$$=\begin{pmatrix}\cos\gamma & \sin\gamma & 0\\ -\sin\gamma & \cos\gamma & 0\\ 0 & 0 & 1\end{pmatrix}\begin{pmatrix}\cos\beta & 0 & -\sin\beta\\ 0 & 1 & 0\\ \sin\beta & 0 & \cos\beta\end{pmatrix}\begin{pmatrix}\cos\alpha & \sin\alpha & 0\\ -\sin\alpha & \cos\alpha & 0\\ 0 & 0 & 1\end{pmatrix}$$

$$=\begin{pmatrix}\cos\gamma\cos\beta\cos\alpha-\sin\gamma\sin\alpha & \cos\gamma\cos\beta\sin\alpha+\sin\gamma\cos\alpha & -\cos\gamma\sin\beta\\ -\sin\gamma\cos\beta\cos\alpha-\cos\gamma\sin\alpha & -\sin\gamma\cos\beta\sin\alpha+\cos\gamma\cos\alpha & \sin\gamma\sin\beta\\ \sin\beta\cos\alpha & \sin\beta\sin\alpha & \cos\beta\end{pmatrix} \quad (8)$$

Substitution of eqns (6) and (8) into eqn (5) yields the explicit representation of σ'_{33} by the Eulerian angles:

$$\sigma'_{ij}=(\mathbf{R}\boldsymbol{\sigma}_\mathrm{p}\mathbf{R}^{-1})_{ij}$$
$$=\sum_{k,l}R_{ik}(\boldsymbol{\sigma}_\mathrm{p})_{kl}(\mathbf{R}^{-1})_{lj}$$
$$=\sum_{k}R_{ik}(\boldsymbol{\sigma}_\mathrm{p})_{kk}(\mathbf{R}^{-1})_{kj} \qquad ((\sigma_\mathrm{p})_{kl}=0 \quad \text{except for} \quad k=l)$$

Accordingly,

$$\sigma'_{33}=R_{31}\sigma_{11}(\mathbf{R}^{-1})_{13}+R_{32}\sigma_{22}(\mathbf{R}^{-1})_{23}+R_{33}\sigma_{33}(R^{-1})_{33}$$
$$=\sigma_{11}\cos^2\alpha\sin^2\beta+\sigma_{22}\sin^2\alpha\sin^2\beta+\sigma_{33}\cos^2\beta \qquad (9)$$

Therefore, the experimentally observable chemical shift σ'_{33} becomes a function of the principal values as well as the Eulerian angles α and β that transform the principal-axes frame to the laboratory frame. As has been mentioned already, the chemical shift $\boldsymbol{\sigma}(\sigma_{11}, \sigma_{22}, \sigma_{33})$ is defined for individual carbon atoms in a molecule depending on the molecular conformation and σ'_{33} depends on Eulerian angles α and β. Therefore, the lineshape is decided by the spatial alignment of the molecule to $\mathbf{B}_0$ during measurement, i.e. by the molecular alignment in the sample as well as by the alignment of the sample with respect to $\mathbf{B}_0$.

Lineshape of Single Crystal or Oriented Sample

In the case of a single crystal, each carbon species yields an individual chemical shift tensor which depends on the molecular conformation.

According to eqn (9), the lineshape of each carbon is determined by the spatial alignment of the sample to $\mathbf{B}_0$ in the laboratory frame. It follows immediately from eqn (9) that the following relations hold.

$$\sigma'_{33} = \sigma_{33} \quad \text{for} \quad \beta = 0$$

$$\sigma'_{33} = \sigma_{11} \quad \text{for} \quad \beta = 90^\circ \quad \text{and} \quad \alpha = 0$$

$$\sigma'_{33} = \sigma_{22} \quad \text{for} \quad \alpha = \beta = 90^\circ$$

Therefore, NMR measurements with different alignments of a single crystal to $\mathbf{B}_0$ using a goniometer give us the principal values of a particular chemical shift according to the above relations. This is the case if the absorption lines from different carbons are distinguishable or can be decomposed. In the case of macromolecules, a sufficiently large single crystal needed for such a measurement is usually not available. However, for most linear-chain molecules, a uniaxially orientated crystalline structure can be obtained. Then one of the principal values of $\boldsymbol{\sigma}$ can be determined if the principal axis (corresponding to the principal value of $\boldsymbol{\sigma}$) is parallel to the molecular chain axis. For example, in linear polyethylene, the σ_{33}-axis of methylene carbons in the *trans–trans* molecular chain sequence coincides with the chain axis. Hence, σ_{33} is detected as a sharp line in the case of a uniaxially oriented sample where the chain axis is parallel to the direction of $\mathbf{B}_0$, as shown later.

Lineshapes of Powdered Crystals

In powdered crystals, a particular molecular conformation may be considered to be stable but oriented at random in space. Let $p(\Omega)\,d\Omega$ be the probability that the principal-axes frame of $\boldsymbol{\sigma}$ is oriented in a solid angle between $\Omega(\alpha, \beta, \gamma)$ and $\Omega + d\Omega$. Then, the absorption intensity $I(\omega)$ at a frequency ω can be correlated with $p(\Omega)\,d\Omega$ as follows.

$$I(\omega)\,d\omega = p(\Omega)\,d\Omega \tag{10}$$

In the case of a polycrystalline sample, $p(\Omega)$ is considered to be constant because of the random orientation of individual crystallites. Then $I(\omega)$ reduces to

$$I(\omega) = |d\Omega/d\omega| \tag{11}$$

where the proportionality coefficient is omitted.

If the crystal structure is axially symmetric with σ_{33} being the symmetry axis and $\sigma_{11} = \sigma_{22}$, then $d\Omega$ is proportional to $\sin\beta\, d\beta$ and independent of α.

Hence

$$I(\omega) = \left|\sin\beta \frac{d\beta}{d\omega}\right| \tag{12}$$

Converting σ'_{33} and σ_{11}, σ_{22}, σ_{33} in eqn (9) to ω, ω_{11}, ω_{22}, ω_{33} (in frequency units), respectively, we have

$$\omega = \omega_{11}\cos^2\alpha\sin^2\beta + \omega_{22}\sin^2\alpha\sin^2\beta + \omega_{33}\cos^2\beta \tag{9a}$$

Substitution of $\omega_{11} = \omega_{22} = \omega_\perp$, $\omega_{33} = \omega_\parallel$, into eqn (9a) yields

$$\omega = (\omega_\parallel - \omega_\perp)\cos^2\beta + \omega_\perp \tag{13}$$

From eqns (9a) and (13) the following equations are obtained.

$$d\beta/d\omega = -\tfrac{1}{2}(\cos\beta\sin\beta(\omega_\parallel - \omega_\perp))^{-1}$$

$$|\cos\beta| = (\omega - \omega_\perp)^{1/2}/(\omega_\parallel - \omega_\perp)^{-1/2}$$

for $\omega_\perp < \omega < \omega_\parallel$. Accordingly, the absorption intensity $I(\omega)$ becomes

$$I(\omega) = \sin\beta\frac{d\beta}{d\omega}$$

$$= \tfrac{1}{2}(\omega - \omega_\perp)^{-1/2}(\omega_\parallel - \omega_\perp)^{-1/2} \tag{14}$$

In the general case where $\omega_{11} < \omega_{22} < \omega_{33}$, treating the absorption intensity $I_\beta(\omega)$ for a temporarily fixed β in a similar way, we have

$$I_\beta(\omega) = \left|\frac{d\alpha}{d\omega}\right|$$

The subsequent averaging over β:

$$I(\omega) = \int_0^\pi I_\beta(\omega)\sin\beta\, d\beta \Big/ \int_0^\pi \sin\beta\, d\beta$$

yields the lineshape. Since this calculation is rather tedious, only the result is given.[5]

For $\omega_{11} \leq \omega < \omega_{22}$, we have

$$I(\omega) = \frac{1}{\pi}(\omega_{33} - \omega)^{-1/2}(\omega_{22} - \omega_{11})^{-1/2}K(m) \tag{15a}$$

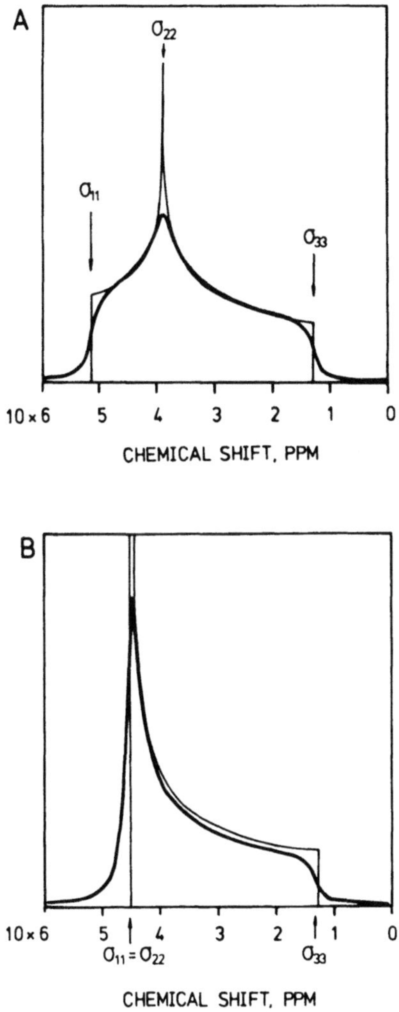

FIG. 2. Schematic representation of powder lineshape. (A) $\sigma_{11} < \sigma_{22} < \sigma_{33}$. The zero point of the chemical shift is arbitrarily fixed. The relative ratios of the principal values are fixed as for the principal values for crystalline polyethylene listed in Table 1. (B) $\sigma_{11} = \sigma_{22} < \sigma_{33}$. The thinner lines indicate the theoretical lineshapes according to eqns (14) and (15) and the solid lines are obtained from the theoretical lines by convoluting the Lorentzian distribution function with a linewidth of 2 ppm.

with

$$m^2 = \frac{(\omega - \omega_{11})(\omega_{33} - \omega_{22})}{(\omega_{33} - \omega)(\omega_{22} - \omega_{11})}$$

For $\omega_{22} < \omega \leqslant \omega_{33}$, we obtain

$$I(\omega) = \frac{1}{\pi}(\omega - \omega_{11})^{-1/2}(\omega_{33} - \omega_{22})^{-1/2}K(m) \tag{15b}$$

with

$$m^2 = \frac{(\omega_{22} - \omega_{11})(\omega_{33} - \omega)}{(\omega_{33} - \omega_{22})(\omega - \omega_{11})}$$

where $K(m)$ is the complete elliptic integral of the first kind:

$$K(m) = \int_0^{\pi/2} (1 - m^2 \sin^2 \phi)^{-1/2}\, \mathrm{d}\phi$$

The absorption intensities $I(\omega)$ calculated from eqns (14) and (15) as functions of ω are schematically shown in Fig. 2.

Sample Rotation and Magic Angle Sample Spinning (MAS)[6,7]
Consider an NMR measurement performed on a sample which is rotated at a high rate around an axis inclined at an angle ζ to B_0. Let a coordinates system (x', y', z') be fixed in the sample, the z'-axis of which coincides with the rotating axis (see Fig. 3). Then the

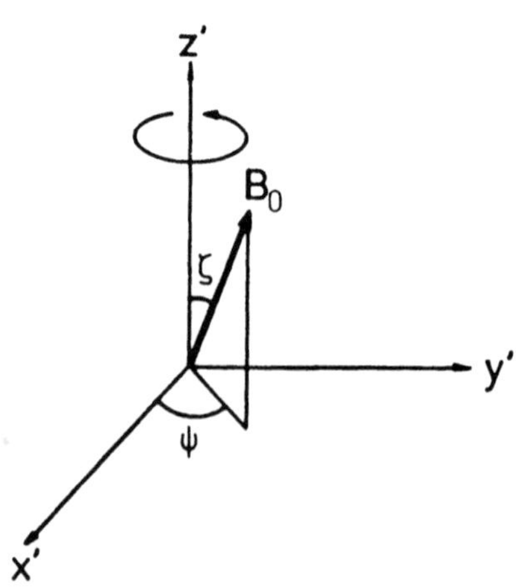

FIG. 3. Geometry for sample spinning.

coordinates (x', y', z') become a rotating frame during the measurement. In this rotating frame, the static field B_0 rotates around the z'-axis. Let $\boldsymbol{\sigma}'$ and $\boldsymbol{\sigma}''$ be the expressions of a chemical shift $\boldsymbol{\sigma}$ for a molecular conformation in the laboratory and rotating frames, respectively. The $\boldsymbol{\sigma}'$ and $\boldsymbol{\sigma}''$ are related to each other by the rotational transformation $\mathbf{R}(\alpha, \zeta, \psi)$ which transforms the laboratory frame to the rotating frame, as follows (cf. eqn (5)).

$$\boldsymbol{\sigma}' = \mathbf{R}^{-1}\boldsymbol{\sigma}''\mathbf{R} \tag{16}$$

Here, $\mathbf{R}(\alpha, \zeta, \psi)$ is expressed similarly to eqn (8),

$$\mathbf{R}(\alpha, \zeta, \psi) = \begin{pmatrix} \cos\psi\cos\zeta\cos\alpha - \sin\psi\sin\alpha & \cos\psi\cos\zeta\sin\alpha + \sin\psi\cos\alpha & -\cos\psi\sin\zeta \\ -\sin\psi\cos\zeta\cos\alpha - \cos\psi\sin\alpha & -\sin\psi\cos\zeta\sin\alpha + \cos\psi\cos\alpha & \sin\psi\sin\zeta \\ \sin\zeta\cos\alpha & \sin\zeta\sin\alpha & \cos\zeta \end{pmatrix} \tag{17}$$

Here, the angle α can be taken arbitrarily and ψ changes at the rate of sample spinning. When the angle α is equal to zero, eqn (17) reduces to

$$\mathbf{R}(0, \zeta, \psi) = \begin{pmatrix} \cos\psi\cos\zeta & \sin\psi & -\cos\psi\sin\zeta \\ -\sin\psi\cos\zeta & \cos\psi & \sin\psi\sin\zeta \\ \sin\zeta & 0 & \cos\zeta \end{pmatrix} \tag{17a}$$

Then the experimentally observable chemical shift σ'_{33} can be derived from eqn (16)

$$\begin{aligned} \sigma'_{33} &= (\mathbf{R}^{-1}\sigma''\mathbf{R})_{33} \\ &= \sum_{i,j} (\mathbf{R}^{-1})_{3i}\sigma''_{ij}R_{j3} \\ &= \sum_{i,j} R_{i3}R_{j3}\sigma''_{ij} \end{aligned}$$

Substitution of the matrix elements R using eqn (17a) yields σ'_{33} as

$$\begin{aligned} \sigma'_{33} = {} & \cos^2\psi\sin^2\zeta\,\sigma''_{11} - \sin\psi\cos\psi\sin^2\zeta\,\sigma''_{21} - \cos\psi\sin\zeta\cos\zeta\,\sigma''_{31} \\ & - \cos\psi\sin\psi\sin^2\zeta\,\sigma''_{12} + \sin^2\psi\sin^2\zeta\,\sigma''_{22} + \sin\psi\cos\zeta\sin\zeta\,\sigma''_{32} \\ & - \cos\psi\sin\zeta\cos\zeta\,\sigma''_{13} + \sin\psi\sin\zeta\cos\zeta\,\sigma''_{23} + \cos^2\zeta\,\sigma''_{33} \end{aligned} \tag{18}\dagger$$

† Since the calculation of σ'_{33} involves only the matrix elements in the third column of $R(\alpha, \zeta, \psi)$ in eqn (17) that does not include the angle α, this relation is valid even if $\alpha \neq 0$.

Since the sample is rotated at a high rate, any function $F(\psi)$ of ψ in the above expression can be averaged as

$$\overline{F(\psi)} = \frac{1}{2\pi}\int_0^{2\pi} F(\psi)\,\mathrm{d}\psi$$

in the first approximation. Then $\overline{\sin^2\psi} = \overline{\cos^2\psi} = \frac{1}{2}$, $\overline{\sin\psi\cos\psi} = 0$, $\overline{\sin\psi} = \overline{\cos\psi} = 0$. Therefore in eqn (18) only the terms with σ''_{11}, σ''_{22} and σ''_{33} remain and we obtain

$$\begin{aligned}\sigma'_{33} &= \tfrac{1}{2}\sin^2\zeta\,\sigma''_{11} + \tfrac{1}{2}\sin^2\zeta\,\sigma''_{22} + \cos^2\zeta\,\sigma''_{33}\\ &= \tfrac{1}{3}(\sigma''_{11}+\sigma''_{22}+\sigma''_{33}) + \tfrac{1}{6}(1-3\cos^2\zeta)(\sigma''_{11}+\sigma''_{22}-2\sigma''_{33})\\ &= \tfrac{1}{3}(\sigma_{11}+\sigma_{22}+\sigma_{33}) + \tfrac{1}{6}(1-3\cos^2\zeta)(\sigma''_{11}+\sigma''_{22}-2\sigma''_{33})\\ &= \tfrac{1}{3}\,\mathrm{tr}\,\sigma + \tfrac{1}{6}(1-3\cos^2\zeta)(\sigma''_{11}+\sigma''_{22}-2\sigma''_{33}) \qquad (19)\end{aligned}$$

In eqn (19), we have used a theorem that the trace of a matrix is unchanged by a rotational transformation of the coordinates in order to derive the third equality from the second one.

Equation (19) provides the principle for the magic angle sample spinning technique in solid-state ^{13}C NMR. If $1-3\cos^2\zeta = 0$, i.e. $\zeta = 54°44'$,

$$\begin{aligned}\sigma'_{33} &= \tfrac{1}{3}(\sigma_{11}+\sigma_{22}+\sigma_{33})\\ &= \tfrac{1}{3}\,\mathrm{tr}\,\boldsymbol{\sigma}\\ &= \sigma_{\mathrm{ave}} \qquad (20)\end{aligned}$$

Therefore, if a sample is rotated at a sufficiently high rate around the axis inclined at 54°44′ to $\mathbf{B}_0$ during the measurement, the average of the principal values of a chemical shift tensor can be observed as a sharp line. We call the angle 54°44′ the 'magic angle' and the technique of rotating a sample at the magic angle is called 'magic angle sample spinning' (MAS). Inasmuch as the molecular conformation is fixed during the measurement, the average of the principal values of the chemical shift tensor definable for individual carbon species is observable in principle even if the sample contains many carbon species in a molecule.[8,9]

In the above discussion, we have put $\overline{\sin\psi} = \overline{\cos\psi} = 0$, $\sin\psi\cos\psi = 0$ in eqn (18). This is valid only for the case where the rate of the sample rotation is infinite. For macromolecules a rate of rotation greater than 3 kHz is sufficient to obtain the high-resolution spectrum in most cases. However, with each spectrum there are associated satellites corresponding to the rotation rate, as is evident from eqn (18). Further discussion of these satellites is beyond the scope of this chapter. It should be noted here that it is necessary for a quantitative analysis of a spectrum to decompose its satellites carefully. These satellites are sometimes useful for assignment of each line.

Cross-Polarization (CP)[1]

In addition to DD and MAS in solid-state ^{13}C NMR, there is one more important technique, so-called cross-polarization or proton-enhancement, based on the use of the cross-magnetization between ^{1}H and ^{13}C. In this technique one enhances the ^{13}C magnetization using the equilibrium magnetization of ^{1}H nuclei. The ^{1}H equilibrium magnetization ${}^{H}M_0$ in the z-axis direction (the direction of static field) is first turned 90° into the y-axis direction by applying a ^{1}H resonant field $\mathbf{B}_{1H}$ oscillating linearly in the x-axis direction; then the phase of this resonant field is shifted 90° so that it oscillates linearly in the y-axis. The linearly oscillating resonance field $\mathbf{B}_{1H}$ at ω is considered to be equivalent in effects to a rotating field with a field strength of $B_{1H}/2$. Then, in a rotating frame at ω, the $\mathbf{B}_{1H}$ becomes a static field and the ^{1}H equilibrium magnetization ${}^{H}M_0$ is fixed in this field. To set the ${}^{H}M_0$ along this static field is referred to as 'locking ${}^{H}M_0$ in $\mathbf{B}_{1H}$'. If one contacts a ^{13}C spin system with this ^{1}H spin system by applying a ^{13}C resonant field $\mathbf{B}_{1C}$ of an adequate amplitude, enhanced ^{13}C magnetization appears (most effectively at the condition: $\gamma_C B_{1C} = \gamma_H B_{1H}$) due to cross relaxation between the ^{1}H and ^{13}C spin systems. By this cross-polarization technique, a ^{13}C magnetization of 16-fold the equilibrium magnetization at the maximum can theoretically be obtained. In principle one can obtain the high-resolution spectrum without this cross-polarization by applying DD with or without MAS, utilizing the ^{13}C equilibrium magnetization. Nevertheless, the natural abundance of ^{13}C is only 1·1% and the gyromagnetic ratio is about one-quarter that of ^{1}H. Accordingly, the ^{13}C equilibrium magnetization which can be obtained is very small and hence it is necessary to repeat the measurement many times in order to obtain a spectrum with a sufficient signal-to-noise ratio. The CP technique could improve this signal-to-noise ratio of the spectrum by equivalent repeating times. In addition, the spin–lattice relaxation time of ^{1}H, T_{1H}, is usually much shorter than that of ^{13}C, T_{1C}. As is evident from eqn (21) it takes at least five times T_1 for a full recovery of the equilibrium magnetization. Hence one must wait at least five times T_{1C} after each pulse in the case of the measurement without CP, whereas waiting only five times T_{1H} (which is much shorter than T_{1C}) is enough in the case of the measurement with CP. Therefore, the CP technique also saves the time necessary for a measurement.

The cross-polarization technique is very important from the technological point of view. Without this technique solid-state high-resolution ^{13}C NMR might not have been developed. However, it has no inherent physical significance. Furthermore, it must be noted that CP efficiency is not necessarily the same for the individual carbon nuclei since it depends on their molecular mobility. Hence sometimes one has to carry out the NMR measurement without CP for a quantitative analysis.

MAGNETIC RELAXATION AND MOLECULAR MOTION[4,10–14]

Introduction

As described above, high-resolution ^{13}C NMR spectra are obtainable also for solid samples, combining techniques such as DD, MAS and CP. It is, then, possible to investigate the molecular chain conformation characteristics of the solid state via the chemical shift of individual carbons. Furthermore, if one measures simultaneously the relaxation of ^{13}C magnetization, one also expects to obtain information on molecular chain dynamics. This section describes the relaxation phenomenon.

A spin system in a static field $\mathbf{B}_0$ generally exhibits a magnetization in the direction parallel to $\mathbf{B}_0$ as the result of energy equilibrium with other degrees of freedom related to the spin system such as the rotation and vibration of atoms and molecules; these degrees of freedom are generally called lattice. This equilibrium magnetization can be disturbed by an appropriate perturbation, but after the perturbation is removed the magnetization tends to return to the equilibrium value. This process is called magnetic relaxation and the relaxation in the directions parallel or perpendicular to $\mathbf{B}_0$ is measured in conventional NMR. The relaxation parallel to $\mathbf{B}_0$ is called longitudinal relaxation, or spin–lattice relaxation since it accompanies energy exchange with the lattice, whereas relaxation perpendicular to $\mathbf{B}_0$ is called transverse relaxation, or spin–spin relaxation because it is carried out by spin exchanges within the spin system. These relaxations generally follow the relations

$$M_z(t) - M_z^0 = [M_z(0) - M_z^0] \exp(-t/T_1) \quad (21)$$

$$M_x(t) = M_x(0) \exp(-t/T_2) \quad (22)$$

Here, $M_z(t)$ and $M_x(t)$ denote the magnetization in the directions parallel and perpendicular to $\mathbf{B}_0$ at time t, respectively, and M_z^0 the equilibrium magnetization. The time constants T_1 and T_2 in the equations which characterize the relaxation rates in the z- and x-axis directions are called spin–lattice (longitudinal) and spin–spin (transverse) relaxation times, respectively. In addition to these relaxations, there are many other kinds of relaxation such as longitudinal relaxation in rotating frame $T_{1\rho}$ or the cross-polarization mentioned in the previous sections.

The magnetic relaxations mentioned above are generally caused by the perturbation of magnetic field in the vicinity of the spins. Any mechanism that initiates the perturbation of magnetic field which includes a frequency component that corresponds to the energy splittings of the spin system can cause these relaxations. For example, since $\mathscr{H}_S$, $\mathscr{H}_J$ and $\mathscr{H}_D$ in eqn (1) are time-dependent terms, they can cause these relaxations. However, the relaxation of magnetization for macromolecules including only nuclei with spin $\frac{1}{2}$ such as ^{13}C and ^{1}H, stems mostly from the time-fluctuation of the dipole–dipole interaction $\mathscr{H}_D$. In ^{13}C NMR, the time fluctuation of the dipole–dipole interaction between chemically bonded ^{13}C and ^{1}H is the primary mechanism for the relaxation because the interaction between ^{13}C nuclei themselves is negligible owing to the very low natural abundance and the dipole–dipole interaction rapidly diminishes with increasing interspin distance (it is inversely proportional to the third power of the distance). Because the time-fluctuation of the interaction, of course, takes place as a result of time-fluctuation of internuclear vectors, examination of the relaxation provides information on the motional mode of the internuclear C–H vectors in the structure. Therefore, in solid-state ^{13}C NMR, the simultaneous interpretation of the chemical shifts of individual carbons and the relaxation parameters provides insight into the chain conformation and dynamics of macromolecules.

Magnetic Relaxations of a Two-Spin System

Consider a spin system of two spins **I** and **S** in a static field $\mathbf{B}_0$ in order to understand the mechanism of the longitudinal and transverse relaxation of ^{13}C magnetization. For simplicity we deal with the case when both spins are $\frac{1}{2}$; i.e. the maximum eigenvalues of one-dimensional components of the spins are $\frac{1}{2}$. Then, four zeroth-order energy eigenstates can be defined for this system; (1) $|++)$: both spins are parallel to $\mathbf{B}_0$; (2) $|+-)$: **I** spin parallel and **S** spin antiparallel; (3) $|-+)$: **I** spin antiparallel and **S** spin parallel; (4) $|--)$: both spins are antiparallel to $\mathbf{B}_0$. The relaxations of magnetization of **I** and **S** spins are conducted by the mutual transitions among these eigenstates. If the rates of the transitions are independent of time, we have the relationships

$$-\frac{dM_{IZ}}{dt} = \rho(M_{IZ} - M_I^0) + \sigma\frac{\gamma_I}{\gamma_S}(M_{SZ} - M_S^0) \tag{23a}$$

$$-\frac{dM_{SZ}}{dt} = \rho'(M_{SZ} - M_S^0) + \sigma\frac{\gamma_S}{\gamma_I}(M_{IZ} - M_I^0) \tag{23b}$$

considering the balance of the occupation probabilities among the four eigenstates. Here M_{IZ}, M_{SZ} denote the expected magnetizations in the z-axis direction ($\mathbf{B}_0$ direction) of **I** and **S** spins, respectively; M_I^0 and M_S^0 denote the equilibrium values of M_{IZ} and M_{SZ}, respectively; γ_I and γ_S are the gyromagnetic ratios of the **I** and **S** spins, respectively. The parameters ρ, σ, ρ' are determined by the rate of transitions among the four eigenstates. If we designate w_0, w_1, w_1' and w_2 as the transition probabilities among the four eigenstates $|++)$, $|+-)$, $|-+)$, $|--)$ as shown in Fig. 4, then

$$\begin{aligned} \rho &= w_0 + 2w_1 + w_2 \\ \rho' &= w_0 + 2w_1' + w_2 \\ \sigma &= w_2 - w_0 \end{aligned} \tag{24}$$

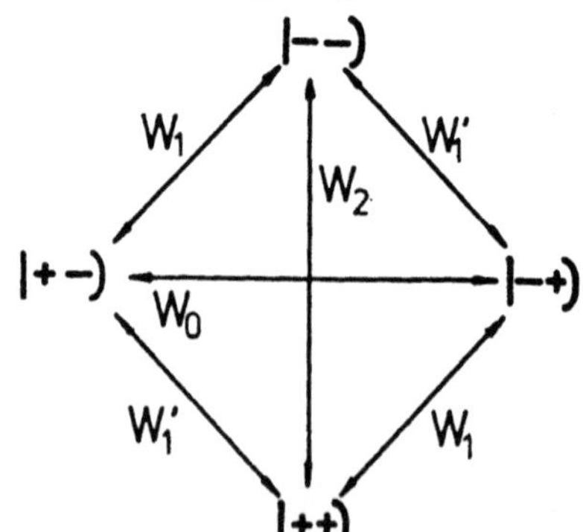

FIG. 4. Energy eigenstates for two-spin system and definitions of transition probabilities.

As is evident from eqns (23), the time-dependencies of the magnetizations of **I** and **S** spins are rather conplicated because they are mutually correlated even if the conjugated differential equations (23) can be generally solved.

However, if the magnetization of one of the two spins is zero or stays unchanged during the relaxation, the conjugated eqn (23) can be easily solved. For example, if M_{IZ} is independent of t, we have

$$M_{SZ}(t) - M_{SZ}(\infty) = [M_{SZ}(0) - M_{SZ}(\infty)]e^{-\rho' t} \tag{25}$$

with

$$M_{SZ}(\infty) = M_S^0 - \frac{\sigma}{\rho'}\frac{\gamma_S}{\gamma_I}(M_{IZ} - M_I^0)$$

In the above, if we consider that the spins **S** and **I** correspond to ^{13}C and ^{1}H spins, respectively, we have the longitudinal relaxation of ^{13}C magnetization in ^{13}C NMR measurements,

$$M_Z(t) - M_Z(\infty) = [M_Z(0) - M_Z(\infty)]e^{-\rho' t} \tag{26}$$

If the magnetization of ^{1}H stays at its equilibrium value during the relaxation, it follows from eqn (26) that

$$M_Z(\infty) = M_Z^0$$

Then, we have

$$M_Z(t) - M_Z^0 = [M_Z(0) - M_Z^0]e^{-\rho' t} \tag{26a}$$

By comparison with eqn (21), the longitudinal relaxation of ^{13}C, T_{1C}, can be defined as

$$\begin{aligned} 1/T_{1C} &= \rho' \\ &\equiv w_0 + 2w_1' + w_2 \end{aligned} \tag{27}$$

If, however, ^{1}H scalar or dipolar decoupling is applied during the relaxation, the magnetization of ^{1}H nuclei can be considered to be zero. Hence, it follows that

$$M_Z(\infty) = M_Z^0 + \frac{\sigma}{\rho'}\frac{\gamma_C}{\gamma_H} M_H^0$$

where M_H^0 denotes the equilibrium magnetization of ^{1}H. Since the equilibrium magnetization of a spin is proportional to the second power of the gyromagnetic ratio, then $M_H^0 = M_C^0\gamma_H^2/\gamma_C^2$. It follows that

$$M_Z(\infty) = M_Z^0\left(1 + \frac{\sigma}{\rho'}\frac{\gamma_H}{\gamma_C}\right) \tag{27a}$$

This means that the ^{13}C magnetization is enhanced by the factor

$$1 + \frac{\sigma}{\rho'}\frac{\gamma_H}{\gamma_C}$$

if ^{1}H decoupling is applied. This factor is referred to as nuclear Overhauser enhancement (NOE) and is widely examined in relation to molecular mobility.

Now we consider the significance of the spin–lattice relaxation time, examining the transition probabilities w_1', w_2 and w_0. The transition between energy eigenstates caused by a perturbation is generally conducted by frequency components corresponding to the energy difference. Mathematically, the transition probability is described as the Fourier transform of the time-correlation function of the perturbation. In the case in which the transitions are carried out by the time-fluctuation of the dipole–dipole interaction between ^{13}C and ^{1}H, the transition probability, $w_{nn'}$, from state n to n' is described as

$$w_{nn'} = \gamma_C^2\gamma_H^2\hbar \sum_m |\langle n'| A_m |n\rangle|^2 \int_{-\infty}^{\infty} \overline{Y_m^*(t+\tau)Y_m(t)} \exp(-i\omega_{nn'}\tau)\, d\tau \tag{28}$$

where $m = 0$, ± 1 and ± 2. The asterisk indicates the complex conjugate and the upper bar the average of the ensemble of the spin system; the ensemble average is independent of the arbitrarily chosen time t. The functions Y_m are the so-called orientation functions that characterize the orientation of the C–H vector in space. Let the C–H vector be $\mathbf{r}$, expressed by (r, θ, ϕ) in polar coordinates as shown in Fig. 5; then

$$\begin{aligned} Y_0 &= r^3(1 - 3\cos^2\theta) \\ Y_{\pm 1} &= r^3 \sin\theta \cos\theta \exp(\pm i\phi) \\ Y_{\pm 2} &= r^3 \sin^2\theta \exp(\pm 2i\phi) \end{aligned} \tag{29}$$

and A_m are time-independent operator terms defined as

$$\begin{aligned} A_0 &= I_Z S_Z - \tfrac{1}{4}(I_+S_- + I_-S_+) \\ A_{\pm 1} &= -\tfrac{3}{2}(I_\mp S_Z + I_Z S_\mp) \\ A_{\pm 2} &= -\tfrac{3}{4} I_\mp S_\mp \end{aligned} \tag{30}$$

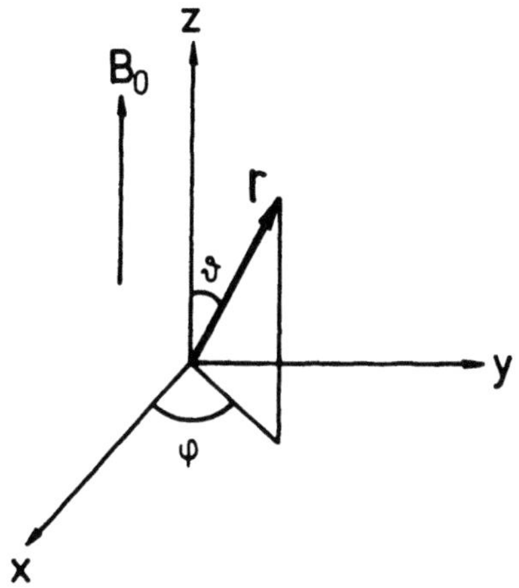

FIG. 5. Rectangular and polar coordinates to define interspin vector **r**.

where I_Z and S_Z denote the Z components of the spin operators of ^{1}H and ^{13}C, respectively. The raising and lowering operators of **I** and **S** are $I_\pm$ and $S_\pm$, given by: $I_\pm = I_x \pm iI_y$, $S_\pm = S_x \pm iS_y$. Then, the terms $|\langle n'|\, A_m\, |n\rangle|^2$, the second power of the matrix element of A_m, are easily calculated as time-independent coefficients for all pairs of four eigenstates of the ^{13}C, ^{1}H spin system. Defining spectral densities as

$$J_m(\omega_{nn'}) = \int_{-\infty}^{\infty} \overline{Y_m^*(t+\tau)Y_m(t)}\exp(-i\omega_{nn'})\,\mathrm{d}\tau / \overline{|Y_m(t)|^2} \tag{31}$$

we have

$$\begin{aligned} w_0 &= \tfrac{1}{20}\,\gamma_C^2\gamma_H^2\hbar^2 r^{-6} J_0(\omega_H - \omega_C) \\ w_1' &= \tfrac{3}{40}\,\gamma_C^2\gamma_H^2\hbar^2 r^{-6} J_1(\omega_C) \\ w_2' &= \tfrac{3}{10}\,\gamma_C^2\gamma_H^2\hbar^2 r^{-6} J_2(\omega_C + \omega_H) \end{aligned} \tag{32}$$

Accordingly, the spin–lattice relaxation time of ^{13}C, T_{1C}, is described in terms of spectral densities as

$$\frac{1}{nT_{1C}} = \tfrac{1}{20}\,\gamma_H^2\gamma_C^2\hbar^2 r^{-6}[J_0(\omega_H - \omega_C) + 3J_1(\omega_C) + 6J_2(\omega_H + \omega_C)] \tag{33}$$

On the other hand, four eigenstates of the spin system in the transverse direction (x-axis direction) are expressed as linear combinations of the four energy eigenstates though they are not themselves energy eigenstates. Hence, the transition probabilities among four eigenstates in the x-axis direction can be similarly formulated, and the spin–spin relaxation time of ^{13}C can be described as

$$\frac{1}{nT_{2C}} = \tfrac{1}{40}\,\gamma_H^2\gamma_C^2\hbar^2 r^{-6}[4J_0(0) + J_0(\omega_H - \omega_C) + 3J_1(\omega_C) + 6J_1(\omega_H) + 6J_2(\omega_H + \omega_C)] \quad (34)$$

In eqns (33) and (34), n denotes the number of protons bonded chemically to the carbon under consideration. The number n has been introduced assuming additivities of the relaxations induced by plural pairs of ^{13}C and ^{1}H.

Single Correlation Time Theory

If the mode of the time-fluctuation of the C–H vector under consideration is designated, the spectral densities appearing in eqns (33) and (34) can be formulated. For example, if the C–H vector **r** undergoes random diffusional rotation while the internuclear length is held constant, i.e. if the vector **r** undergoes rotational diffusional motions around the x, y, and z-axes with a common rotational diffusion constant R, all of the correlation functions of the orientation functions of **r** can be described by a correlation time τ_c as

$$\overline{Y_m^*(t+\tau)Y_m(t)} = \overline{|Y_m(t)|^2}\exp(-\tau/\tau_c) \quad (35)$$

where $\tau_c = 1/(6R)$. Then, the spectral densities defined by eqn (31) are described as

$$J_m(\omega_{nn'}) = \frac{2\tau_c}{1+\tau_c^2\omega_{nn'}^2} \quad (36)$$

for all m.

Therefore T_1 and T_2 are described explicitly as a function of τ_c and Larmor frequencies of ^{1}H and ^{13}C under $\mathbf{B}_0$, using eqns (33), (34).

$$\frac{1}{nT_{1C}} = \tfrac{1}{10}\,\gamma_H^2\gamma_C^2\hbar^2 r^{-6}\left[\frac{\tau_c}{1+\tau_c^2(\omega_H-\omega_C)^2} + \frac{3\tau_c}{1+\tau_c^2\omega_C^2} + \frac{6\tau_c}{1+\tau_c^2(\omega_H+\omega_C)^2}\right] \quad (37)$$

$$\frac{1}{nT_{2C}} = \tfrac{1}{20}\,\gamma_H^2\gamma_C^2\hbar^2 r^{-6}\left[4\tau_c + \frac{\tau_c}{\tau_c^2(\omega_H-\omega_C)^2} + \frac{3\tau_c}{1+\tau_c^2\omega_C^2} + \frac{6\tau_c}{1+\tau_c^2\omega_H^2} + \frac{6\tau_c}{1+\tau_c^2(\omega_H+\omega_C)^2}\right] \quad (38)$$

In the extreme case where $\tau_c^2\,\omega_{nn'}^2 \ll 1$, the relations reduce to

$$\frac{1}{nT_{1C}} = \frac{1}{nT_{2C}} = \gamma_H^2\gamma_C^2\hbar^2 r^{-6}\tau_c \tag{39}$$

These relations are together generally referred to as single correlation time theory and used to correlate the relaxation phenomena for monomeric substances in solution to their molecular motion. Nevertheless, in the case of macromolecules, the C–H vectors in the molecular structure are not thought to undergo such isotropic spherical diffusional rotation. In fact, the relaxation phenomena of macromolecules seldom follow these relations and particular modes of motion must be assumed for the internuclear vectors considering the detailed molecular structure.

Three Correlation Times Theory

Many models for the motional mode of the internuclear vector (C–H vector) to describe the magnetic relaxation of macromolecules have been proposed. Examples are the ellipsoidal or spherical rotational models where an ellipsoidal or spherical molecule undergoes independent diffusional rotations around the long and short axes and the C–H vectors are embedded in the ellipsoidal or spherical molecule.[15,16] Here consider one model including three independent motions as a pertinent model for long-chain molecules (referred to as the 3τ model). In this model, schematically depicted in Fig. 6, the

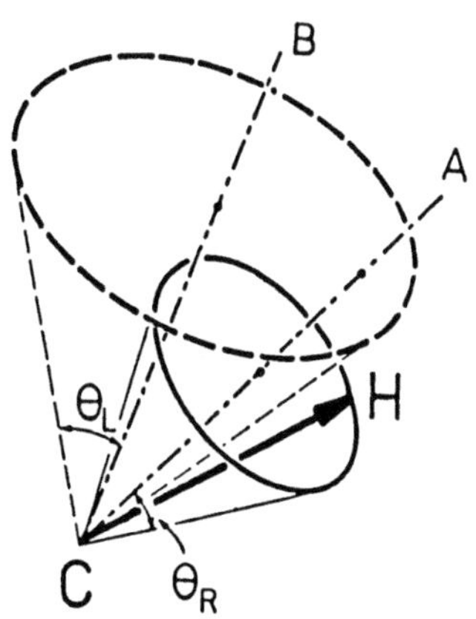

FIG. 6. Schematic diagram of 3τ model for the motion of the internuclear C–H vector.

C–H vector first undergoes a random rotational motion around an axis A with an angle θ_R and a correlation time τ_R, the axis A undergoes a librational motion with a correlation time τ_L within a cone making an angle θ_L to its axis B, and the axis B finally undergoes random spherical rotation with a correlation time τ_I. This is the model proposed by Howarth for long-chain molecules.[16] Although he derived a mathematical formula for the spectral densities, intuitively connecting the spectral density for Woessner's 2τ rotational model[15] with that for his 2τ librational model,[16] the strict formula for the spectral densities in this model is as follows.[17]

$$\begin{aligned} J_m(\omega_{nn'}) = {} & A_R A_L \frac{2\tau_I}{1+\omega_{nn'}^2\tau_I^2} + A_R(1-A_L)\frac{2\tau_1}{1+\omega_{nn'}^2\tau_1^2} \\ & + B_R B_L \frac{2\tau_2}{1+\omega_{nn'}^2\tau_2^2} + B_R(1-B_L)\frac{2\tau_3^2}{1+\omega_{nn'}^2\tau_3^2} \\ & + C_R C_L \frac{3\tau_4^2}{1+\omega_{nn'}^2\tau_4^2} + C_R(1-C_L)\frac{2\tau_5^2}{1+\omega_{nn'}^2\tau_5^2} \end{aligned} \tag{40}$$

Here, A_R, B_R, C_R, A_L, B_L, C_L are the parameters designated by the angles θ_R and θ_L, and

$$\tau_1 = (\tau_L^{-1} + \tau_I^{-1})^{-1}$$

$$\tau_2 = (\tau_R^{-1} + \tau_I^{-1})^{-1}$$

$$\tau_3 = (\tau_R^{-1} + \tau_L^{-1} + \tau_I^{-1})^{-1}$$

$$\tau_4 = (4\tau_R^{-1} + \tau_I^{-1})^{-1}$$

$$\tau_5 = (4\tau_R^{-1} + \tau_L^{-1} + \tau_I^{-1})^{-1}$$

Substitution into eqns (33) and (34) yields T_{1C} and T_{2C} as explicit functions of τ_I, τ_L, τ_R and θ_R, θ_L. Equation (40) is fairly complex, but the condition $\tau_R \ll \tau_L \ll \tau_I$ is assumed for most macromolecules in solution or in the rubbery state. In this case eqn (40) is simplified to

$$\begin{aligned} J_m(\omega_{nn'}) = {} & A_R A_L \frac{2\tau_I}{1+\omega_{nn'}^2\tau_I^2} + A_R(1-A_L)\frac{2\tau_L}{1+\omega_{nn'}^2\tau_L^2} \\ & + B_R \frac{2\tau_R}{1+\omega_{nn'}^2\tau_R^2} + C_R \frac{2(\tau_R/4)}{1+\omega_{nn'}^2(\tau_R/4)^2} \end{aligned} \tag{41}$$

This relation can be further approximated by neglecting the fourth term as

$$J_m(\omega_{nn'}) = A\frac{2\tau_{\mathrm{I}}}{1+\omega_{nn'}^2\tau_{\mathrm{I}}^2} + B\frac{2\tau_{\mathrm{L}}}{1+\omega_{nn'}^2\tau_{\mathrm{L}}} + C\frac{2\tau_{\mathrm{R}}}{1+\omega_{nn'}^2\tau_{\mathrm{R}}^2} \tag{42}$$

Although we have started with a particular 3τ model, the result becomes equivalent to the general case that three independent random motions with sufficiently different correlation times are involved, in spite of the model adopted.

As mentioned above, the relaxation phenomena of macromolecules seldom follow the single correlation time theory dictated by eqn (36). In such cases, a wide distribution is usually introduced in the correlation time. However, as discussed elsewhere,[18] the distribution of correlation time not only fails to explain the temperature dependencies of T_1, T_2 and the NOE of the non-crystalline components observed by scalar decoupled ^{13}C NMR on linear polyesters and polyethylene, but also overlooks the intrinsic motion of long-chain molecules. On the contrary, the 3τ theory dictated by eqn (41) was found to be very effective to describe such temperature dependencies of the relaxation parameters. Irrespectively of whether the motional mode assumed in the 3τ model for the C–H vector is really true, the concept that the C–H vector in macromolecules involves plural independent diffusional motions with discretely different correlation times is very useful to explain the magnetic relaxation phenomena of macromolecules, as will be shown later.

To this end, it is generally to be pointed out that the spin–lattice relaxation time T_1 is related to the molecular motion corresponding to the ^{13}C and ^{1}H Larmor frequencies under $\mathbf{B}_0$, as is evident from eqn (33). On the other hand, the spin–spin relaxation time T_2 contains zero-frequency spectral density, as is evident from eqn (34). Then the T_2 relaxation is related to much slower molecular motion than is the T_1 relaxation. The Larmor frequencies ω_{H} and ω_{C} are of the order of $(1–5) \times 10^8$ Hz for usual NMR conducted with a static field of 2–10 T. Therefore, the T_1 relaxation can detect molecular motion in this frequency range, whereas the T_2 relaxation detects molecular motion of the order of *ca* 10^6 Hz. According to the single correlation time theory dictated by eqns (37) and (38), both T_1 and T_2 decrease with increasing correlation time τ_c when the molecular motion is enhanced. However, it is to be noted that T_1 increases again after showing a minimum at τ_c corresponding to the reciprocal of the Larmor frequencies of ^{13}C or ^{1}H, whereas T_2 decreases monotonically with increasing τ_c. Thus, T_1 becomes very long and T_2 very short for rigid material.

In addition to T_1 and T_2, there is one more relaxation time, referred to as $T_{1\rho}$, which is widely measured in ^{13}C NMR. This is the spin–lattice relaxation time in relation to the resonant oscillating magnetic field $\mathbf{B}_1$ in the rotating frame where $\mathbf{B}_1$ can be considered as a static field (cf. the section on cross-polarization). Since $\mathbf{B}_1$ is a few millitesla in usual solid-state ^{13}C NMR, $T_{1\rho}$ detects molecular motions in the region of several kilohertz. Since macromolecules involve chain motions at different frequencies, concomitant measurements of T_1, T_2 and $T_{1\rho}$ are desirable. The measurement of T_2 in solid-state ^{13}C NMR is rather difficult at present because T_2 generally is shorter than several microseconds. Nevertheless, interpretation of both longitudinal and transverse relaxation phenomena are very useful in studying the phase structure of crystalline polymers as the following example of linear polyethylene shows.

STUDIES OF LINEAR POLYETHYLENE (LPE) BY SOLID-STATE ^{13}C NMR

Principal Values and Axes of Chemical Shift of *trans–trans* Methylene Chains in the Orthorhombic Crystal Form

Solid-state ^{13}C NMR is, of course, applicable to many crystalline polymers whose only nuclei possessing spin are ^{13}C and ^{1}H. As the first example this section introduces mainly recent studies of LPE. The carbon species of this polymer is considered to be methylene carbons, except for methyl end-groups. All observed chemical shifts are assignable to methylene carbons for samples with typical molecular weight. Hence, the spectrum reflects the principal values of the chemical shift tensor for a particular chain conformation or chemical shifts for different chain conformations of methylene sequences. The examination of the spectrum provides information on the phase structure of samples, in relation to both crystalline and non-crystalline phases, in terms of their chain conformation. Figure 7 shows the DD/CP ^{13}C spectrum of highly crystalline PE taken at $\mathbf{B}_0 = 4{\cdot}6\,\mathrm{T}$ at room temperature by a JEOL JNM-FX200 spectrometer. Because of the high degree of crystallinity of the sample, no appreciable contribution from non-crystalline material in the structure is recognized. This is a typical powder crystalline pattern as is dictated from eqns (15). Since

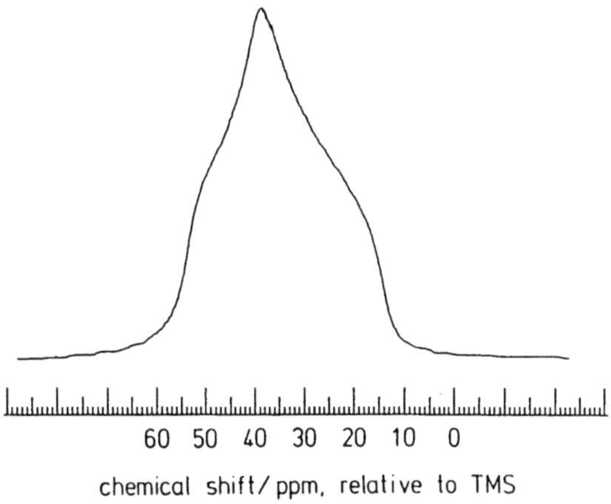

FIG. 7. 50 MHz powder crystal spectrum of high-crystalline polyethylene at room temperature.

the crystalline form of this sample was confirmed by X-ray analyses as being orthorhombic, the observed chemical shift tensor is assigned to methylene carbons in *trans–trans* chain conformation in the orthorhombic crystal form. The principal values (σ_{11}, σ_{22}, σ_{33}) of this chemical shift tensor and their average (σ_{ave}) can be easily determined from the spectrum as listed in Table 1.[19] On the other hand, the

TABLE 1
PRINCIPAL VALUES AND DIRECTIONS OF PRINCIPAL AXES OF CHEMICAL SHIFT TENSORS FOR POLYETHYLENE[19]

Tensor element	*Value (ppm)*[a]	*Direction*	
σ_{11}	51·4 ± 1·3	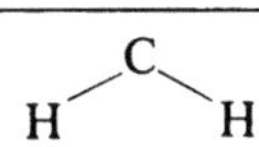	proton–proton vector
σ_{22}	38·8 ± 1·3	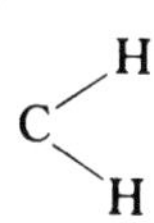	CH_2 angle bisector
σ_{33}	12·9 ± 1·3	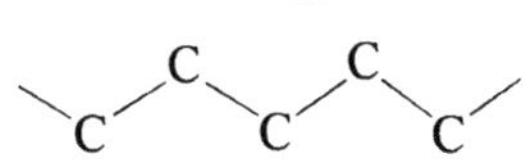	Chain axis
σ_{av}	34·4 ± 1·3		
σ_{iso}	31·0, 29·67		

[a] Chemical shift from tetramethylsilane (TMS).

directions of the principal axes cannot be determined directly from the spectrum. Nevertheless, the assignment is of foremost importance not only in relation to discovering the electronic structure of molecular chains but also to characterizing their anisotropic chain motion. If a single crystal sufficiently large for NMR measurement is available and the crystallographic directions are macroscopically known, it is fairly easy to sort out all principal axes according to the suggestion from eqn (9). Nevertheless, this is almost always hopeless for macromolecules: then other methods must be sought.

The determination of the principal axes of the chemical shift of polyethylene was carried out by VanderHart[19] as follows. First, the DD/CP ^{13}C NMR spectrum taken on a highly drawn polyethylene sample with the draw direction (molecular chain direction) parallel to $\mathbf{B}_0$ shows a single sharp line at *ca* 13 ppm. Then the σ_{33} axis can be assigned to the chain axis as indicated in the table. The other two axes were determined from the DD/CP ^{13}C NMR of an *n*-eicosane ($C_{20}H_{42}$) single crystal. VanderHart measured the rate of cross-polarization at 10° intervals rotating the eicosane single crystal around the chain axis (σ_{33}-axis) which was fixed perpendicular to $\mathbf{B}_0$, and observed a minimum CP rate when the σ_{22}-axis was parallel to $\mathbf{B}_0$ (when $\alpha = \beta = 90°$, cf. eqn (9); i.e. when only the 38·8 ppm line appeared). On the basis of this result he assigned the σ_{22}-axis as the bisector of the H–C–H bond in the plane of these atoms according to the following steps.

(1) The σ_{11} and σ_{22} axes were assumed to be either parallel to the line joining two protons on the same CH_2 unit or the bisector of the H–C–H bond, considering the symmetry of the methylene chains.
(2) The CP rate is proportional to the dipolar second moment M^2_{CH} for the ^{13}C resonance line due to the C–H dipolar interaction.
(3) M^2_{CH} is proportional to $(3\cos^2\theta - 1)^2$,[20] where θ is the angle between the C–H vector and $\mathbf{B}_0$, and hence it should be minimum when $3\cos^2\theta - 1 = 0$, i.e. $\theta = 54°44'$.
(4) The condition $\theta = 54°44'$ corresponds to the orientation of the single crystal when the chain axis is perpendicular to $\mathbf{B}_0$ while the bisector of the H–C–H bond is parallel, because the tetrahedral angle H–C–H is twice the magic angle 54°44′. Therefore, M^2_{CH} is expected to be minimum at this orientation and hence the CP rate will be minimum.

Thus all directions of the principal axes could be determined as listed in Table 1.

On the other hand, in ^{13}C NMR spectra of solutions of this polymer, the chemical shift of about 29–30 ppm is reported as being assignable to inner methylene chains. This is the motionally conformation-averaged chemical shift of methylene chains. Corresponding to this, the non-crystalline absorption line at *ca* 31 ppm is always recognized to some extent even in the solid-state ^{13}C NMR spectrum, as listed in Table 1 as σ_{iso}; the difference of *ca* 1 ppm is at present thought to be due to some packing effects of molecules in the solid state. These principal values as well as the motionally averaged chemical shift are reflected in the absorption spectra, depending on the conditions for measurements and the nature of the samples. Examples are shown in the following sections.

DD/CP ^{13}C NMR of Uniaxially Oriented PE

Figure 8 shows the CP/DD ^{13}C NMR spectrum taken at 70 and 100°C for highly stretched PE film with the tensile axis parallel to $\mathbf{B}_0$.[21] Here, two sharp peaks are evidently recognized at 11·7 and 32·4 ppm which are assignable to the crystalline σ_{33} and non-crystalline σ_{iso} line, respectively. This stretched PE sample was made as follows. The film of a molecular-weight fraction ($\bar{M}_v = 3{\cdot}8 \times 10^5$) of LPE was irradiated to 8·5 Mrad with γ-rays from a ^{60}Co source at room temperature *in vacuo*. After removal of the sol by extraction, the film was uniaxially stretched 600% in the molten, rubbery state at 160°C (elastic stretching) and crystallized by cooling while the stretched length was held. The density of intermolecular cross-links of this gel sample was estimated to be *ca* $2{\cdot}4 \times 10^{-4}$ from the sol–gel distribution curve for the irradiation, and the degree of crystallinity to be 0·64 from density measurements. The previous X-ray and birefringence analyses with ^{1}H broad-line NMR showed a very unusual phase structure for this sample with a crystalline phase of large crystallites oriented almost completely parallel to the stretching direction such that the orientation factor was 0·997, whereas the molecular chains of the non-crystalline component were almost unoriented in a relaxed state.[22]

In response to this phase structure the non-crystalline σ_{iso} line is recognizably enhanced in intensity and sharpness in comparison with usual drawn PE samples. The spin–lattice relaxation of this line in both laboratory and rotating frames showed simple exponential decay. Then it was concluded that the non-crystalline material in this sample

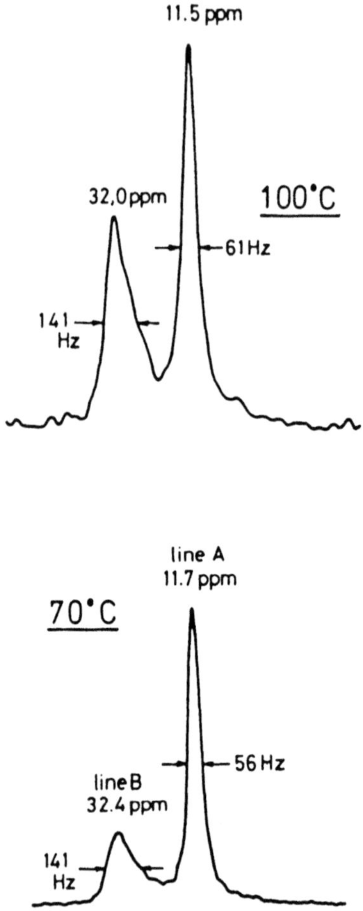

FIG. 8. 15 MHz CP/DD ^{13}C NMR spectra of highly oriented polyethylene films at 70 and 100°C. The drawing direction is parallel to B_0.

was composed of a single component, at least in respect to the longitudinal relaxation. Nevertheless, even if the longitudinal relaxation phenomenon showed single-component behaviour, the transverse relaxation may show plural-component behaviour, as will be shown in the DD/MAS experiment on PE.

On the other hand, the σ_{33} line involved two relaxation times in both laboratory and rotating frames. Two T_1 values, 1·6 and 80 s, were recognized at 70°C and two $T_{1\rho}$ values, 3·5 and 39 s, were observed at 40°C, respectively. Such plural T_1s have been sometimes observed for

CP/DD/MAS ^{13}C NMR but their origin has not yet been clarified. This problem will be discussed elsewhere but it should be noted here that even at 70°C, corresponding to the α-transition temperature where some molecular mobilities must originate in the crystalline region, T_1 is as long as 80 s. This implies a very stable crystalline structure of this sample, reflecting its very ordered crystalline phase of large crystallites.

Studies of Phase Structure of LPE Samples with Lamellar Crystalline Structure by Magic Angle Sample Spinning

In the case of magic angle sample spinning, the average (σ_{ave}) of the principal values and the motionally conformation-averaged chemical shift (σ_{iso}) listed in Table 1 are detectable. Examination of these σ_{ave} and σ_{iso} values, as well as the relaxation phenomena associated with these chemical shifts provides information on the phase structure of samples in terms of molecular conformation and dynamics. As is known, the phase structure of LPE samples differs widely depending on the mode of crystallization. For example, when crystallized isothermally from dilute solution or from the melt, LPE usually exhibits a lamellar crystalline structure; the thicknesses of lamellae are relatively limited, say *ca* 100–500 Å, in comparison with the lateral dimensions of over several microns, and the crystalline methylene chains orients perpendicularly to the wide face of the lamellae with molecular chains inclined to lamellar normals at angles of up to 38°.[23] These lamellar samples, crystallized either from the melt or dilute solution, contain a non-crystalline component. The problem of whether this non-crystalline component is in an isotropically disordered state or in a particular state due to the coexistence of lamellar crystallites has been extensively studied for a few decades.

As an approach to investigating the conformational characteristics of such non-crystalline material in the structure, the observation and calculation of chemical shifts of different carbon species of cyclic alkanes[24–26] or *n*-alkanes[27,28] have been carried out. However, the results obtained have not been correlated with the phase structure of LPE samples, although such researches are interesting in themselves. On the other hand, DD/MAS ^{13}C NMR spectroscopy can provide detailed information on this problem.[29,30]

For simplicity, we deal with two LPE samples with lamellar crystalline structure.[29] One, referred to as the bulk-crystal, is an unfractionated LPE, $\bar{M}_v$ 3×10^6, crystallized isothermally at 129°C from the melt. The other, referred to as the solution-crystal, is a molecular-weight fraction of LPE with $\bar{M}_v$ 91 000 crystallized at 85°C from 0·08% toluene solution. The previous ^{1}H broad-line NMR analyses show that the bulk-crystal consists of lamellar crystalline region together with a non-crystalline interfacial region with limited molecular mobility and non-crystalline interzonal regions with liquid-like molecular mobility, whereas the solution-crystal consists of lamellar crystals and non-crystalline overlayer without or with a negligible amount of non-crystalline interzonal material possessing pronounced molecular mobility.[31–33] As has already been pointed out in the first section (p. 180), since ^{1}H broad-line NMR can analyse the samples only via dipolar interactions among protons, it cannot provide direct information on the molecular chain conformation of each component. On the other hand, the DD/MAS ^{13}C NMR analyses of these samples not only support this conclusion but also provide information on the chain conformation of each component via their chemical shifts.

Figure 9 shows the DD/MAS spectrum of the bulk-crystal (A) and the solution-crystal (B) taken by a JEOL JNM-FX200 spectrometer at room temperature under the following conditions: static field strength 4·7 T, ^{1}H decoupling field 1·4 mT, magic angle spinning at 3·3–3·5 kHz. First, in spectrum A at least two separable absorption lines are recognized at *ca* 33 and 31 ppm, corresponding to the σ_{av} and the σ_{iso} shifts, respectively. It evidently shows that this sample comprises both orthorhombic crystalline and motionally conformation-averaged non-crystalline components. On the other hand, in spectrum B of the solution-crystal the σ_{av} line at *ca* 33 ppm is also clearly recognized but the existence of the non-crystalline line at *ca* 31 ppm is a little vague, partly because of the high degree of crystallinity. These spectra were obtained with pulse programs, not including the CP process, with very long waiting times before each pulse to ensure complete recovery of the ^{13}C equilibrium magnetizations of all components in the structure. Hence, these spectra reproduce the contributions from all components. Nevertheless, it is very difficult to seek out detailed information on the individual phases in the structure except for the coexistence of the crystalline and non-crystalline components.

However, the different components in the structure may differ not only in their chemical shifts but also in their magnetic relaxation times. We have examined the longitudinally and transversely relaxed spectra of these samples. First of all, we examined the longitudinally relaxed

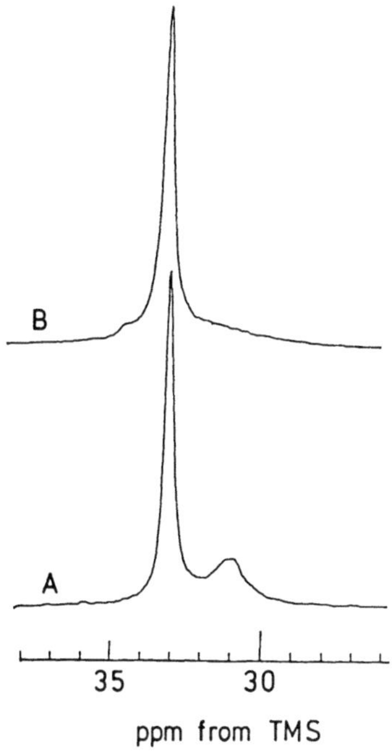

FIG. 9. 50 MHz DD/MAS ^{13}C spectra of PE samples obtained using pulse sequences [(90°–FID)–τ] where ^{1}H dipolar decoupling was applied only for (90°–FID). (A) Bulk-crystal, $\tau = 17\,000$ s; (B) solution-crystal, $\tau = 1500$ s.

spectra using a usual inversion recovery pulse sequence $(\pi\text{–}t\text{–}\pi/2\text{–}10\text{ s})_n$. The waiting time for every pulse sequence is only 10 s as is described. Then the contribution from components with T_1 longer than several tens of seconds cannot appear because of the lack of time for recovery of the equilibrium magnetization. Hence, we can see preferentially the relaxation process of the non-crystalline components, with a diminished contribution from the crystalline component. Figure 10 shows the partially relaxed spectrum of the bulk-crystal by the pulse sequence $(\pi\text{–}t\text{–}\pi/2\text{–}10\text{ s})_n$ with $t = 0{\cdot}40$ s. It is recognized that the remaining 33 ppm crystalline line† is still in the negative direction while the 31 ppm line has already appeared in the

† The 33 ppm line involves plural T_1s,[30,34] several thousand and several hundred seconds and a few seconds. All of these components are assigned to the methylene sequences in *trans–trans* conformation in the orthorhombic crystal form.

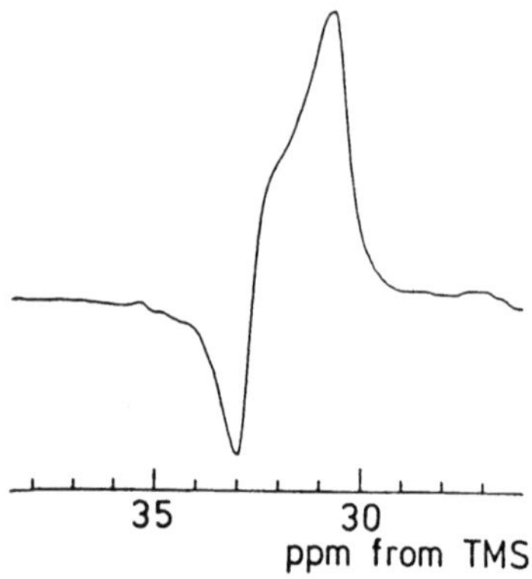

FIG. 10. Partially relaxed 50 MHz DD/MAS ^{13}C spectrum of bulk-crystal. $(\pi-t-\pi/2-10\text{ s})_n$ with $t = 0{\cdot}40$ s.

positive direction with a downfield shoulder. This behaviour must imply the presence of two kinds of non-crystalline components with an equivalent T_1 but different chemical shifts, one at 31 ppm and the other a little downfield. On the other hand, for the solution-crystal we could not obtain such evidence suggesting the existence of two different non-crystalline components. We have concluded that the bulk-crystal contains two different non-crystalline components whereas the solution-crystal contains only one non-crystalline component, at least by examining the longitudinal relaxation.

Next, we examined the partially relaxed spectra in the transverse direction. As was explained in the theoretical sections, the relaxation of ^{13}C magnetization of macromolecules whose only nuclei possessing spin are ^{13}C and ^{1}H is predominantly conducted by the time-fluctuation of the dipole–dipole interaction between ^{13}C and neighbouring ^{1}H nuclei. However, the mechanism is somewhat different depending on the direction of the relaxation. The longitudinal relaxation is carried out by the ω_C and $\omega_C \pm \omega_H$ frequency components (see eqn (33)) of the fluctuation arising from random reorientation of chemically bonded ^{13}C–^{1}H internuclear vectors via molecular motion, where ω_C and ω_H denote the Larmor frequencies of ^{13}C and ^{1}H under the static magnetic field, respectively. On the other hand, the transverse relaxation contains a contribution from the zero-frequency component of the fluctuation (eqns (34) and (38)). Accordingly, the transverse relaxation may distinguish the different non-crystalline components via a slower molecular chain motion even

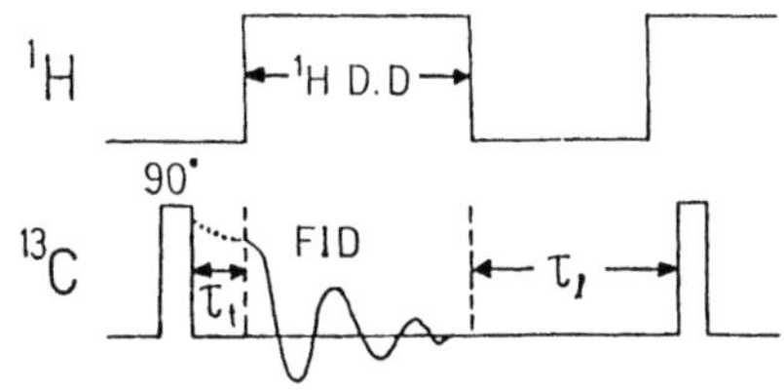

FIG. 11. The pulse sequence for the partially relaxed spectra in the transverse direction.

when they involve an equivalent longitudinal relaxation time due to an equivalent high-frequency chain motion.

The pulse sequence used for obtaining the transversely relaxed spectra is shown schematically in Fig. 11. This is devised to observe the free induction decay under ^{1}H dipolar decoupling conditions, after relaxing the magnetization in the transverse direction for a chosen period τ_t without ^{1}H decoupling. It is possible by choosing the time τ_t appropriately to sort out the contribution from different components via difference in spin–spin relaxation time T_2, even when the components involve equal T_1s. Figure 12 shows the spectra for the

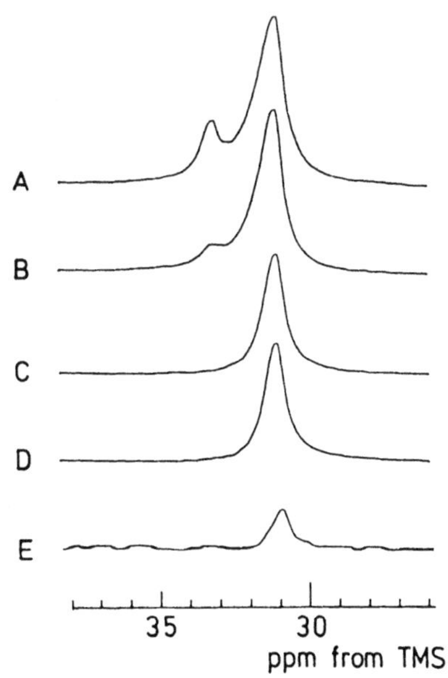

FIG. 12. Partially relaxed spectra in the transverse direction of the bulk-crystal with $\bar{M}_v$ 3×10^6, taken by the pulse sequence shown in Fig. 11 with $\tau_l = 3{\cdot}5$ s. The τ_t values are (A) 0·5, (B) 20, (C) 100, (D) 140, (E) 4000 μs.

bulk-crystal with $\tau_l = 3{\cdot}5$ s and different τ_t values. Here, the contribution from the rigid crystalline component with $T_1 \geqslant 10$ s is eliminated because of the short τ_l adopted. The spectrum at $\tau_t = 0{\cdot}5\ \mu$s is thought to represent the contribution from all non-crystalline components and a crystalline component with T_1 shorter than a few seconds. With increasing τ_t, however, it is seen that the contribution from the crystalline and non-crystalline components with shorter transverse relaxation times T_2 rapidly disappears leaving only the contribution from a non-crystalline component with a longer T_2. Thus, at $\tau_t = 100\ \mu$s only a single peak remains at 31·0 ppm. With further increasing τ_t, this single line gradually disappears while holding the lineshape essentially unchanged. It is to be noted here that each remaining line can be well approximated by a Lorentzian distribution function and that the peak position is kept at 31.0 ppm. These results evidently suggest the existence of two non-crystalline components associated with an equivalent T_1 but different T_2s. Furthermore, one of the two is characterized as being non-crystalline material forming a monophase involving a pronounced rubbery chain motion with T_2 longer than several hundred microseconds. The lineshape of the other non-crystalline component with shorter T_2 will be obtained by comparing the partially relaxed spectra (at $\tau_t > 100\ \mu$s) representing the rubbery non-crystalline phase with the spectrum (at $\tau_t = 0{\cdot}5\ \mu$s) representing the contribution from all non-crystalline components.

In Fig. 13, a difference spectrum which was obtained by subtracting the spectrum at $\tau_t = 140\ \mu$s from the spectrum at $\tau_t = 0{\cdot}5\ \mu$s is shown. The difference spectrum (C) is considered to represent the other non-crystalline component with the contribution from the crystalline component with shorter T_1. By subtracting the crystalline contribution, we have a rather broad but symmetric lineshape centring at 31·3 ppm, as is indicated by the broken line in curve C. Since the lineshape can be well approximated by a single Lorentzian distribution function, this non-crystalline component is also assumed to comprise a monophase. The very short transverse relaxation time (which disappeared completely within 100 μs) and the downfield chemical shift characterize this non-crystalline phase as comprising somewhat rigid methylene sequences in a *trans*-rich conformation.

As was pointed out previously, the spin–lattice relaxation is related to the molecular motion corresponding to ^{13}C and ^{1}H Larmor frequencies usually in the range of 10^8–10^9 Hz, whereas the spin–spin relaxation is related to much slower molecular motion. Then the equivalent T_1 must suggest the equivalency in the high-frequency mode

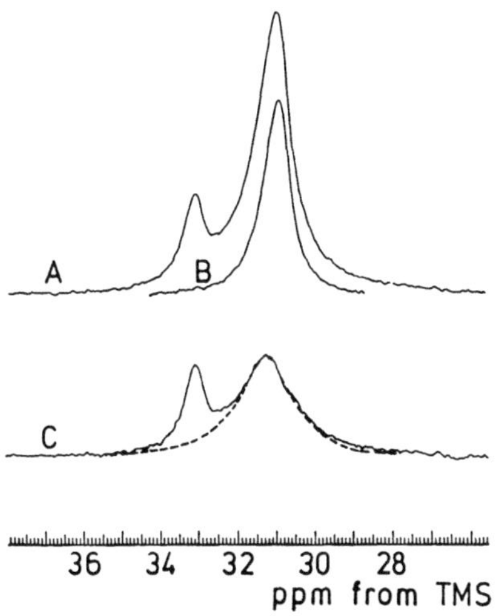

FIG. 13. Difference spectrum demonstrating the existence of two non-crystalline phases with different T_2s. A and B correspond to A and D in Fig. 12. C = A − B.

of chain motion (probably very local chain motion) in the order of the Larmor frequencies for both non-crystalline phases. On the other hand, the very short T_2 for the 31·3 ppm non-crystalline phase must imply the immobility of the molecular chains in the lower-frequency mode of motion in this phase. The fact that two non-crystalline phases involve an equivalent T_1 but quite different T_2s cannot be explained by the single correlation time theory dictated from eqns (35) and (36). It can be explained, however, by a motional mode of the C–H vectors including three correlation times, τ_I, τ_L, τ_R ($\tau_I \gg \tau_L \gg \tau_R$), as shown in Fig. 6. It is usually assumed that τ_L is in the range of the reciprocal of the Larmor frequencies of ^{13}C and ^{1}H and hence τ_I is much longer than those. Hence, the T_1 relaxation is mainly determined by τ_L and τ_R whereas the T_2 relaxation is determined only by τ_I. Therefore, if it is assumed that the τ_L and τ_R are the same for both components and the τ_I of the 31·3 ppm component is much longer than that of the 31·0 ppm component, the above experimental relaxation behaviour can be easily understood.†

In consideration of the morphological phase structure of this

† In order to describe the above-observed relaxation behaviour of two components only at room temperature, a motional mode including only two correlation times may be sufficient. However, in order to describe the temperature dependencies of the relaxation phenomena the assumption of three correlation times is thought to be inevitable.

bulk-crystal, the two non-crystalline phases with distinctly separable chemical shifts at 31·3 and 31·0 ppm as well as quite different spin-spin relaxation times T_2 with equivalent spin–lattice relaxation time T_1 are assumed to consist of the crystalline–amorphous interface and the isotropic amorphous phase, respectively. The necessity for the existence of such a crystalline–amorphous interface involving molecular chains emerging from the 001 face of crystalline lamellae and re-entering the same lamellae from which they emerge or dissipating in an isotropic amorphous phase was pointed out[35] and the chain conformation was statistically treated with a lattice theory.[36] However, it is interesting that such a crystalline–amorphous interface has been experimentally detected in terms of chemical shift and relaxation times.

In the previous ^{1}H broad-line NMR spectrum analyses, we treated the non-crystalline material as two components which contributed to the medium and narrow components of the broad-line spectrum, and obtained the conclusion that the bulk-crystals consist of two non-crystalline components, i.e. interfacial and interzonal material between lamellar crystallites. The present result not only evidences the rationality of the broad-line analysis but also responds to a criticism raised that the contribution in the broad-line spectrum from non-crystalline material in semicrystalline PE should not be treated as two components.

On the contrary, in the case of the solution-crystal, as mentioned already, no evidence that suggests the existence of two non-crystalline components with different chemical shifts has been obtained by examining the longitudinal relaxation of ^{13}C magnetization. Moreover, we could not obtain any evidence by examining the transverse relaxation phenomena in a similar manner to the case of the bulk-crystal. In the spectral series obtained by the pulse sequence with different τ_ts (shown in Fig. 11) all the non-crystalline components at *ca* 31 ppm quickly disappeared within 50 μs without leaving a line, suggesting the presence of a mobile rubbery component with longer T_2s. It was confirmed that the non-crystalline contribution could be presented by a broad line shape centred at 31·3 ppm which resembled the lineshape for the crystalline–amorphous interface in the bulk-crystal. Therefore, it is concluded that the non-crystalline component of the solution-crystal comprises the overlayer of the basal surface of crystallites in the lamellar structure without an appreciable amount of isotropically amorphous material.

Here some comments should be made in relation to the previously reported scalar decoupled ^{13}C NMR work of LPE samples.[37,38] The non-crystalline material in these bulk- and solution-crystals can also be detected by the usual scalar decoupled ^{13}C NMR as it is associated with rubbery molecular chain motion. It was found that T_1 was almost equivalent for the non-crystalline components of both crystals whereas T_2 of the solution-crystals was very much shorter than that of the bulk-crystals. Referring to the above-mentioned results from solid-state ^{13}C NMR, it is assumed that the 31·0 ppm non-crystalline component in the bulk-crystals and the 31·3 ppm non-crystalline overlayer in the solution-crystals were detected by scalar decoupled ^{13}C NMR.

Thus, in addition to the crystalline phase, two different non-crystalline phases and one non-crystalline phase are identified respectively for the bulk-crystal and the solution-crystal. The characterization of each phase of these samples is summarized in Table 2. Here, the spin–lattice and spin–spin relaxation times of ^{13}C magnetization (T_{1C} and T_{2C}, respectively) and the spin–lattice relaxation time of ^{1}H magnetization (T_{1H}) are listed for individual absorption lines. The T_{1C} values longer than several tens of seconds were obtained by use of the pulse sequence with CP developed by Torchia[39] and those shorter than a few seconds were obtained by the usual saturation recovery pulse sequence without CP. The T_{2C}s of the non-crystalline components were estimated from the semilogarithmic plot versus time of the peak height at 31·0 and 31·3 ppm after decomposing. The T_{1H}s were estimated by a pulse sequence [^{1}H(180°–τ–90°)–CP–^{13}C(FID)], measuring ^{13}C magnetization which appeared by CP after relaxing ^{1}H equilibrium magnetization for time τ. The phase structures discussed above of the samples are characterized here by actual NMR parameters of individual component phases.

In relation to the T_{1H}, several points are to be noted. First, although the crystalline 33·0 ppm line involves three different T_{1C}s, it is associated with only one T_{1H} of 1·87 s. This implies that proton spin diffusion occurs throughout the crystalline region. On the other hand, two evidently different T_{1H}s were obtained for the non-crystalline 31·3 and 31·0 ppm lines. This implies that the spin diffusion does not occur between the two non-crystalline components. This shows again that the two non-crystalline components comprise two separate phases. Furthermore, the T_{1H} of the non-crystalline 31·3 ppm line is roughly equivalent to that of the crystalline 33·0 ppm line for all samples. This

TABLE 2
SAMPLES AND CHARACTERIZATION[29]

Sample	Property	Crystalline		Non-crystalline	
		Triclinic 34·4 *ppm*	*Orthorhombic* 33·0 *ppm*	31·3 *ppm*	31·0 *ppm*
(1) Bulk-crystal					
$\bar{M}_v = 3 \times 10^6$, $(1-\lambda)_{BL} = 0{\cdot}62^a$	Mass fraction	≈0	0·66	0·18	0·16
	Half-width, $\nu_{1/2}$ (Hz)		18	85	38
	T_{1H} (s)		1·87	1·61	0·39
	T_{1C} (s)		2560, 263, 1·7	0·37	
	T_{2C} (ms)		—	0·044	2·4
(2) Solution-crystal					
$\bar{M}_v = 91\,000$, $(1-\lambda)_{BL} = 0{\cdot}78^a$	Mass fraction	0·09	0·68	0·23	≈0
	Half-width, $\nu_{1/2}$ (Hz)	60	20	133	
	T_{1H} (s)	—	1·90	1·90	
	T_{1C} (s)	—	220, 21, 2·0	0·46	
	T_{2C} (s)	—	—	—	

[a] Crystalline fraction from ^{1}H broad-line NMR analyses.

implies that the ^{1}H spin diffusion occurs appreciably between the crystalline component and the non-crystalline 31·3 ppm component, verifying the assignment of the latter component to be the crystalline–amorphous interface material.

In the above discussion, the result from an LPE sample with $\bar{M}_v$ 3×10^6 was the only example shown of samples which are crystallized from the melt (bulk-crystal). However, we have actually examined many bulk-crystals with different molecular weights, and have elucidated the phase structure of each sample. Of particular interest is the evaluation of the thickness ζ_i of the crystalline–amorphous interface. If the lamellar crystal thickness ζ_c of each sample is known, it is possible to evaluate ζ_i from the data obtained by this work, using the relation $\zeta_i = \zeta_c x_i / 2x_c$ where x_c and x_i are the fractions of the crystalline and the interfacial components, respectively. We have no ζ_c value for our samples. Nevertheless, Voigt-Martin and Mandelkern[40] and Bassett and Hodge[23] have reported some ζ_c data for LPE samples crystallized under the almost same conditions as ours. We have evaluated the ζ_i of our samples with different molecular weight using their reported ζ_c values, and found that χ_i is about 34 Å for samples with molecular weights of 20 000–110 000. This value corresponds well to the thickness 15–20 Å of the crystalline–amorphous interface estimated by Flory *et al.* from their theoretical calculation based on the lattice model[36] as well as from analysis of small-angle neutron scattering results.[41,42] It is to be noted that we could actually not only detect the crystalline–amorphous interface, the presence of which has been predicted as mentioned above, but also the experimentally estimated thickness was in rough accordance with the theoretical expectation.

In this section, our recent investigation of LPE by solid–state ^{13}C NMR has been briefly reviewed. Through these studies solid–state ^{13}C NMR has been found to be very useful to studying the crystalline structure of polymers. Although we have been concerned only with LPE of very simple chemical structure, by use of different magnetic relaxation phenomena the developed spectroscopic technique will be applicable to other polymers of complex chemical structure.

STUDY OF SOLID CELLULOSES BY SOLID–STATE ^{13}C NMR

As further examples of the application of ^{13}C NMR to crystalline polymers, this section reviews recent studies of various celluloses,

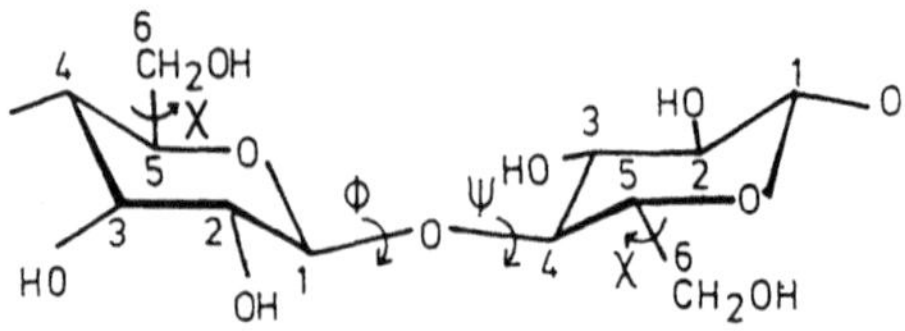

FIG. 14. Chemical structure of cellulose and the definition of three torsional angles.

including cotton, ramie, bacterial and valonia cellulose, and regenerated cellulose. The cellulose molecule includes six kinds of carbon species, as shown in Fig. 14. The ^{13}C NMR spectrum (for polycrystalline samples) without the magic angle sample spinning process is usually unresolved for individual carbons. However, the MAS process enables the resolution of the absorption signals for most of the individual carbons as shown below. Furthermore, since the molecular chains are fairly rigid, the rate of cross-polarization is almost equivalent for the crystalline and non-crystalline states for all carbon species. The CP process can be used efficiently in most cases.

Assignments of Absorption Lines of Individual Carbons in the Crystalline and Non-crystalline Components

Figure 15 shows a CP/MAS ^{13}C NMR spectrum of ramie cellulose with a stick-type spectrum[43] of low molecular weight cellulose (DP ~ 10) in deuterated dimethyl sulphoxide solution. The signals of C1, C4 and C6, involving the β-1,4-glycosidic linkage and the exocyclic C5–C6 bond, have been assigned as indicated in the figure, whilst the signals for C2, C3 and C5 are not resolved so well.[44,45] It can be seen that the resonance lines of C1, C4 and C6 are shifted downfield by 2·3–9·6 ppm in comparison with those of the low molecular weight oligomer in solution. The cause of such downfield shifts is mainly attributed to intrinsic chain conformation in the solid state as discussed below. Furthermore the lines of C4 and C6 split into two components, sharp downfield and broad upfield. Similar splittings are observed also for the C4 and C6 resonance lines of cotton, bacterial and valonia cellulose, and regenerated cellulose such as cupra rayon fibres. In relation to the C4 lines, it was noted for samples with different crystallinities over a wide range that the downfield intensity decreases and the upfield intensity increases with decreasing crystallinity. An assumption was then made that the sharp downfield and broad upfield signals were contributed from the crystalline and non-crystalline

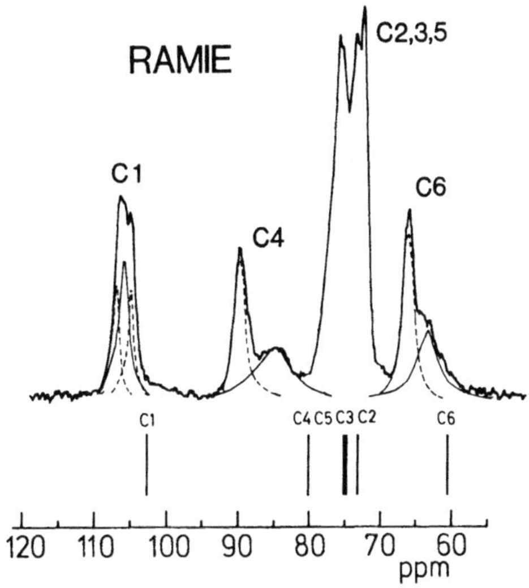

FIG. 15. 50 MHz CP/MAS ^{13}C NMR spectrum of ramie, and stick-type scalar-decoupled ^{13}C NMR spectrum of low molecular weight cellulose in DMSO-d_6 solution. Broken and thin solid lines in the CP/MAS spectrum are for the crystalline and non-crystalline components, respectively.

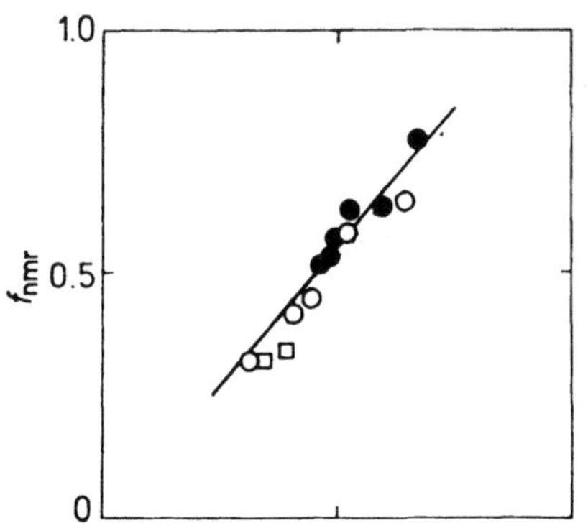

FIG. 16. Integrated fraction f_{nmr} of the downfield component of the C4 resonance line versus degree of crystallinity, f_x, determined by X-ray analysis. ●, Native cellulose; ○, regenerated cellulose; □, mercerized cotton and ramie. (The data for regenerated cellulose samples, except for cupra rayon fibres, were reproduced from ref. 45).

components, respectively.[46] We have decomposed the C4 signals into two Lorentzian downfield and upfield components for different cellulose samples with different crystallinities in order to confirm this assumption.[47] In Fig. 16, the integrated intensity of the downfield component is plotted against the degree of crystallinity estimated from X-ray analyses. It is evident that a linear relationship exists between the integrated fraction of the downfield component and the degree of crystallinity. A similar result was obtained for the C6 signals. It is then concluded that the sharp downfield and broad upfield components of the C4 and C6 carbons of different cellulose samples, including native and regenerated celluloses, are contributed from the crystalline and non-crystalline components, respectively.

Separable Detection of Crystalline and Non-crystalline Spectra of Various Cellulose Samples

As mentioned above, the C4 and C6 signals consist of downfield crystalline and upfield non-crystalline components. Hence, the signals of all carbon species are thought to consist of the crystalline and non-crystalline components even if they cannot be seen explicitly in the spectra. However, if the spin–lattice relaxation times are sufficiently different between the crystalline and non-crystalline lines for all carbon species, the crystalline and non-crystalline spectra can be obtained separately as mentioned below.

Table 3 summarizes the T_1 values of some samples;[48] they were obtained by Torchia's pulse sequence,[39] which is shown schematically in Fig. 17. It is seen that for all samples examined each carbon contains two distinctly different (long and short) T_1 values although the shorter T_1 of bacterial cellulose could not be discerned except for

TABLE 3

^{13}C SPIN–LATTICE RELAXATION TIMES OF THE CARBONS OF DIFFERENT CELLULOSE SAMPLES[47]

	T_1 (*s*)			
Sample	*C1*	*C4*	*C6*	*C2, C3, C5*
Ramie	76 8·8	130 16	65 4·6	82 8·3
Cotton	78 ~7	87 11	89 ~7	67 ~8
Bacterial	87 —	107 —	86 —	104 5·7
Rayon fibres	58 6·8	82 9·9	53 4·2	37 4·7

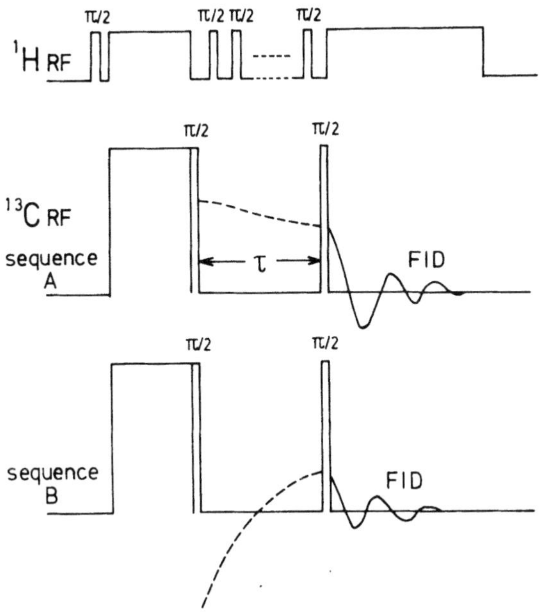

FIG. 17. Pulse sequence for T_1 measurement developed by Torchia.

C2, C3 and C5 because of the high crystallinity. The longer and shorter T_1s for each carbon are assignable to the crystalline and non-crystalline components respectively. The comparison of the longer and shorter T_1s among the samples provides some information on the molecular chain mobility of both crystalline and non-crystalline components. However, we only attempt here to obtain separately the crystalline and non-crystalline spectra by use of the difference in T_1.

Torchia's pulse sequence for T_1 measurements comprise two pulse sequences A and B as shown in Fig. 17. In sequence A, the ^{13}C magnetization which appears by the CP process in the direction parallel to $\mathbf{B}_{1C}$ is measured after turning to $\mathbf{B}_0$ and relaxing for a chosen time t. Contrarily, in sequence B, the ^{13}C magnetization which appears by CP in the direction antiparallel to $\mathbf{B}_{1C}$ is measured after turning to $-\mathbf{B}_0$ and relaxing for time t. In the pulse sequences A and B the longitudinal magnetization of ^{13}C follows the relations, respectively.

$$M_Z^A(t) = M_Z^{CP} \exp(-t/T_1) + M_Z(\infty)[1 - \exp(-t/T_1)] \qquad (43)$$

$$M_Z^B(t) = -M_Z^{CP} \exp(-t/T_1) + M_Z(\infty)[1 - \exp(-t/T_1)] \qquad (44)$$

which are derived from eqn (26) by inserting the initial conditions: $M_Z(0) = M_Z^{CP}$ for sequence A and $M_Z(0) = -M_Z^{CP}$ for sequence B. Here, $M_Z^A(t)$ and $M_Z^B(t)$ denote the ^{13}C magnetization at time t for the sequences A and B, respectively. $\pm M_Z^{CP}$ denotes the magnetization which appears by the CP processes. Subtraction of eqn (44) from eqn (43) yields

$$M_Z^-(t) = 2M_Z^{CP} \exp(-t/T_1) \tag{45}$$

Equation (45) suggests that, if there are two T_1 values, shorter (non-crystalline) and longer (crystalline), for each carbon, the magnetization of the non-crystalline component with the shorter T_1 disappears after a time t which is longer than about five times the shorter T_1, and only the magnetization of the crystalline component with longer T_1 remains. Therefore, the observation and subsequent Fourier transformation of the free induction decay of this remaining magnetization provide the crystalline spectrum. Figure 18 shows the crystalline spectrum of ramie cellulose obtained by this method. Similar crystalline spectra could be obtained for all samples examined. Although the relative line intensities of different carbon species are not quantitatively valid because of the difference in T_1s of individual carbons (see eqn (45)), the chemical shifts and linewidths of C1, C4 and C6 could be determined accurately, eliminating the non-crystalline contributions. In Table 4, the chemical shifts obtained from such

TABLE 4

^{13}C CHEMICAL SHIFTS OF THE CRYSTALLINE COMPONENTS OF NATIVE AND REGENERATED CELLULOSE[47]

Sample	Chemical shift (ppm)					
	C1			*C4*		*C6*
Native cellulose						
Valonia	107·0	106·4	105·3	91·3	90·1	66·5
Bacterial	107·3	106·3	105·3	91·0	90·1	66·4
Cotton	107·1		105·4	89·9		66·5
Ramie	107·2		105·0	89·7		66·0
Regenerated cellulose						
Cupra rayon	108·0		106·3	89·5	88·2	64·1
High-crystalline cellulose II	108·1		106·3	89·6	88·4	63·7

crystalline spectra of native and regenerated cellulose samples are listed.[47] We will consider these data later in relation to the molecular chain conformation of the crystalline component of the various samples.

On the other hand, the non-crystalline spectrum can be obtained by a standard single $\pi/2$ pulse sequence with a short waiting time before each pulse, by decreasing the crystalline contribution with use of the difference in T_1 between the crystalline and non-crystalline components. Alternatively it can be obtained by subtracting appropriately the crystalline spectrum from the corresponding whole spectrum including both crystalline and non-crystalline contributions. An example is given in Fig. 18, where the non-crystalline spectrum of ramie obtained by a single $\pi/2$ pulse sequence is shown. The intensities of the crystalline lines are much decreased and the non-crystalline signals of individual carbons can be observed predominantly. The chemical shifts and linewidths of C1, C4 and C6 estimated from such non-crystalline spectra of different samples are listed in Table 5. These data will also be discussed later in relation to the chain conformation of the non-crystalline components.

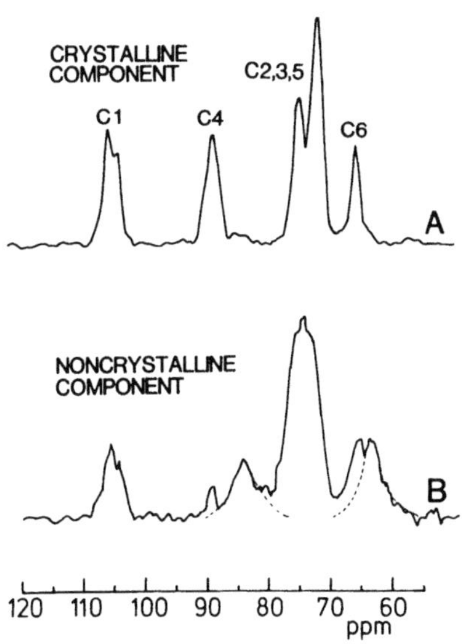

FIG. 18. 25 MHz ^{13}C spectra of crystalline and non-crystalline ramie.

TABLE 5
^{13}C CHEMICAL SHIFTS AND LINEWIDTHS OF THE NON-CRYSTALLINE COMPONENTS OF NATIVE AND REGENERATED CELLULOSE[47]

	C1		*C4*		*C6*	
Sample	$\delta(ppm)$	$\Delta\nu(Hz)$	$\delta(ppm)$	$\Delta(Hz)$	$\delta(ppm)$	$\Delta\nu(Hz)$
Native cellulose						
Valonia	106·2	114	85·2	249	63·4	185
Bacterial	106·6	120	85·0	242	63·4	163
Cotton	106·0	143	85·1	256	63·7	185
Ramie	105·8	128	84·8	284	63·0	171
Regenerated cellulose						
Cupra rayon	105·3	284	85·1	327	62·8	185
Amorphous	105·2	284	81·6	284	62·8	284

Molecular Chain Conformation and Chemical Shifts of C1, C4 and C6

The chemical shift observed by the CP/MAS NMR technique is the average value (σ_{av}) of the principal values of the chemical shift tensor. This chemical shift σ_{av} reflects, of course, the electronic environment around the nucleus concerned. The electronic environment is thought to depend on the packing state of molecules, the hydrogen-bonding effects, and their molecular chain conformation. Only a little is known about the relationship between the chemical shifts and these parameters for different substances at present. In relation to the packing effects, for example, VanderHart[49] has found that the resonance line of CH_2 carbons of triclinic crystals of n-$C_{20}H_{42}$ is shifted $1{\cdot}3 \pm 0{\cdot}4$ ppm downfield in comparison with that of n-paraffins of other crystal forms. As for the hydrogen-bonding effects, Terao *et al.*[50,51] have reported that large downfield shifts appear for the carbons bonded chemically to hydroxyl groups which form strong hydrogen bonds. Firstly, we have examined these two effects for the crystals of monosaccharides and disaccharides, which are monomeric compounds related to cellulose.

Table 6 compares the ^{13}C CP/MAS chemical shifts of these components in the crystalline state and in D_2O solution. The comparison is made only for carbons which are not associated with the exocyclic C–C bond and the β-1,4-glycosidic linkage. The respective resonance lines of the solid-state spectra of α- and β-glucose and those of the solution spectra of all samples were assigned according to the

TABLE 6

COMPARISON OF ^{13}C CHEMICAL SHIFTS OF MONOSACCHARIDES AND DISACCHARIDES IN THE SOLID STATE AND IN SOLUTION[47]

Sample	*Carbon*	$\delta(ppm)$ *Solid*	$\delta(ppm)$ *D_2O solution*	$\Delta\delta(ppm)$	O...O *distance*(Å)
α-D-Glucose	1	93·54	93·30	0·24	2·847[a]
	2	71·22	72·73	−1·51	2·776[a]
	3	73·75	74·05	−0·30	2·707[a]
	4	73·36	70·93	2·43	2·773[a]
β-D-Glucose	1	97·34	97·15	0·19	2·666[b]
	2	76·39	75·46	0·93	2·686, 2·708[b]
	3	76·39	77·17	−0·78	2·772[b]
	4	70·34	70·93	−0·59	no[b]
β-D-Cellobiose	1′	98·02	97·05	0·97	2·776[b]
α-D-Lactose monohydrate	1′	93·50	93·15	0·35	2·790[c]
β-D-Lactose	1′	99·00	97·10	1·90	2·772[d]

[a] Ref. 54.
[b] Ref. 55.
[c] Ref. 56.
[d] Ref. 57.
no = hydrogen bonding was not formed.

results of Pfeffer *et al.*[52,53] In the Table, the O . . . O distances of the hydrogen bonds in the solid state are also shown. The difference $\Delta\delta$ in chemical shift between the solid and solution ranges from −1·51 ppm to 2·43 pm. Nevertheless, it is difficult to find any dependence of the chemical shifts on the O . . . O distances. It is then assumed that the hydrogen bonding does not significantly affect the chemical shifts in these carbohydrates, as in cellulose. However, rather large shifts may originate as the result of the overall intermolecular effects, i.e. the packing effects together with the hydrogen-bonding effects.

Next, we have examined the relationship between the chemical shifts and the molecular chain conformation. The versatility of chain conformation of cellulose molecules is expressed in three torsion angles as shown in Fig. 14: ϕ and ψ, rotations around the β-1,4-glycosidic linkages and χ, the rotation of methylol side groups around the C5–C6 bonds. In order to find out the relationship between the chemical shifts of cellulose samples and the torsion angles, the

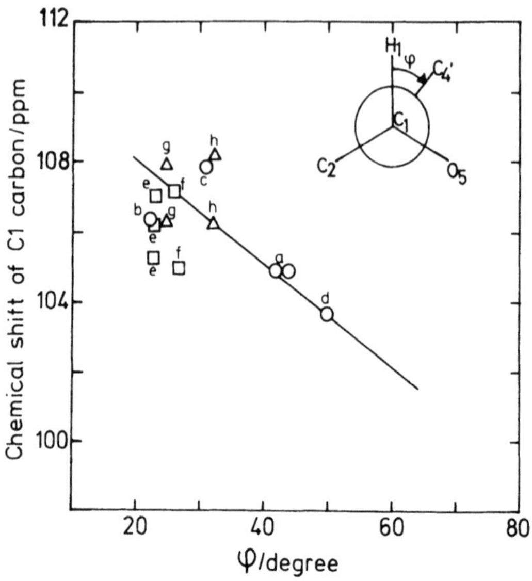

FIG. 19. ^{13}C chemical shifts of the C1 carbons versus torsion angles ϕ. (a) β-D-cellobiose; (b) β-D-methylcellobioside·CH_3OH; (c) α-D-lactose monohydrate; (d) β-D-lactose; (e), (f) cellulose I; (g), (h) cellulose II. (Data for a–d and g–h from ref. 58.)

chemical shifts were examined of different disaccharides with β-1,4-glycosidic linkages, the torsion angles of which are known from X-ray diffraction analyses. In Fig. 19, the chemical shifts of the C1 carbon are plotted against the torsion angle ϕ. The chemical shifts of the crystalline C1 carbon of cellulose samples which are shown in Table 4 are also plotted in the figure using the ϕ values reported. In the case where two or three different chemical shifts are observed and only one ϕ is reported, all chemical shifts are plotted against the same ϕ. Although the data are somewhat scattered, the C1 chemical shift seems to be correlated with the torsion angle ϕ.

Similar plots of the C4 and C6 chemical shifts against the torsion angles ψ and χ obtained from X-ray analyses are shown in Figs 20 and 21. It is evident that the C4 and C6 chemical shifts depend primarily on these torsion angles, ψ and χ, respectively. According to a recent survey of the rotation about the C5–C6 bonds in low molecular weight glucosides, *gauche–gauche* and *gauche–trans*† are the two favourable

† The C6–O6 bond is *gauche* in relation to O5–C5–C6 and *trans* in relation to C4–C5–C6.

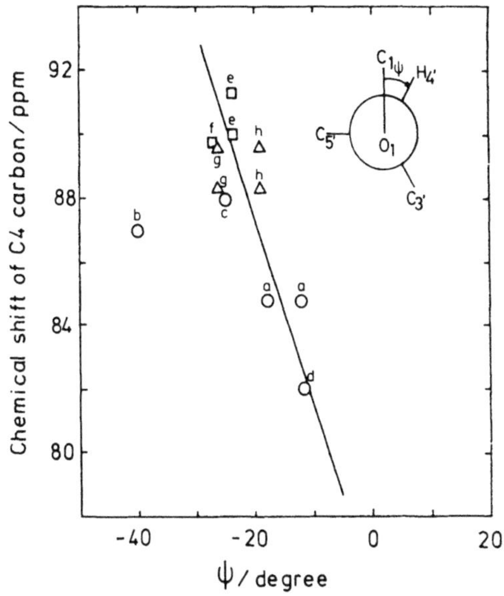

FIG. 20. ^{13}C chemical shifts of the C4 carbons versus torsion angle ψ. (a) β-D-cellobiose; (b) β-D-methylcellobioside·CH_3OH; (c) α-D-lactose monohydrate; (d) β-D-lactose; (e), (f) cellulose I; (g), (h) cellulose II. (Data for a–e and g–h from ref. 4.)

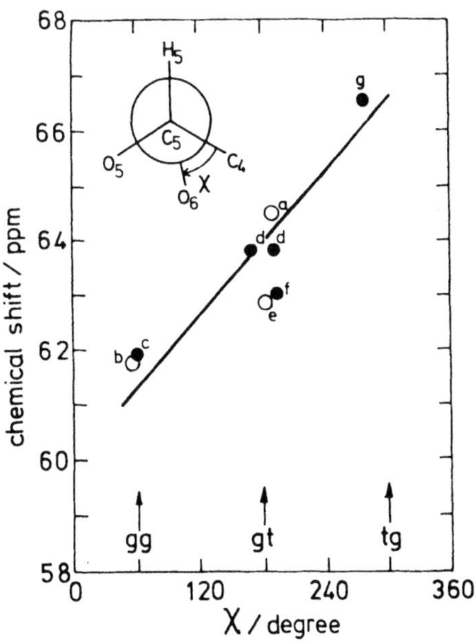

FIG. 21. ^{13}C chemical shifts of the C6 carbons versus torsion angles χ. (a) α-D-glucose; (b) α-D-glucose monohydrate; (c) β-D-glucose; (d) β-D-cellobiose; (e) α-D-lactose monohydrate; (f) β-D-lactose; (g) cellulose I. (Data for a–f from ref. 46.)

conformers. In agreement with this survey, no data for monosaccharides and disaccharides are included in Fig. 21 which correspond to the χ of *ca* 300° (*trans–gauche*). On the other hand, the conformation of the CH_2OH side-groups of the crystalline component of native cellulose is found to be *trans–gauche,* against the expectation from the data of monomeric glucosides. It is found from Fig. 21 that three preferred conformations: *gauche–gauche, gauche–trans,* and *trans–gauche,* correspond to the chemical shifts of approx. 62, 62·7–64·5 and 66 ppm, respectively.

It is concluded that the solid–state chemical shifts of the C1, C4 and C6 carbons are mainly correlated with the torsion angles ϕ, ψ and χ, respectively. However, intermolecular effects such as packing and hydrogen bonding produce the total downfield or upfield shift of 0·2–2·4 ppm for the carbohydrates studied here. Therefore, when a downfield or upfield shift of such an order appears even in the solid-state ^{13}C NMR spectra, intermolecular effects must be also considered.

Now we are in a position to be able to discuss the crystalline and non-crystalline chemical shifts for different cellulose samples. First, consider the chemical shifts of the crystalline components shown in Table 4. The C1 and C4 resonance lines of valonia and bacterial celluloses split, respectively, into three and two lines, in good accord with the results reported by others.[59,60] On the other hand, in cotton and ramie the splitting into two lines appears only in the C1 line. This may be caused by the disappearance of the central line of the C1 carbon of valonia and bacterial celluloses. If these splittings are associated with the different torsional angles in the β-1,4-glycosidic linkages, the splittings of 0·6–2·2 ppm recognized in the C1 carbon and those of 0·9–1·2 ppm in the C4 carbon correspond to the difference of 4·0–14·7° and 1·5–2·0° in the torsion angles ϕ and ψ, respectively (see Figs 19, 20). The validity of the assumption and the estimation will be examined for ramie by an X-ray crystal refinement because in this sample a relatively large splitting appears only for the C1 carbon.

An alternative cause of the splittings may be the intermolecular effects, because their extent is in the order of these effects, as described before. In that case molecular chains must be packed in such a manner that at least two different intermolecular environments surrounding the C1 and C4 carbons are defined. The effects of hydrogen bonding associated with O5 and O1 oxygens might be somewhat significant. These effects are very difficult to estimate precisely at present. Nevertheless, it can be concluded here that the crystal structures of valonia and bacterial celluloses are different from those of cotton and ramie, although the same crystal structure (of cellulose I) has been assumed for all these samples in the X-ray crystal analyses.

On the other hand, the results for regenerated cellulose are different from those for both types of native cellulose; both C1 and C4 carbons split into two lines. The chemical shift of the downfield line of the C1 carbon is the highest value among the C1 resonance lines. On the contrary, that of the upfield line of the C4 carbon is the lowest value for the C1 resonance lines. These facts may be related to the crystal structure of cellulose II, but the detailed structure must be also re-examined by considering two kinds of values for both ϕ and ψ or two different kinds of intermolecular environments for C1 and C4 carbons. In addition, the chemical shifts of the C6 carbon are 63·7–64·1 ppm for regenerated cellulose, which correspond to the *gauche–trans* conformation as is seen in Fig. 21. This conflicts with the result of X-ray crystal analyses, where both *gauche–trans* and *trans–gauche* conformations are assumed. In an X-ray crystal refinement this fact must be also considered.

Table 5 shows that the non-crystalline components of native cellulose have no significant difference in chemical shift and linewidth from each other. The chemical shifts of the regenerated cellulose fibres, except for the amorphous cellulose, are also almost the same as those of native cellulose. This may suggest that the most probable conformation of the regenerated fibres is very similar to that of native cellulose. However, there is a marked difference in linewidth $\Delta\nu$ of the C1 carbon between native and regenerated celluloses; the linewidth of the C1 carbon of native cellulose is about half that of the regenerated cellulose. Although the cause of the line broadening in CP/MAS spectra is not clear as yet, it is most likely that the linewidth of the C1 carbon is primarily dependent on the distribution of the torsion angle ϕ. Thus, it is suggested that the distribution in the angle ϕ is relatively narrow for the non-crystalline component of native cellulose compared with that of regenerated cellulose. This means that the conformation of the non-crystalline component is comparatively confined for native cellulose, whereas in the regenerated cellulose fibres it is fairly versatile. The result obtained here is in good accord with our previous ^{1}H broad-line NMR analyses which indicate that the non-crystalline component of regenerated cellulose fibres is much enhanced in molecular mobility by a swelling agent such as DMSO, whereas no significant effect appears for native cellulose.[61]

The regenerated cellulose fibres were produced by spinning from the concentrated solution in which molecular chains were coiled and entangled with each other. Therefore, it is thought that this coiled and entangled state of molecular chains has been brought into the non-crystalline region of the final products. On the contrary, there is no necessity in native celluloses for such an entangled and coiled state, involving a wide distribution in the torsional angle ϕ, to remain in the non-crystalline region.

In the amorphous cellulose sample the chemical shift of the C4 carbon is as low as 81·6 ppm and very close to the corresponding value of the low molecular weight cellulose in DMSO solution (see Table 4). This sample was prepared by dissolution of Whatman cellulose powder CF-1 in DMSO–paraformaldehyde followed by precipitation in ethanol. Therefore, the molecular chains of this sample must be fully disordered in comparison with those of the regenerated cellulose fibres and native cellulose. The more detailed structure of the non-crystalline components of different cellulose samples will be discussed elsewhere.[47]

REFERENCES

1. Pines, A., Gibby, M. G. and Waugh, J. S., *J. Chem. Phys.*, 1972, **56,** 1776; 1973, **59,** 569; *Chem. Phys. Lett.*, 1972, **15,** 373.
2. Mehring, M. in *Principles of High Resolution NMR in Solids,* 1976 and 1983, Springer-Verlag, Berlin.
3. Waugh, J. S., Huber, L. M. and Haeberlen, U., *Phys. Rev. Lett.*, 1968, **20,** 180.
4. Kubo, R. and Tomita, K., *J. Phys. Soc. Japan,* 1954, **9,** 888.
5. Bloembergen, N. and Rowland, J. A., *Acta Met.*, 1953, **1,** 731.
6. Andrew, E. R., Bradbury, A. and Eades, R. G., *Nature* (*London*) 1958, **182,** 1659; 1959, **183,** 1802.
7. Lowe, I. J., *Phys. Rev. Lett.*, 1959, **2,** 285.
8. Schaefer, J., Stejskal, E. O. and Buchdahl, R., *Macromolecules,* 1977, **10,** 384.
9. Stejskal, E. O., Schaefer, J. and McKay, R. A., *J. Magn. Res.*, 1977, **25,** 569.
10. Bloemberger, N., Purcell, E. M. and Pound, R. V., *Phys. Rev.*, 1948, **73,** 679.
11. Solomon, I., *Phys. Rev.*, 1955, **99,** 559.
12. Abragam, A., *The Principle of Nuclear Magnetism,* 1961, Oxford University Press, Oxford.

13. Slichter, C. P., *Principles of Magnetic Resonance,* 1978, Springer-Verlag, Weinheim.
14. Kitamaru, R., in *Applications of NMR Spectroscopy to Problems in Stereochemistry and Conformational Analysis,* Y. Takeuchi and A. P. Marchand (Eds), 1986, Verlag Chemie, Deerfield Beach, FL, USA.
15. Woessner, D. E., *J. Chem. Phys.*, 1962, **36,** 1; 1962, **37,** 647.
16. Howarth, O. W., *J. Chem. Soc. Faraday Trans. II,* 1978, **74,** 1031; 1979, **75,** 863.
17. Murayama, K., Horii, F. and Kitamaru, R., *Bull. Inst. Chem. Res. Kyoto Univ.*, 1983, **61,** 229.
18. Horii, F., Hirai, A., Murayama, K. and Kitamaru, R., *Macromolecules,* 1983, **16,** 273.
19. VanderHart, D. L., *J. Chem. Phys.*, 1976, **64,** 830.
20. Demco, D. E., Tegenfeldt, J. and Waugh, J. S., *Phys. Rev.*, 1975, **B11,** 4133.
21. Horii, F., Kitamaru, R., Maeda, S., Saika, A. and Terao, T., *Polym. Bull.*, 1985, **13,** 179.
22. Kitamaru, R. and Hyon, S.-H., *J. Polym. Sci., Macromol. Rev.*, 1979, **14,** 207.
23. Bassett, D. C. and Hodge, A. M., *Proc. R. Soc., London,* 1978, **A359,** 121; 1981, **A377,** 25.
24. Möller, M., Cantow, H.-J., Krüger, J. K. and Höcker, H., *Polym. Bull.*, 1981, **5,** 124.
25. Möller, M., Gronski, W., Cantow, H.-J. and Höcker, H., *J. Amer. Chem. Soc.*, 1984, **106,** 5093.
26. Ando, I., Yamanobe, T., Sorita, T., Komoto, T., Sato, H., Deguchi, K. and Imanari, M., *Macromolecules,* 1984, **17,** 1955.
27. Yamanobe, T., Chujo, R. and Ando, I., *Mol. Phys.*, 1983, **50,** 1231.
28. Sorita, T., Yamanobe, T., Komomoto, T. and Ando, I., *Makromol. Chem. Rapid Commun.*, 1984, **5,** 657.
29. Kitamaru, R., Horii, F. and Murayama, K., *Macromolecules,* 1986, **19,** 6360.
30. Axelson, D. E., Mandelkern, L., Popli, R. and Mathien, P., *J. Polym. Sci., Polym. Phys. Ed.*, 1983, **21,** 2319.
31. Kitamaru, R., Horii, R. and Hyon, S.-H., *J. Polym. Sci., Polym. Phys. Ed.*, 1977, **15,** 821.
32. Horii, F. and Kitamaru, R., *J. Polym. Sci., Polym. Phys. Ed.*, 1978, **16,** 265.
33. Kitamaru, R. and Horii, F., *Advances in Polymer Science,* Vol. 26, 1978, Springer-Verlag, Berlin, pp. 137–78.
34. Schröter, B. and Posern, A., *Makromol. Chem. Rapid Commun.*, 1982, **3,** 623.
35. Flory, P. J., *J. Amer. Chem. Soc.*, 1962, **84,** 2857.
36. Flory, P. J., Yoon, D. Y. and Dil, K. A., *Macromolecules,* 1984, **17,** 862.
37. Komoroski, R. A., Maxfield, J., Sakaguchi, F. and Mandelkern, L., *Macromolecules,* 1977, **10,** 550.

38. Horii, F., Murayama, K. and Kitamaru, R., *ACS Polym. Prepr.*, 1983, **24,** 384.
39. Torchia, D. A., *J. Magn. Reson.*, 1978, **30,** 613.
40. Voigt-Martin, I. G. and Mandelkern, L., *J. Polym. Sci., Polym. Phys. Ed.*, 1981, **19,** 1769.
41. Flory, P. J. and Yoon, D. Y., *Nature* (*London*) 1978, **272,** 226.
42. Yoon, D. Y., *J. Appl. Cryst.*, 1978, **11,** 531.
43. Gast, J. C., Atalla, R. H. and McKelvey, R. D., *Carbohydr. Res.*, 1980, **84,** 137.
44. Atalla, R. H., Gast, J. C., Sindorf, D. W., Bartuska, V. J. and Maciel, G. E., *J. Amer. Chem. Soc.*, 1980, **102,** 3249.
45. Earl, W. L. and VanderHart, D. L., *J. Amer. Chem. Soc.*, 1980, **102,** 3251.
46. Horii, F., Hirai, A. and Kitamaru, R., *Polym. Bull.*, 1982, **8,** 163.
47. Horii, F., Hirai, A. and Kitamaru, R., *ACS Symposium Series*, 1983, **260,** 28.
48. Horii, F., Hirai, A. and Kitamaru, R., *J. Carbohydr. Chem.*, 1984, **3,** 641.
49. VanderHart, D. L., *J. Magn. Reson.*, 1981, **44,** 117.
50. Imashiro, F., Maeda, S., Takegoshi, K., Terao, T. and Saika, A., *Chem. Phys. Lett.*, 1982, **92,** 642; 1983, **99,** 189.
51. Terao, T., Maeda, S. and Saika, S., *Macromolecules*, 1983, **16,** 1535.
52. Pfeffer, P. E., Valentine, K. M., and Parrish, F. W., *J. Amer. Chem. Soc.*, 1979, **101,** 1265.
53. Pfeffer, P. E., Hicks, K. B., Frey, M. H. and Opella, S. J., *J. Carbohydr. Chem.*, 1984, **3,** 197.
54. Brown, G. M. and Levy, H. A., *Science*, 1965, **147,** 1038.
55. Chu, S. S. and Jeffrey, G. A., *Acta Cryst.*, 1968, **B24,** 830.
56. Fries, D. C., Rao, S. T. and Sundaralingam, M., *Acta Cryst.*, 1971, **B27,** 994.
57. Hirotsu, K. and Shimada, A., *Bull. Chem. Soc. Japan*, 1974, **47,** 1872.
58. Horii, F., Hirai, A. and Kitamaru, R., *Bull. Magn. Reson.*, 1983, **5,** 190.
59. Earl, W. L. and VanderHart, D. L., *Macromolecules*, 1981, **14,** 570.
60. Kwoh, D. L. W., Bhattacharjee, S. S., Cael, J. J. and Patt, S. L., *TAPPI Proc.*, 1982, 114.
61. Hirai, A., Horii, F. and Kitamaru, R., *J. Polym. Sci., Polym. Phys. Ed.*, 1980, **18,** 1801.

Chapter 5

GELATION

M. J. MILES
AFRC Institute of Food Research, Norwich, UK

SUMMARY

The phenomenon of polymer gelation is discussed in terms of the interactions and mechanisms which could result in the formation of a gel network. The possible mechanisms leading to network formation are first discussed. These include chemical cross-linking, specific polymer–polymer interactions between identical and between chemically different polymers, phase separation of flexible and rigid polymers, crystallization of homopolymers and copolymers, and topological cross-linking. Examples of particular polymer gelling systems for each of the proposed mechanisms are then described. The aim is to illustrate the proposed mechanisms rather than give an exhaustive list of examples. In some cases, several mechanisms have, over time, been proposed for one polymer–solvent system. This has the advantage of illustrating how the ideas of gelation have developed and how, in fact, features of many of the mechanisms proposed may be involved. Although the emphasis is on crystalline polymers, the exclusion of what are usually classified as non-crystalline polymers would lead to a distorted view of polymer gelation. The breadth of the variety of polymers which exhibit gelation is emphasized. Synthetic-polymer and biopolymer gels are included as well as gels formed in biological systems. The similarities in gelation behaviour across this range are highlighted in the hope of revealing new insights.

INTRODUCTION

Definition and Approach

Definition
The bulk characteristics of a gel are that it exhibits solid-like mechanical behaviour, with a high compliance and a high elastic-strain region, while consisting mostly of liquid. Astonishingly, some gels containing in excess of 99·9% liquid by weight behave as solids under static loading. A structural characterization is that a gel consists of a microscopic, three-dimensional network which exists over macroscopic distances and which holds or entraps the liquid component, with the osmotic force balanced by the entropic (and free-energy) retractive force of the network. A more precise definition of a gel is difficult to formulate concisely, although intuitively a gel can usually be recognized.

Approach
A discussion of polymer gelation could proceed by many different routes depending on the particular aspect of the gel that was to be considered. Polymer gels could be categorized by their chemical structure, their application or properties, the origin of the polymer, and so on. The approach adopted here is based on the mechanisms for gelation.

Firstly general mechanisms for gelation are discussed, and then examples for such mechanisms presented. In some cases, the polymer system chosen is the best example for a particular mechanism, but it may nevertheless be currently only a minority view for this polymer. Examples are drawn from both biopolymer and synthetic-polymer systems, and although the emphasis is on crystalline polymers, other systems are considered for perspective and completeness. Indeed, there are crystalline-polymer systems for which the proposed mechanism of gelation does not involve crystallization, and, conversely, there are polymers which would generally be classified as non-crystalline for which the gelation mechanism is based on limited crystallization. It is hoped that by its inherent interdisciplinary nature, this approach may provide new insights.

Within the bounds of this chapter it is not possible to review the developments in gelation theories, both 'classical'[1,2] and 'modern', involving scaling-theory corrections[3–5] to mean-field exponents, the inclusion of solvent effects[6,7] and association enthalpy,[8] growth kinetics in percolation,[9] the development of fractal-percolation[10,11] and renormalized-group[12,13] methods. Unfortunately, the impact of such theories on the experimental field of physically cross-linked polymer gels remains small. Gordon[14] recently demonstrated that data on casein gelation could be fitted by both 'classical' and 'modern' theories, although the former implicitly assume that excluded-volume effects cancel and the latter assume that they dominate. There have also been developments in the understanding of the interaction between the gel network and free molecules including two-component liquids,[15,16] rigid polyions,[17] and an 'ant' (the jumping and capture of a Brownian particle).[18] It is also worth noting the topoanalogous transformation principle,[19] which, by considering networks of identical topology but different chemical structure, has potential to give insight into some unresolved problems on the application of rubber elasticity to gel networks.

The Role of Gels

The importance of gelation can be judged by the prevalence of gels in many far-flung fields, both scientific and technological. The generality of gelation is evident from the disparate materials which form gels. These range, for example, from the inorganic silicates found in clays through to complex biopolymers. However, most studies of gelation on a given class of material have been developed in isolation and have not benefited from an interdisciplinary approach. Similarly, developments in network and gelation theory have found application in few experimental studies of physically crosslinked gels.

Polymer gels and gelation are of importance in three respects:

(1) *For the gelation process itself,* bringing about a change in viscosity. Perhaps the most vital example of this is that of blood clotting in which the first stage of network formation reduces flow.[20] Another example in which the reversibility of the gelation process is apparently important is in the sol–gel transitions in the cytoplasm of the amoeba accompanying the extension of pseudopods responsible for the organism's movement.[21]

(2) *For the gel per se.* Here the emphasis is on the properties of the gel itself, i.e. its mechanical, optical and diffusion properties. Examples are far-ranging but can encompass food gels[22–26] (such

as starch or gelatin gels), the vitreous humour of the eye,[27] drug encapsulation,[28] gel-permeation chromatography,[29] and mucus,[30] and might arguably include plasticized PVC,[31] one of the most common synthetic materials. The gel structure of extracellular matrices (involving fibronectin and collagen) has even been implicated in the important problem of embryogenesis and cell differentiation in a recent study on the movement of non-motile cells.[32] The above examples represent only a few fields of the many in which gels play a major role.

(3) *As an intermediary state* in the fabrication of specialized materials. Non-polymer examples include the production of high-quality glasses[33] and ceramics.[34] The polymer parallel is the manufacture of rubber via the gel state. Microporous foams can be manufactured by the processes of gelation and crystallization followed by solvent extraction. An example of the use of such materials can be found in the current laser nuclear-fusion research in which the potential of such foams to damp out hydrodynamic instabilities in laser-induced nuclear-fusion reactions is being studied.[35] A final example is the production of high-stiffness, high-strength fibres by ultra-drawing of the gel structure to produce crystalline fibres[36–38] with elastic moduli (in the fibre direction) comparable with steel.

GELATION MECHANISMS

Network Structure

The classical theories of gelation[39,40] and network formation are concerned in general with networks formed by point junctions. A more extensive variety of junctions and network structures can, of course, be imagined, and indeed, corresponding examples of actual gels discovered. A logical expansion from point junctions is to increase the dimensionality to produce one-, two- and three-dimensional junction zones (Fig. 1a).

As the dimensionality is increased the probable functionality of the junctions also increases. Functionality and density of junction zones are important in gel stiffness and porosity, as is the spatial extent of the junction zones relative to the polymer in solution between them. An alternative network structure is composed not of individual polymer molecules interlinking the junctions but of discrete regions of higher polymer concentration, such that the network is now a

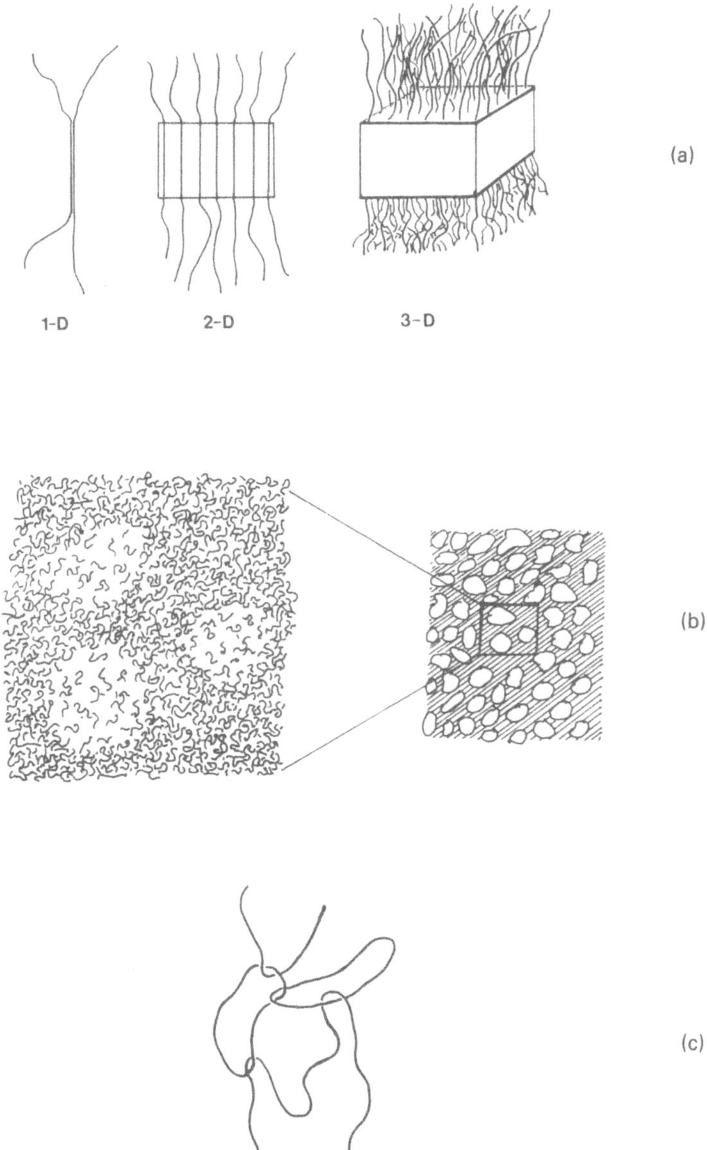

FIG. 1. (a) Diagrammatic representation of one-, two- and three-dimensional junction zones in a polymer gel. (b) A network formed by a continuous polymer-rich phase. (c) Diagrammatic representation of topological cross-links or knots.

continuous phase (Fig. 1b). Another class of network is that formed by geometric interlinking of polymer molecules. The cross-linking is neither chemical nor physical but topological (Fig. 1c).

There are two major considerations in the formation of a network: firstly the nature of the interaction and the conditions which favour it, and secondly the conditions which lead to network formation rather than precipitation of the aggregating species. Recently, the emphasis has been on studies into the nature of the junction zones with less emphasis on the factors influencing the overall network structure. The process of aggregation, which could result in a network junction, may involve a phase transition, but gelation itself is not a thermodynamic transition; it cannot be uniquely described by or distinguished from a non-gelling phase transition on the basis of free energy alone. The essential feature of gelation is the connectivity of the chains as a result of cross-linking. The formation of a continuous network extending throughout the solution rather than a disconnected precipitate is a topological constraint and a percolation problem. One perhaps obvious but necessary condition for the formation of a molecular network, rather than precipitation, is that the molecules should be in contact. In many cases, the minimum concentration for gelation, c_0, follows the overlap concentration, c^*, i.e. the concentration at which the volumes occupied or swept out by neighbouring molecules overlap. However, the value of c^* can usually only be determined experimentally away from the gelling conditions. Therefore, the average volume encompassed by a molecule during gelation may be somewhat different from such an equilibrium measurement. Consider, for example, the effect on molecular size and shape of a change in molecular conformation at the gelling conditions such as a coil→helix transition, or molecular extension in an elongational flow field, or denaturation. Similarly, a change in the quality of the solvent on reaching the gelling conditions would cause a change in the expansion of the polymer molecule. All of these would result in an overlap concentration very different from the equilibrium value.

Non-equilibrium State

In the case of a flexible-coil molecule, even under quiescent conditions, there is no abrupt cut-off of overlap with decreasing concentration, owing to the Gaussian probability distribution of a segment occupying a volume at some distance from the origin.

The situation is dynamic. The position of any segment is continuously changing so that a characteristic disentanglement time can be associated with a given overlap of the Gaussian distributions. Whether such an overlap is detected will depend on the timescale of the experiment. A novel method for studying overlap and disentanglement has been reported[41] recently. If there are weak interactions then their lifetime will contribute to the disentanglement time, and short-timescale measurements (e.g. high strain-rate deformation) would register entanglements as cross-links.

A distinction should be made between a gel network involving linear homopolymer molecules and copolymers or polymer molecules having 'interruptions' along their length. In general, a linear-homopolymer network which involves phase-separated regions, including crystallites, is also in a non-equilibrium state, as the system could lower its free energy by the growth of the larger phase regions at the expense of the smaller, reducing the total interfacial energy. If unhindered, this would eventually lead to a loss of connectivity and result in a precipitate. That such an ageing process is under way is often made manifest by a decrease in gel volume and limited syneresis (the expulsion of liquid from the gel) or by an increase in opacity as the dimensions of the phase-separated regions become comparable with the wavelength of light. This ageing of the system is in most cases kinetically hindered by the geometric constraints imposed by any one polymer molecule being pinned in a number of phase-separated regions, and a quasi-equilibrium is established. However, if the aggregating species is limited to sections only of the polymer molecule, as would be the case for a copolymer, then growth of aggregates will be severely restricted. With only limited ageing possible, such a gel network could rapidly approach a structure which may be regarded as at equilibrium. Various interruptions to the aggregating sequence have been proposed to limit the size of the interacting region in order to explain the relative stability of a particular gel. However, it should be emphasized that such interruptions may not be essential for the formation of an apparently stable gel as there are many examples of gelation occurring from solutions of linear homopolymers. For these cases, kinetic hindrance due to the presence of any one chain in several junction zones coupled with the low diffusion or reptation rates through these denser regions are invoked to account for the quasi-stability, and in some cases the very existence, of a gel. The non-equilibrium nature of gelation resulting from partial demixing suggests that the degree of supercooling and rate of quenching will affect the nucleation density, and therefore the number of cross-links per polymer molecule. This indicates the possibility of manipulating gel properties by controlling the conditions during network function. In the case of decompositions occurring by nucleation and growth, it can be expected that the initial stage will be dominated by the nucleation rate, but that the subsequent growth will be arrested by the geometric hindrances described above.

Gelation can occur either (1) from solution as a result of some change in solution conditions or as a result of chemical reaction, or (2) from the solid state by swelling of an existing network on the addition of a solvent. The latter route to gelation will not be pursued further here. Mechanisms that could lead to the formation from solution of the gel-network structures described above are discussed in the immediately following subsections. Examples of actual polymer–solvent systems in which gelation has been attributed to such mechanisms are considered in the last section.

Covalent Junctions

The ways in which a covalently linked gel network can form, in principle, are outlined below. Some examples are given in the subsection on Chemical Cross-Linking Gelation (pp. 256–8).

A distinction should be made between 'tight' networks formed in multifunctional polymerization, e.g. phenolic resins, and 'loose' networks, such as vulcanized rubber, involving the occasional cross-linking of linear chains. The 'tight' networks will not be discussed further.

Crosslinking During Polymerization

Condensation Polymerization. Networks can form as a result of non-linear condensation polymerization. For example, if bifunctional units, A, are present in solution with A_f, polyfunctional units of functionality f, then the following structure may result:

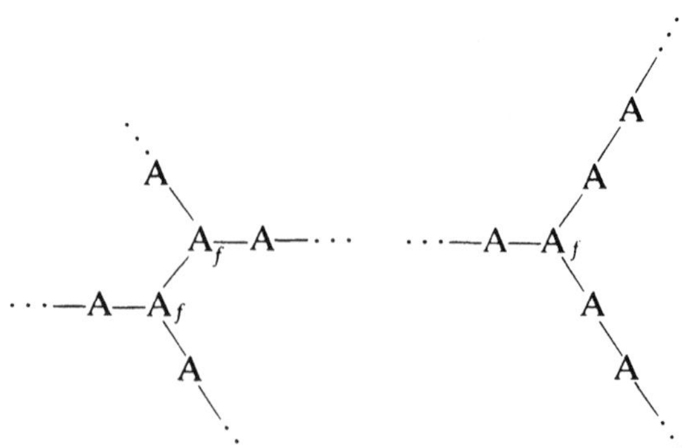

Addition Polymerization. Gelation by addition polymerization involves a chain reaction in which a free radical is capable of reacting to open the double bond of a vinyl monomer resulting in polymerization by linear addition of these units.

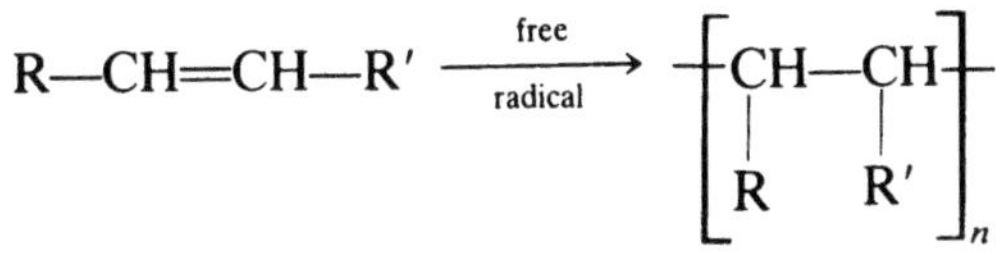

The presence of some tetrafunctional vinyl derivative, i.e. one possessing two double bonds,

$$CH_2{=}CH{-}R''{-}CH{=}CH_2$$

can lead to the bridging of two polymer chains by its incorporation in each where n_1, n_2, n_3 and n_4 are the degrees of polymerization within the brackets.

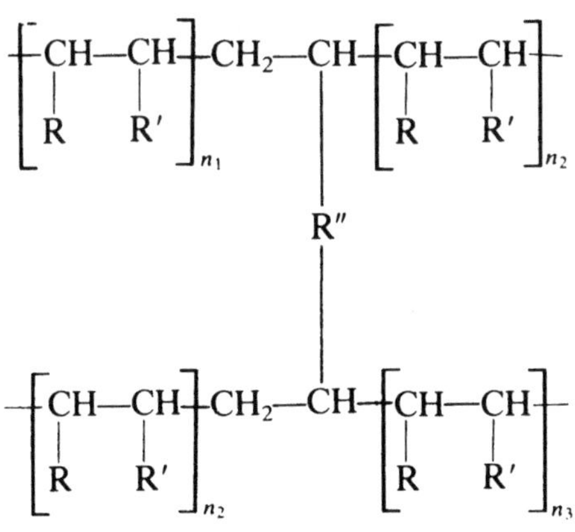

Cross-Linking of Pre-synthesized Polymers

A polymer solution can be cross-linked by the triggering of reactive sites incorporated in the chain as a result of a change in solution conditions to favour this reaction, e.g. changes in pH or temperature. Another possibility is the introduction of a cross-linking agent. This could be a chemical agent, but might also include radiation. Ionizing radiation[42] is usually used in the solid state, but UV-promoted photochemical reactions can occur in solution (see Fig. 2a).

One variation on the cross-linking of polymers is by the bridging of chains by oligomers or polymers of a different species (see Fig. 2b).

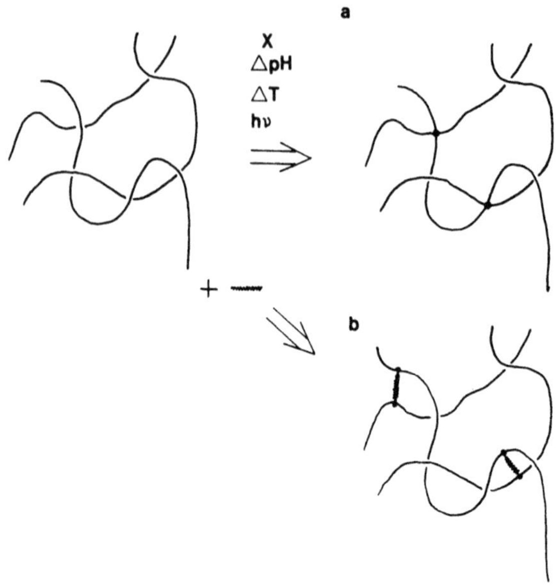

FIG. 2. (a) Cross-linking of pre-synthesized polymers by chemical agents (X), changes in pH (ΔpH) or temperature (ΔT), or radiation ($h\nu$). (b) Cross-linking of pre-synthesized polymers through bridging by oligomers or polymers of a different species.

Physical Cross-Links

The formation of a network in which the junctions are physical cross-links can be brought about by a change in solution conditions such that the polymer solubility is reduced and demixing occurs. Polymer–polymer interactions may then be sufficiently favourable to allow the formation of relatively stable junction-zones and a network. This association or aggregation of polymer molecules can in general terms be described as a phase separation, and must clearly occur in such a manner that the morphology of the separated phases favours network formation rather than precipitation. Such phase separations from solution may result from liquid–liquid or liquid–solid phase transitions. This is clearly central to an understanding of gelation from solution and will form the main thrust of this subsection.

In the following subsections, physical interactions which might result in network formation and gelation are briefly discussed. Later (from p. 258) some examples of polymer systems in which gelation has been ascribed to processes involving these interactions will be described.

Specific Interactions

Perhaps the most easily visualized physical cross-link is that of a specific interaction between two polymer molecules. This is probably the closest physical parallel to the chemical (covalent) cross-link. Although solubility remains a factor, the formation of such cross-links cannot strictly be described as a phase separation, as a dimeric junction zone cannot be regarded as a new phase. To be stable, strong anisotropic interactions are usually involved, resulting in one-dimensional periodic junctions. The length of the interacting sequence limits the length of the junction, but there will usually be a minimum length for stability, analogous to the critical-nucleus in crystallization. For the junctions to be limited to two molecules, the interaction must be specific and anisotropic; thus enantiomorphic complexes present themselves as strong candidates, as, in general, do molecules of complementary shape.

Although perhaps more easily visualized, a block structure along the polymer molecule for such interacting regions is not essential for gelation (see subsection on Network Structure, pp. 236–8).

Phase Separation

Liquid–Liquid Phase Separation. Partial demixing of a polymer solution results in the formation of a polymer-rich and a polymer-deficient phase. The polymer-rich phase could, in principle, result directly in the formation of a network either by forming a continuous network-like region (Fig. 1b) or by forming the junction regions of the network (Fig. 1a). Alternatively, liquid–liquid phase separation may provide the concentration conditions that subsequently favour more specific molecular interactions or transitions.

The free energy of mixing is given by:

$$\Delta G_{\mathrm{m}} = \Delta H_{\mathrm{m}} - T\,\Delta S_{\mathrm{m}} \tag{1}$$

For demixing to be favoured, $\Delta G_{\mathrm{m}} > 0$, where ΔH_{m} and ΔS_{m} are the enthalpy and entropy of mixing. In an ideal solution, the free energy depends only on the entropy of mixing. The molecules do not interact ($\Delta H_{\mathrm{m}} = 0$) and the components are completely miscible under all conditions. Using the Boltzmann equation for configurational entropy and the lattice model, Flory[43] and Huggins[44] derived the expression for the entropy of mixing of polymer and solvent, the so-called Flory–Huggins expression:

$$S = k(N_1 \ln \phi_1 + N_2 \ln \phi_2)$$

where ϕ_1 and ϕ_2 are the volume fractions and N_1 and N_2 are the numbers of polymer and solvent molecules, respectively.

For a system in which interactions occur, $\Delta H_m > 0$, a temperature exists below which complete miscibility is unfavourable. For weak interactions thermal motion may still sufficiently randomize the positions of the components on the lattice to allow the use of the ideal entropy of mixing. This is the basis for the concept of a regular solution.[45] The enthalpy of mixing can be expressed in terms of an exchange energy, ω, given by:

$$\omega = \tfrac{1}{2}\varepsilon_{11} + \tfrac{1}{2}\varepsilon_{22} - \varepsilon_{12}$$

where ε_{ij} are the energies of interaction per contact between components i and j.[46] The enthalpy of mixing is then given by:

$$\Delta H_m = \left(N_1 + \frac{N_2 V_2}{V_1}\right) z\omega\phi_1\phi_2$$

where z is the coordination number, and V_2/V_1 is the volume ratio of sizes of the polymer to solvent molecules. The free energy of mixing/unit volume becomes:

$$\frac{\Delta G}{V} = kT\left(\frac{\phi_1}{V_1}\ln\phi_1 + \frac{\phi_2}{V_2}\ln\phi_2 + \frac{z\omega}{kT}\frac{\phi_1\phi_2}{V_1}\right)$$

where $V = V_1N_1 + V_2N_2$.

The quantity $z\omega/kT$ is the Flory interaction parameter, χ. Another quantity, which should be mentioned because of its frequent use in the prediction of solubilities in polymer systems, is the solubility parameter,[45,46] δ. Optimum miscibility occurs when polymer and solvent have the same value of δ. δ^2 is defined as the cohesive energy density of the material. Its value will change with temperature and pressure, but, in its simplest form, the approach gives the enthalpy of mixing as:

$$\frac{\Delta H_m}{V} = (\delta_1 - \delta_2)^2\phi_1\phi_2$$

where δ_1, ϕ_1 and δ_2, ϕ_2 are the solubility parameters and volume fractions of the polymer and solvent, respectively.

Thus, the smaller the difference in the solubility parameters, the greater will be the mutual solubility of components 1 and 2.

Below the critical solution temperature, T_c, which is associated with a critical composition, the free energy of mixing, ΔG_m, will be positive, and demixing will be favoured. In the limit of infinite molecular weight, T_c equals the θ temperature, i.e. the temperature at which the polymer assumes the dimensions of the idealized random walk. In a binary system consisting of polymer and solvent, this results in polymer-rich and polymer-deficient phases (Fig. 3a). Below T_c, a region of concave (downwards) curvature appears in the otherwise convex free energy versus composition plot (Fig. 3b). In the phase diagram, the binodal or coexistence line separating the one-phase from the two-phase region can be shown to correspond to the compositions at the points of contact of a common tangent to the two concave regions in the free energy versus composition plots.[46,47]

Immediately below the binodal, the appearance of the new phases will be associated with an interfacial energy. The phase separation will be an activated process requiring the formation of a critical nucleus from which the new phase can grow. However, further into the two-phase region, the decrease in free energy on phase separation effectively reduces the size of the critical nucleus to zero. This is the spinodal region,[48] which is represented in the phase diagram (Fig. 3a) by a broken line, and corresponds to the point of inflection separating the convex and concave regions in the free-energy plots (Fig. 3b). Spinodal decomposition or phase separation is associated with a negative diffusion coefficient such that any composition fluctuation is amplified and the resulting phase separation is sinusoidally periodic in composition with a characteristic wavelength.[48] The network-like character of the spinodal phase separation, apparent even in the original simulations[48] (see Fig. 4) makes the spinodal mechanism a particularly attractive idea as a mechanism for the formation of a phase-separated network. Van Aartsen[49] has adapted the Cahn theory to spinodal decomposition of polymer solutions. Simulations of binodal and spinodal phase separations have been performed for various high-polymer mixtures.[50] Derham *et al.*[51] have developed the pulse-induced critical scattering (PICS) technique to study the kinetics of phase separation in polymer solutions. In particular, PICS allows a better estimation of the spinodal line.

For polymer–solvent systems, the critical composition occurs at low polymer concentrations. This results from the large volume of the polymer molecule compared with the solvent molecule. One consequence is that even the polymer-rich phase is relatively dilute.

The phase separation has so far been assumed to occur as the temperature is decreased to below the critical solution temperature,

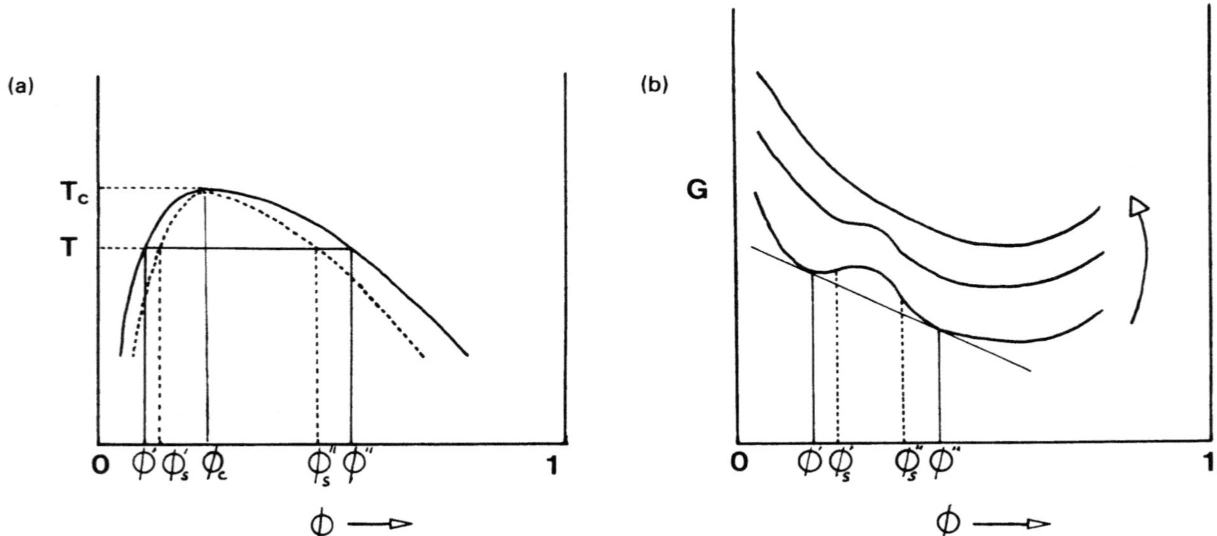

FIG. 3. (a) A schematic phase diagram showing the coexistence or binodal (—) and spinodal (---) lines for composition, ϕ, and temperature, T. T_c and ϕ_c are the critical temperature and composition, respectively. Cooling to T results in phase regions of compositions ϕ' and ϕ'' or ϕ_s' and ϕ_s'' for binodal and spinodal decompositions, respectively. (b) A schematic representation of the corresponding free energy, G, versus composition, ϕ, plot.

FIG. 4. Computer simulations of spinodally phase-separated regions. (Reprinted with permission, from J. W. Cahn, *J. Chem. Phys.*, 1965, **42,** 93.)

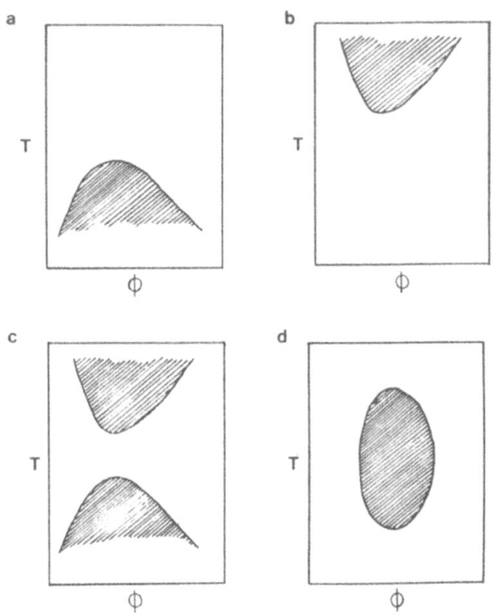

FIG. 5. Schematic phase diagrams for polymer–solvent systems. The shaded area represents two-phase regions. T is temperature and ϕ composition.

which, in this case where solubility decreases with decreasing temperature, is also known as the upper critical solution temperature, ucst (Fig. 5a). Some systems exhibit decreasing solubility with increasing temperature, and have a lower critical solubility temperature, lcst (Fig. 5b), with phase separation occurring on heating. Such behaviour can occur when there is an attractive specific interaction between the polymer and the solvent. With increasing temperature, dissociation increases with an accompanying decrease in solubility. Many polymers in poor solvents exhibit two regions (Fig. 5c) of immiscibility with the ucst being attributed to the difference in the chemical nature of the components and incorporated in the classical Flory–Huggins theory via the interaction parameter, and the lcst being attributed, in the absence of a specific interaction, to a difference in free volume between the components.[52] Free-volume effects are not taken into account in the classical lattice theory. The more recent equation-of-state theories[53,54] are more suitable, and also include volume changes on, and the pressure-dependence of, mixing and demixing. Suzuki *et al.*[55] have studied the effect of pressure on the sol–gel transformation of several polymers. They found that the changes in enthalpy and volume on gelation were both negative for gelatin and for poly(vinyl alcohol) (PVA) but positive for methylcellulose. This was attributed to hydrogen bonding being the main cross-linking interaction in gelatin and PVA, and hydrophobic interactions in the case of methylcellulose. Figure 5d shows an uncommon but possible binary phase diagram that can occur for polymer–solvent systems.

Of at least equal importance to gelation are the ternary systems. The third component may be a low molecular weight molecule which could be either neutral or ionic and either a solvent or a non-solvent for the polymer. The calculation of such ternary phase diagrams is, of course, more complicated than in the binary case, particularly when electrolytes and polyelectrolytes are involved. Indeed, the role of water in gelation[56] is far from being fully resolved, and is a major issue in its own right.[57] A schematic representation of the triangular phase diagram for a solvent–non-solvent–polymer system at a particular temperature is shown in Fig. 6(a). There is a parallel between the solvent/non-solvent ratio and the role of temperature in the solvent–polymer system.

The third component in a ternary system may also be a polymer. A schematic phase diagram for a polymer–solvent–polymer system is shown in Fig. 6(b) for the case where the polymer species are of

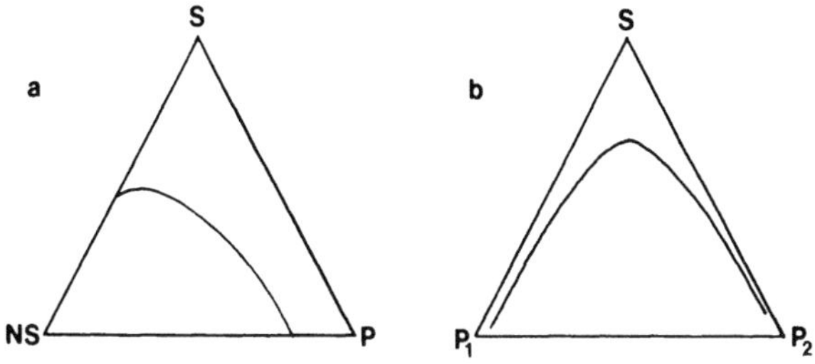

FIG. 6. Schematic representation of ternary phase diagrams for the systems (a) polymer–solvent–non-solvent and (b) polymer–solvent–polymer.

similar molecular weight. In general, polymer–polymer phase separation only occurs at concentrations greater than the overlap concentration.[58]

Quaternary and higher systems have been investigated.[59] In such cases, many phases may coexist. Indeed, since the phase behaviour is dependent on the molecular weight of the polymer, polydispersity itself results in a multicomponent system.

Changes in the temperature or composition of the solution may result in a conformational transition of one of the components. For example, intramolecular forces may stabilize a helical conformation. Such an ordered structure may also be favoured because of the associated decrease in the volume occupied.[60] A helical conformation can affect the phase-separation behaviour in several ways, for example by presenting attractive or repulsive groups on its surface, promoting either polymer–polymer or polymer–solvent interactions, or, if the helical molecule is sufficiently stiff and rod-like, by an isotropic–liquid crystal phase separation. Flory[61] has predicted the general features of the corresponding phase diagram (Fig. 7).

Phase-Separated Networks. There are two basic morphologies of phase separation which could, at least in principle, produce a network structure. Firstly, a phase separation may occur which results in a localized polymer-rich region, possibly by a nucleation and growth process. This could act as a junction zone. The growth of this rich phase may be limited by geometric constraints, such as entanglements or the copolymeric nature of the molecule. Secondly, the polymer-rich phase may not be localized but form a continuous network. This is more likely to be the case when the nucleation density is high or when spinodal phase separation occurs.

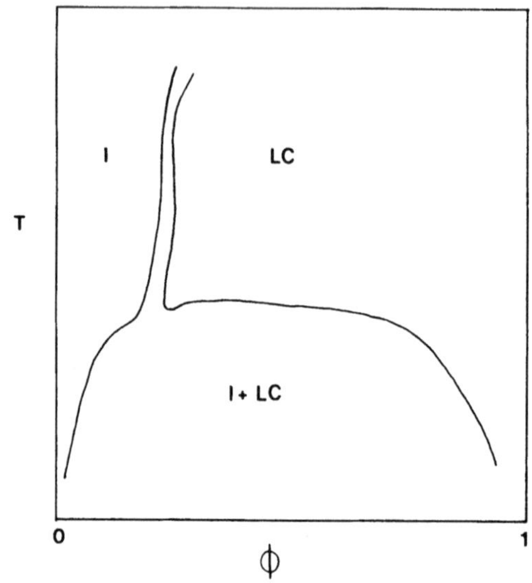

FIG. 7. Schematic representation of a liquid crystal–isotropic phase diagram for rigid rods in solution.

In polymer–solvent systems even the polymer-rich phase is relatively dilute and contains much solvent. This is clearly unsuitable for forming a permanent junction in the network, as it would seem that even a low stress would cause the junction to break up. One suggestion to overcome this objection to phase separation as a prime mechanism in gelation is that the increased concentration in the polymer-rich phase is sufficient, at the given temperature, for the polymer to be in a glassy state.[62] Alternatively, it can be argued that even above the glass transition temperature, the viscosity of the polymer-rich region is sufficiently high, compared with the dilute region, to confer on the junction a lifetime which is sufficient for a gel to be detected rather than a viscous liquid.

Another possibility, perhaps more appropriate in the case of a continuous phase-separated network, is that the retractive force of the gel arises from a minimization of interfacial energy. For example, if the gel structure is imagined to be cellular, consisting of approximately spherical holes in the polymer-rich phase entrapping the dilute phase, then small deformations at constant volume (illustrated in two dimensions in Fig. 8) would result in an increase in interfacial area and, thus, an increase in free energy. The retractive force in such a

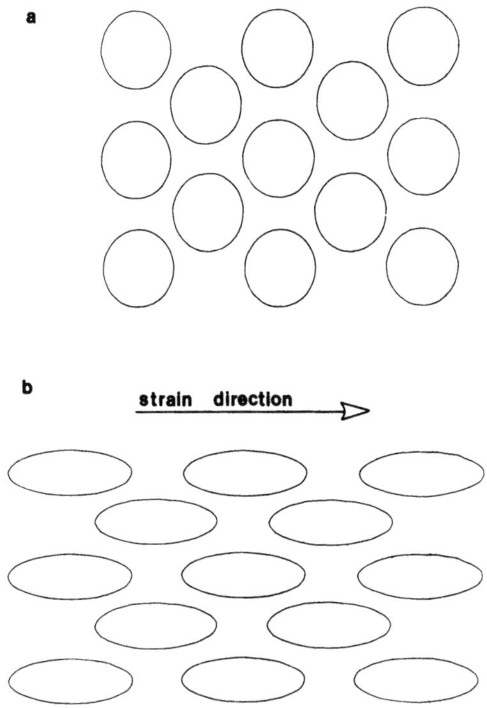

FIG. 8. A two-dimensional representation to illustrate the increase in surface area (which corresponds to line length in this diagram) on deformation of an array of spherical pores.

model would have a negative component to its temperature coefficient. Since the polymer-rich regions still contain much solvent, diffusion of small molecules through these regions could still occur relatively easily.

Crystallization

Crystallization is the major mechanism that has been proposed for the formation of the physical cross-links in gel networks. There are indications that any of the principal polymer-crystal morphologies, i.e. chainfolded, fringed-micellar, and extended-chain (Fig. 9), can be responsible for gelation. Some polymers exhibit all three crystal morphologies depending on the crystallization conditions. In general, of course, the Gibbs free energy of the crystal phase must be lower than the corresponding solution for crystallization to be favourable.

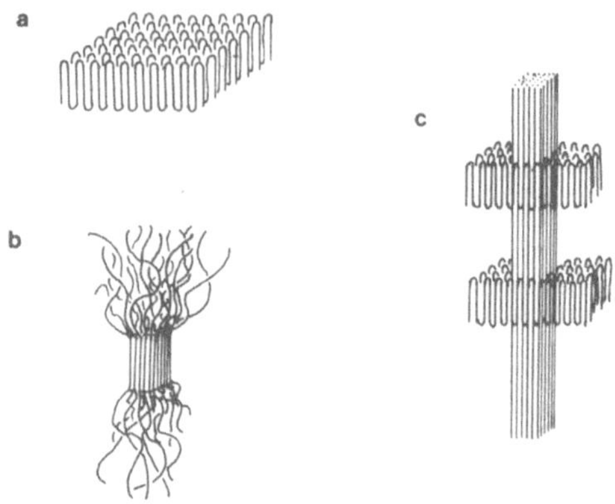

FIG. 9. The basic polymer-crystal morphologies of (a) folded-chain, (b) fringed-micellar and (c) extended-chain crystals. The last is shown with fold-chain crystal overgrowth which results in the so-called shish-kebab structure.

However, as with the liquid–liquid separation, the interfacial energy contributes a positive term, and thus there is a minimum, or critical, size for a stable nucleus which can subsequently grow. At low supercooling and for flexible polymers, the nucleus is chain folded.[63] Subsequent crystal growth is also chainfolded, and at low concentrations, platelet-like single-crystal suspensions form. Clearly, the molecule must be sufficiently flexible to allow a fold to form readily. With increasing concentration, entanglements will also lead to the disruption of the fold surface, with the possibility of interlocked loose loops from neighbouring crystals. At suitable polymer concentration and crystallization temperature, a network can thus form with interlocked loops linking chainfolded crystals (Fig. 10).

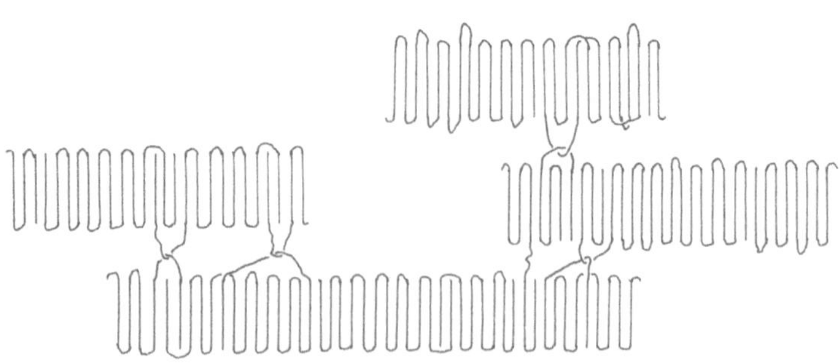

FIG. 10. Interlocking of loops in irregular fold surfaces following folded-chain crystallization from suitably high molecular weight polymer solutions over a limited concentration range to form junctions in a network.

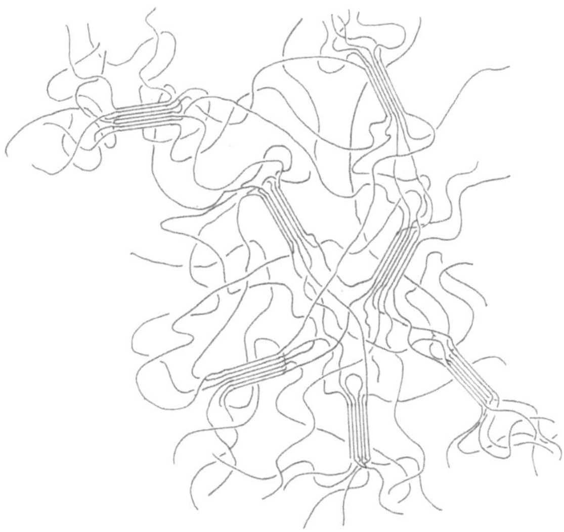

FIG. 11. Schematic representation of fringed-micellar crystals as junctions in a network.

The fringed-micellar crystal offers the most suitable morphology for an intermolecular junction zone (see Fig. 11). At larger supercoolings, there is evidence that fringed-micellar crystals are stable.[64] The melting endotherm for crystals formed at large supercoolings occurs at a much lower temperature than that associated with folded-chain crystals. This low melting point itself suggests smaller, less perfect crystals. That these are not just imperfect folded-chain crystals is deduced from the observation that both the high- and low-temperature endotherms can be observed in the same sample if a suitably intermediate crystallization temperature has been chosen.[64] The crystals formed at large supercoolings have a shape (derived from X-ray line-broadening measurements) which is consistent with the fringed-micellar morphology,[64] i.e. the dimension in the chain direction is larger than the lateral dimension. The lateral dimensions of such crystals are limited by the lattice-strain energy at the crystal–amorphous interface at either end of the crystal, resulting from the difference in density between the two regions.[65] Fringed-micellar crystallization[66] occurs more readily in the case of stiffer, less flexible polymers and for high quenching rates which are sufficiently rapid not to allow the disentanglement necessary for the other crystallization modes. High molecular weight polymer molecules can enter more than one micellar nucleus and, thus, form the basis of a network. Longitudinal crystal growth may be inhibited by the geometric constraints of polymer molecules in any given crystal being 'pinned' in a number of other crystals. However, longitudinal growth also stops even in very dilute solutions. This is not understood and remains a major unresolved issue. In some cases the crystallizable sequence is limited to block-like regions along the molecule. These blocks may be a crystalline polymer in a regular block copolymer or they may be iso- or syndio-tactic regions of an otherwise atactic homopolymer. The role of tacticity and other isomeric irregularities has been carefully considered with particular reference to isotactic polystyrene.[64] Another factor which can limit the extent of a crystalline region[67] is branching of the polymer.

For extended-chain crystallization to occur, the polymer molecule must either be inherently stiff and extended in solution, or be extended by an elongational flow field immediately prior to crystallization, usually at low supercooling. Pennings and coworkers[68] have studied flow-induced crystallization. The extended chain crystals could form a fibrous network in which the junctions are simply common extended-chain crystalline regions (Fig. 12a), or regions of common chainfolded crystal overgrowth (Fig. 12b). Hoffman has developed a

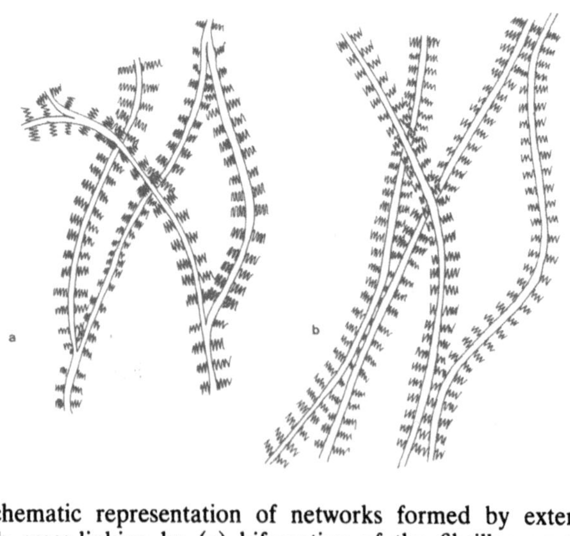

FIG. 12. Schematic representation of networks formed by extended-chain crystals with cross-linking by (a) bifurcation of the fibrillar crystals or (b) common folded-chain crystals.

theory for extended-chain crystallization.[69] The extension of the molecule by an elongation flow field reduces the overlap concentration. This would still be the case if, instead of extended-chain crystallization, some localized, intermolecular interaction occurred in the extended state. It is likely that such a gel network, without extended-chain crystallization, would undergo considerable syneresis on removal of the flow field.

In general, crystallization will be enhanced by the increased concentration of the polymer-rich regions following a phase separation (subsection on Liquid–Liquid Phase Separation, pp. 243–9). Thus, in some cases, crystallization can occur subsequently to a liquid–liquid or an isotropic–liquid crystal phase separation rather than directly from solution as a liquid–solid transition.

Crystallization usually occurs on cooling. A conformational transition may be an *a priori* requirement. The conformational transition can occur on cooling or heating, and could then be followed by crystallization. In some cases, the limiting process is the phase separation or conformational transition rather than the crystallization itself, although it is often difficult to determine whether crystallization is the result of a conformational transition, or whether the new conformation is stabilized by the crystal structure.

Colloidal Aggregation

The aggregation of solid-like colloidal particles, brought about by a decrease in their solubility, can result in the formation of flocs or of a network. The colloidal particles need not be spherical, and may be crystalline or non-crystalline. The interparticle interaction can range from van der Waals forces to coalescence or recrystallization of particles in contact. A detailed theory of structure formation in colloidal systems has been developed and applied to clays, concretes, and synthetic and natural polymers.[70] In the case of polymers, the argument for using this approach is that as the solvent becomes poorer, the polymer becomes less swollen and more rigid such that gelation becomes more colloidal in nature. Considering the viscoelastic nature of the polymer-rich region in a liquid–liquid separation, such a distinction must, in practice, be arbitrary.

Topological Cross-Linking

Two basic structures have been suggested for geometric cross-linking. Firstly, cross-linking can occur by the interlinking of looped or cyclic

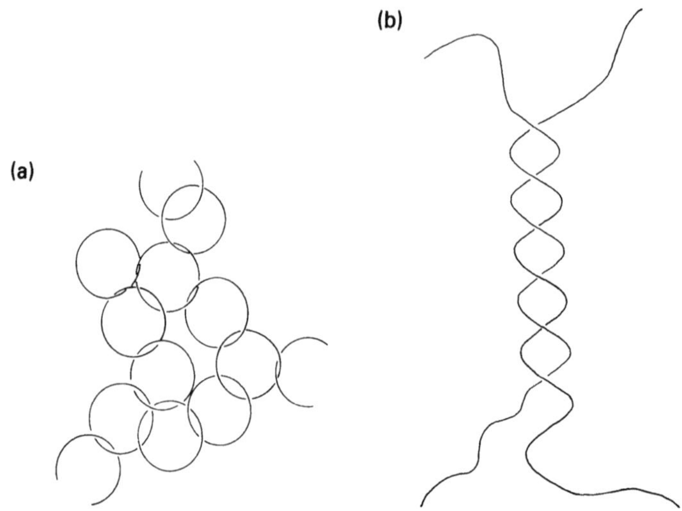

FIG. 13. Topologically cross-linked networks formed by (a) the interlocking of cyclic or ring polymers and (b) the intertwining of linear polymer molecules.

molecules (Fig. 13a). A simple molecular entanglement can also be regarded as a network with geometric cross-links, if the timescale of the measurement does not permit the molecules to disentangle (see subsection on Non-equilibrium State, pp. 238–40). Secondly, there is the localized intertwining of polymer molecules to form regions of double-, triple- or more complex helical structure (Fig. 13b).

POLYMER EXAMPLES OF THE MECHANISMS

Chemical Cross-Linking Gelation

The formation of chemically cross-linked gels from synthetic polymers has been comprehensively reported,[71] including comparison with theoretical predictions[72] and study of the thermodynamic behaviour of such systems after gelation.[73] Therefore, the examples described here are in a different vein and are based on a cross-linking mechanism which occurs in biological gel-network formation in both animals and plants and also forms the basis of a potential industrial process. As described in the subsection on Covalent Junctions (pp. 240–1), there are two basic methods by which chemically cross-linked gelation can occur: (1) copolymerization of monomers, some fraction of which are bifunctional; and (2) bridge formation between existing polymer molecules brought about by vulcanization, peroxidation, etc. The example chosen follows the latter method, although it is interesting to note in passing that network formation in such systems as blood clotting[74] may involve both the polymerization and bridge-formation methods.

Phenolic Cross-Linking

Oxidative phenolic cross-linking of polymer molecules at pre-existing reactive sites is a mechanism common to many biological systems. The two types of cross-links which form are shown in Fig. 14. The symbol R represents the connection between the phenolic group and the polymer which carries it.

In the case of the rubbery material resilin, which is a chemically cross-linked protein gel found in some elastic ligaments of the animal phylum *Arthropoda,* the phenolic group is coupled to tyrosine in the protein sequence.[75] Isodityrosine and isotrityrosine (linking two and three polymers, respectively) have been isolated.[76] Similarly, in plants the protein extensin, which contains ~10% tyrosine, is secreted into

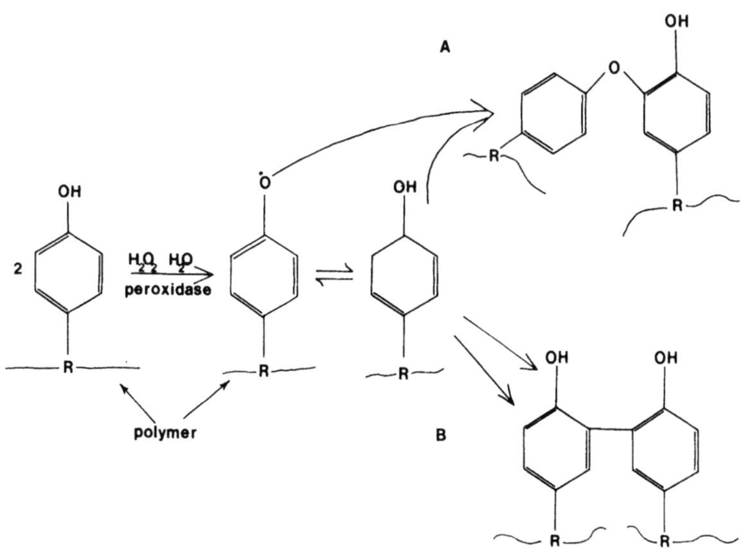

FIG. 14. The oxidative phenolic cross-linking reactions which can be responsible for gel-network formation. The dot · indicates the formation of a free radical.

the primary cell wall, where it is associated with acidic polysaccharides. Phenolic cross-linking (route A) gradually occurs[77,78] and the network becomes insoluble. The purpose of this network formation is probably to suppress growth and/or exclude microorganisms. Some cell-wall polysaccharides carry phenolic groups directly.[79] These can be cross-linked *in vitro* via peroxidase $+ H_2O_2$ (route B) to form a gel,[80,81] but these cross-linked phenolic groups have also been detected in many grass cell walls,[81,82] indicating that an analogous peroxidase-catalysed reaction probably occurs *in vivo,* to retard growth and/or microbial attack.[79] This mechanism, employing *in vitro* oxidation by peroxidase, has been proposed[83,84] as the basis of a commercial process for the gelation of sugarbeet pectin for industrial application.

Cross-Linking by a Chemically Different Oligomeric Bridge

This type of covalent cross-linking (see subsection on Cross-Linking of Pre-synthesized Polymers, p. 241) is quite rare. An interesting example is the dense three-dimensional mucopeptide network found in the cell walls of bacteria.[85] The oligomeric bridge is a pentapeptide (see Fig. 15) which, during biosynthesis, is linked to the polysaccharide molecules via their peptide side-arms.

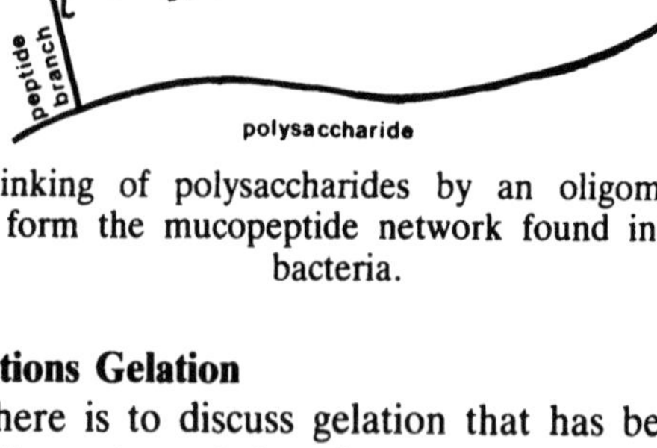

FIG. 15. Cross-linking of polysaccharides by an oligomeric bridge of a pentapeptide to form the mucopeptide network found in the cell walls of bacteria.

Specific-Interactions Gelation

The intention here is to discuss gelation that has been suggested to occur by the formation of junction zones in which the polymer molecules are bound together by specific interactions between complementary regions in each molecule, in contrast to simple clustering via, say, dipole–dipole interactions. An ordered structure for the junction zone is implied, with the detailed physical shape of the interacting molecules of prime importance. The two molecules involved in the junction may be chemically different, similar or identical.

Identical Molecules

The mechanism proposed for the gelation of the polysaccharide alginate provides an example of dipolymeric junction zones involving only one chemical species. The alginates are the salts of alginic acid, which is a linear copolymer of $(1\rightarrow4)$ linked β-D-mannuronic acid (M) and α-L-guluronic acid (G) occurring in blocks of M, G, and randomly mixed M and G. Alginate occurs as the major structural component of brown seaweed (*Phaeophyceae*) and as an extracellular polysaccharide of the bacteria *Azotobacter vinelandii* and *Pseudomonas aeruginosa.*[86] Addition of a divalent cation (typically Ca^{2+}) to an alginate solution reduces the solubility of the polymer resulting in precipitation or, if the concentration and molecular weight are sufficiently high, gelation. The gel, which is not thermally reversible, can be redissolved by ion-exchanging the divalent for a monovalent cation. Monomeric M and G sugars show the expected affinity for Ca^{2+} cations[87] but, with increasingly molecular weight, blocks of $G > 18$ residues show a sudden decrease in Ca^{2+} activity with accompanying polymer aggregation.[88] X-ray and infrared studies on partially dried gels[89,90] and oriented films of the acids of M and G[91–94] suggest a 2_1-helical, sheet-like structure stabilized by hydrogen bonding for homopolymers of both M and G. The diaxial $(1\rightarrow4)$ linkage of the latter leads to a more buckled sheet resulting in a cavity in the structure which is envisaged as holding a Ca^{2+} ion. This is the 'eggbox' model (Fig. 16) for the gelation of alginate.[95] Although the junction zones in the gel network were originally proposed as many-chain aggregates or crystallites,[95,96] chiroptical and stoichiometric evidence[97] and light-scattering evidence[98] suggest that only two polymer molecules are involved initially in forming the one-dimensional periodic junction as shown with a minimum length for stability, equivalent to the critical nucleus in a crystal. The mechanical properties of the gel depend on the ratio of M to G blocks and their respective lengths. The gel stiffness increases with the length and the fraction of G. In the seaweed, this ratio depends on both the location and nature of the tissue and its age and environment. It is suggested that the biosynthesis involves a mannuronan-C5-epimerase converting mannuronic to guluronic acid, thus controlling the nature of the copolymer and

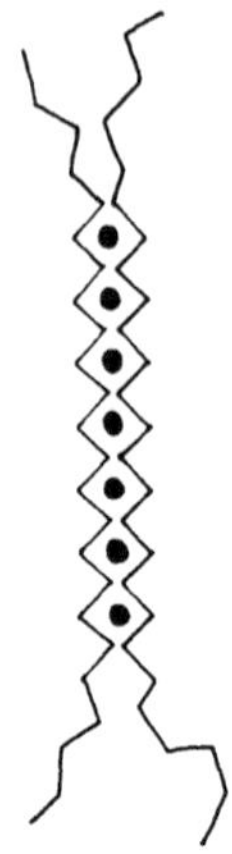

FIG. 16. The 'eggbox' model for the formation of junctions between alginate molecules. Calcium ions occupy the cavities formed by the opposing two-fold helical structures of the alginate molecules.

resulting gel structure. The block structure of the alginate from *Azotobacter vinelandii* is similar to that of the seaweed (algal) alginate but has a high M content and is acetylated. Alginate from *Pseudomonas aeruginosa* is avoided commercially as it is associated with a pathogenic condition in humans. A mucoid strain produces alginate under the conditions imposed by cystic fibrosis; the resulting gel interferes severely with lung function. The mucoid strain is unstable, but by the cloning of the genes controlling alginate biosynthesis,[99] it is anticipated that a better understanding of its synthesis will indicate a means for suppression, possibly by enzymatic modification of the polysaccharide. Characterization by infrared and chromatographic techniques[100] suggests this alginate is highly acetylated and consists of only M blocks and random M–G blocks despite the elevated levels of Ca^{2+} in the lungs. Here the prevention of gelation is of vital importance, but enzymatic modification could also have a role in controlling gel properties in industrial applications.

Complementary Molecules

Alginate-pectin gelation[101] will be presented as an example for which the formation of junction zones has been proposed to involve two chemically different but geometrically complementary polymer molecules. Pectin is the structural polysaccharide in plants equivalent to alginate in brown seaweed. The term 'pectin' covers a wide range of chemical structures. The essential features are a backbone of (1→4)-α-D-galacturonic acid with an occasional rhamnose residue, complete or partial methyl esterification, and some side-branches, typically of arabinose or galactose. For a detailed review of pectin gelation see ref. 24. Highly-methyl-esterified pectins gel under acid conditions and on the addition of a co-solute, such as sucrose. These conditions probably decrease the solubility of the polymer by decreasing dissociation and increasing the competition for water, respectively. Pectin of low methyl esterification gels on the addition of a divalent cation, again typically Ca^{2+}. A minimum of 15 galacturonic acid residues[88] are apparently required for the change in conformation associated with gelation on the addition of Ca^{2+}. The circular-dichroic changes are similar but opposite in sign to those for alginate.[88,95] There is also some evidence that low methyl-esterified pectin can form a 2_1 helix[102,103] (although dried gels exhibit three-fold symmetry[104,105]) and that junction zones are dipolymeric.[106] The similarities in the gelation behaviour of alginate and pectin reflect the fact that polygalacturonic acid and polyguluronic acid are near mirror images (see Fig. 17), with the mirror plane parallel to the chain direction and perpendicular to the plane containing the chain directions of both polymer molecules. This mirror-image property has been proposed to account for the gelation of solutions containing both alginate and pectin at concentrations below which neither component forms a gel on its own. The mixtures gel under acid conditions and in the absence of Ca^{2+}, but its addition after gelation increases the strength of the gel. An interpretation of circular-dichroic measurements[107] suggests an interaction between alginate and pectin. The strongest gels form from mixtures of high-methoxy pectin and high G-content alginate, suggesting hydrogen bonding between the methyl group of the pectin and the hydroxyl groups on C1 and C2 of the guluronic acid.

In all these models an important feature is the reduction of the net charge on the carboxyl group, allowing closer approach of the molecules and the formation of hydrogen bonds.

Specific interactions based on this mirror-image property of the interacting polymers have been observed in other enantiomorphic polymer mixtures.[108] Precipitation has been reported when enantiomorphic solutions of the polypeptide poly(γ-methyl D- and L-glutamate) (known to be 'α'-helical) at 1% concentration in dimethylformamide (DP = 1700) were mixed at 100°C. Mixing in different proportions gave a precipitate, the weight of which corresponded to twice that of

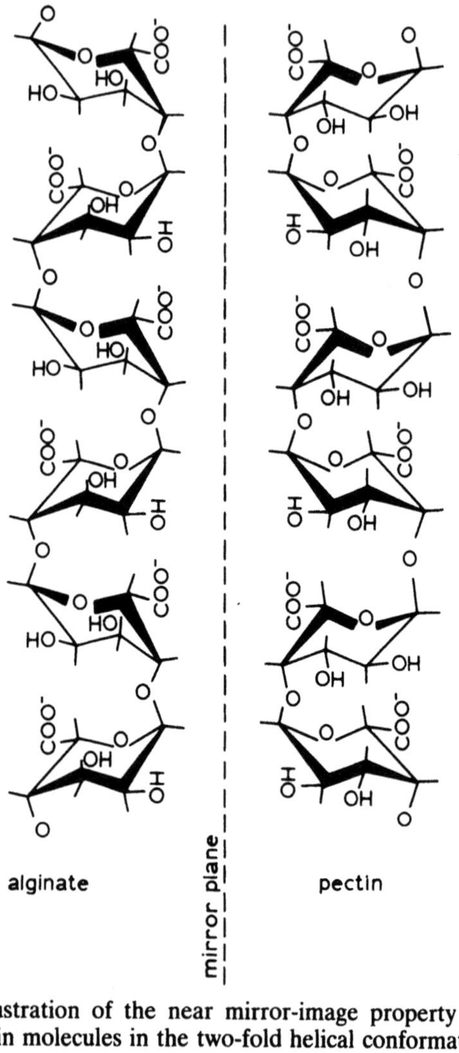

FIG. 17. An illustration of the near mirror-image property of alginate and pectin molecules in the two-fold helical conformation.

the lesser component in the mixture, indicating the formation of 1:1 racemic interaction of D- and L-PMG. At higher concentration and molecular weight (8·5% in 2:1 chloroform–dioxane and DP of 2900). mixing of the enantiomorphic solutions at 37°C resulted in a rigid gel, with specific interactions between the enantiomorphs suggested to result in a racemic-compound junction zone.[108]

Stereo-complex formation between polyesters of opposite chiralities has been reported for poly-(*t*-butylthiirane),[109] poly(α-methylbenzyl methacrylate),[110] and poly(α-methyl-α-ethyl-β-propiolactone).[111] Complex formation has also been observed in solutions of stereoisomers. Isotactic and syndiotactic poly(methyl methacrylate), PMMA, form stereo complexes from solutions in polar solvents. The complex is proposed to consist of isotactic and syndiotactic polymer in the ratio of 2:1, and a molecular structure to account for this has been proposed.[112] More recently, Bosscher *et al.*[113] have concluded from X-ray diffraction data and conformational energy calculations that the complex consists of a double-stranded isotactic 30/4 helix surrounded by a syndiotactic 60/4 helix, and ten Brinke *et al.*[114] have considered the theoretical aspects of such a complexation. On mixing solutions of isotactic and syndiotactic PMMA (i- and s-PMMA), gelation can occur. The gels are transparent and a stereo-complex junction zone has been proposed. The extent of complex formation between the isotactic and syndiotactic sequences depends on concentration and the thermal history. All possibilities from dipolymeric interactions to mixed-isomer microcrystallites apparently occur. Schwarz and Rehage,[115] who have studied the role of complex formation in i–sPMMA gelation, have suggested that it may be pertinent to specific complex formations that occur in biopolymers.

Phase-Separation Gelation

Phase separation in polymer solutions[51,59,116–118] and polymer blends[46,119–122] has been greatly studied from both the theoretical and experimental aspect. The observation that turbidity sometimes accompanies polymer gelation has led to the suggestion, which can be traced back to Hardy,[123] that phase separation could be the mechanism for gelation of many polymers. A liquid–liquid or a liquid–solid transition may be involved. Although the latter includes polymer crystallization from solution, this is usually considered as a separate gelation mechanism.

Colloidal Gelation

Platelet- or needle-shaped particles, often inorganic crystals, can associate via van der Waals forces to form a framework interpenetrated by liquid.[124–126] The possibility that large, folded-chain lamellar crystals could fulfil a similar role and form a colloidal 'framework' has not been reported, but the formation of other types of network involving folded-chain crystals has been studied (see subsection on

Folded-Chain Crystallite Gelation, pp. 284–5). Typical examples are clays,[127] where opposite charges on faces and edges of platelet particles may lead to the association.[128,129] Rebinder and others[130–132] have developed a detailed theory for the formation of such structures in colloidal systems, applying it also to supersaturated solutions of synthetic polymers and biopolymers. The argument is that if the solvent becomes sufficiently poor then the macromolecule will no longer be swollen and will collapse into what may be regarded as a colloidal particle. Such particles, assumed to be roughly spherical, could then associate to form either a framework (see Fig. 18) or a precipitate, depending on the concentration and kinetics of association. The application of this colloidal mechanism to the gelation of macromolecules has, in general, received little support, a possible exception being the gelation of globular proteins. The conformational change associated with partial denaturation results in a decrease in the solubility of the protein. It was earlier thought that denaturation

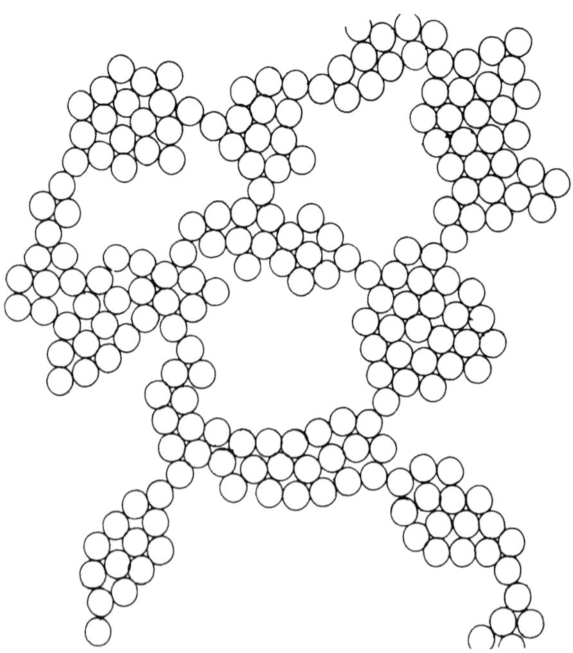

FIG. 18. Association of spherical colloid particles forming a 'framework' or network.

resulted in the unfolding and extension of the protein molecules, which then formed fibrous structures and a network.[133–135] Today the accepted view,[136] first proposed by Barbu and Joly,[137] is that the denaturation associated with gelation of globular proteins maintains the approximately spherical or 'corpuscular' shape of the protein but moves to the surface some of the more hydrophobic regions of the molecule, in particular regions of β-sheet. This reduction in solubility could lead to the colloidal-type gel structure. In fact, two types of aggregation occur depending on the charge carried by the native protein. At high charge and large electrostatic repulsion, the aggregates are linear, having an overall fibrous appearance (see Fig. 19) but still consisting of essentially globular molecules. As the isoelectric pH is approached, the aggregation is less ordered, the overall shape of the aggregate tending to be globular (see Fig. 18). If the aggregation, particularly in the linear case, is regarded as a direct association of β-sheet regions acting as 'sticky' patches on the protein surface, then the gelation mechanism may be better categorized as a specific interaction (see subsection on Specific-Interactions Gelation).

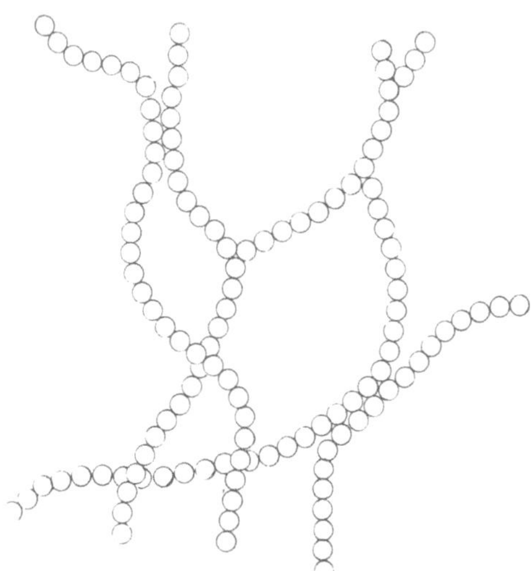

FIG. 19. Linear association of essentially spherical colloid particles which occurs in globular-protein gelation under conditions resulting in high net charge on the molecules.

Copolymer Phase Separation

Perhaps the most intuitively acceptable gelation mechanisms involve copolymers. These also feature in the section on gelation involving crystallization (see subsection on Fringed-Micellar Crystallite Junctions, pp. 275–81). It is easy to visualize a gel network in which one constituent of the copolymer is less soluble in a given solvent than the other constituent. A decrease in temperature, for example, could result in the precipitation of the less-soluble constituent to form junction zones between the more soluble constituents, which remain in solution (see Fig. 20). The number, length and distribution of the constituents along the polymer molecule are clearly of importance to the formation of a network. Liquid–liquid phase separation in block copolymers has been reported. Virtually one-dimensional[138,139] (nematic) and two-dimensional[140] (smectic) phase regions as well as three-dimensional regions have been observed.[141,142] A practical example of such systems is the ABA block copolymer, Pluronic® F127, which has been proposed as a potential drug-delivery system.[143] Aqueous solutions form thermoreversible gels on heating from 5°C to room temperature.[144] The water-insoluble regions (B) tend to aggregate into micellar regions, each containing about 40 molecules. It has been

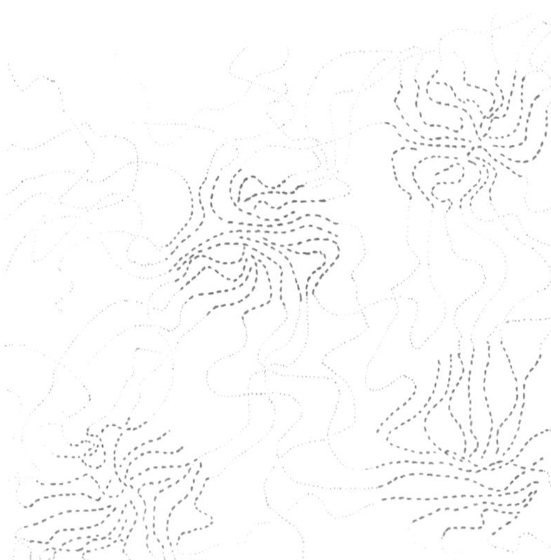

FIG. 20. Schematic representation of network structure in which one constituent of a copolymer has formed phase-separated regions while the other constituent has remained in solution.

proposed that, on heating, the A blocks become less solvated and that the resulting higher 'friction' between these regions is responsible for the reverse thermal gelation behaviour.[145]

Ionomers provide an interesting example of what can be regarded as a random copolymer in which the two constituents have different solubility behaviour. An ionomer is an organic polymer, such as polystyrene, which has been partially substituted with an ionic group, e.g. sulphate or carboxylate, and has an associated metallic cation. In a polar solvent, dissociation of the polar groups occurs and modified polyelectrolyte behaviour dominates. In non-polar solvents, dissociation does not occur and the polar groups are now electric dipoles. Clustering of these dipoles via dipole–dipole interactions has been observed to occur both in solution and in melt. Such interactions can act as junction zones in gelation.[146] Joanny[147] has discussed the global phase diagram (see Fig. 21) to be expected, In dilute solution, intramolecular associations dominate, and the radius of gyration of the polymer decreases with increasing degree of substitution by polar groups, eventually leading to the collapse of the coil and phase separation or precipitation. In the semi-dilute concentration region where the chains become entangled, there is competition between

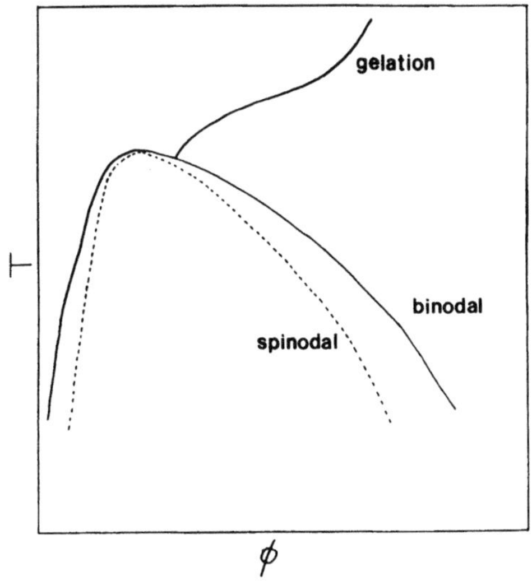

FIG. 21. Schematic phase diagram expected when there are strong interactions limited to regions along the molecule as in the case of ionomers, segment–segment interactions, or chemical cross-linking during polymerization.

intra- and inter-molecular associations. Gelation competes with precipitation. Similar phase diagrams must, of course, exist for other strong attractive interactions including crystallization and chemical cross-linking.[148]

A similar clustering behaviour resulting in gelation may also occur from aqueous solution for polymer molecules with a distribution of hydrophobic groups along their lengths. For example, methylcellulose (degree of substitution 1·3–2·0) is soluble in cold water and can gel on heating. It exhibits a lower critical solution temperature (lcst) (see Fig. 5b). The gels are thermoreversible and melt on cooling. The decrease in solubility with increasing temperature is ascribed to the formation of a methylcellulose–water complex which is unstable at higher temperatures.[149] Other examples of systems exhibiting a lcst and forming thermoreversible gels at high temperature are casein in alkaline solution,[150] PVA in dimethylformamide,[151] and cellulose nitrate in ethanol.[152]

Liquid–Liquid Phase Separation

Liquid–liquid phase separation as described on pp. 243–9, and in particular spinodal decomposition,[48] has been suggested as a primary gelation mechanism for many homopolymers. The periodic and network-like character of the phase separation, apparent even in the original simulations[48] (see Fig. 4), makes the spinodal mechanism a particularly attractive idea. However, the question usually arises as to whether liquid–liquid phase separation alone can result in a stable gel capable of sustaining a static load or whether an additional specific interaction or crystallization is required. It should be mentioned that from thermodynamic consideration, crystallization cannot begin without first a phase separation once the solution has been quenched into the two-phase region of the phase diagram.[153] Thus, phase separation could form the initial network structure which may subsequently be stabilized. However, it is worth noting that two-phase liquid systems, such as emulsions and foams, can exhibit solid-like characteristics which suggest that a microscopically phase-separated network could result in a gel, the increase in interfacial energy providing the restoring force on deformation. In the solid state, however, studies on 'hard' elastic fibres suggest that polymer surface energy can provide a macroscopic restoring force.[154]

In a series of papers using light-scattering techniques, Prins and coworkers[155–157] have interpreted their data in terms of spinodal decomposition as a gelation mechanism for the polysaccharide agarose under certain conditions. For the 1% concentration used, the sol–gel transition occurred at about 42°C.[156] Quenching to temperatures between 32·5°C and 19°C resulted in a maximum in the Rayleigh ratio versus scattering vector plot[156] (see Fig. 22). This maximum broadened and shifted to higher angles with increasing supercooling. Quenching to 41°C resulted in no maximum being observed. This was also the case for all supercoolings of dilute, non-gelling solutions.[155] The maximum was interpreted as being due to Bragg scattering. Subsequent studies[157] on the development of the scattering curve with time from quenching confirmed that the maximum resulted from the formation of a supramolecular structure and that this and the shift and broadening of the maximum with increasing supercooling were consistent with spinodal decomposition. At smaller supercoolings phase separation proceeded by nucleation and growth, i.e. the quench

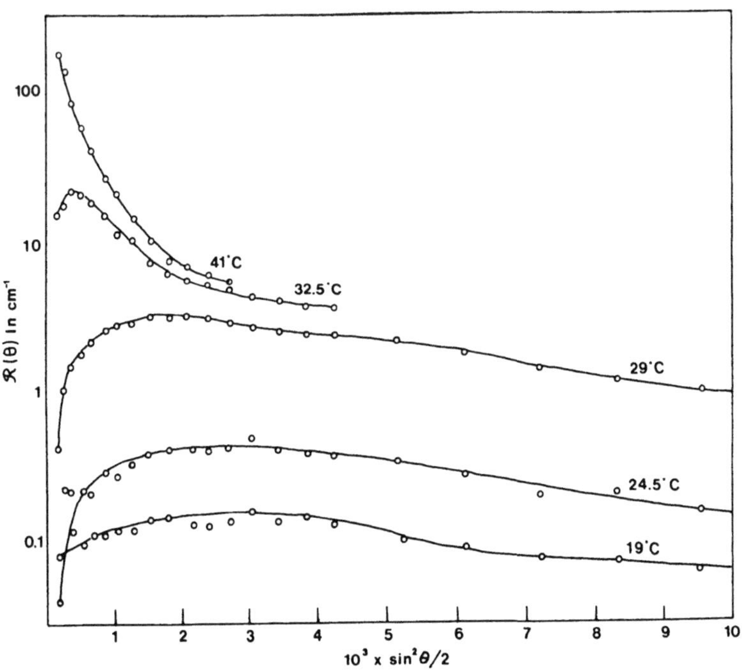

FIG. 22. Plot of Rayleigh ratio versus scattering vector showing a maximum following quenching of a 1% agarose solution to between 32·5 and 19°C. (Reprinted with permission from G. T. Feke and W. Prins, *Macromolecules*, 1974, **7**, 527.)

temperature was above the spinodal line (see sub-section on Liquid–Liquid Phase Separation, pp. 243–9).

In the field of synthetic polymers, Baer and coworkers[158–160] have studied the gelation of atactic polystyrene, aPS. The gelation of aPS in CS_2 was initially considered to result from a spinodal phase separation with the formation of a fine polymer-rich network cross-linked by glassy domains.[158] Localized crystallization of isotactic or syndiotactic sequences was ruled out by DSC and FTIR measurements. Subsequently, clear gels were also obtained above ucst.[159] These gels become turbid when cooled below ucst. Phase diagrams of reduced temperature versus concentration were plotted to include the gelation line. These plots were very similar to those obtained by Joanny[147] (Fig. 21) when dipole–dipole interactions dominated the gelation. Later, for the case of aPS, segment–segment contact was proposed for the cross-linking interaction[160] and to explain the observation that the gelation temperature, T_{gel}, and gel melting temperature were equal. The value of T_{gel}, which in this later work was found to be greater than T_g, had a minimum when the solubility parameters of the polymer and solvent were equal. This is consistent with a similar observation involving solubility parameter and indentation modulus of PVC gels in various solvents (see subsection on Fringed-Micellar Crystalline Junctions, pp. 275–81). However, Gan *et al.*[161] interpret their enhanced low-angle light scattering data in contradiction to this hypothesis. They conclude that highly solvating solvents promote gelation by forming 'microcrystals' in which the solvent forms a complex with the polymer in the crystal lattice in a manner similar to that proposed by Tadokoro *et al.*[162] for syndiotactic methyl methacrylate.

The gelation of isotactic polystyrene, iPS, is intimately associated with crystallization (see subsection on Crystallization, pp. 275–85 and ref. 64), but phase separation also plays a role. Challa and coworkers[163,164] observed gelation to occur at a temperature much lower than that at which folded-chain crystallization occurred. Girolamo *et al.*[64] found that for solutions in decalin with polymer concentrations greater than about 5%, gelation occurred on cooling to 20°C with turbidity subsequently appearing over a period of days. Above 30°C a turbid precipitate without gelation was observed. In the region around 25°C both processes occurred simultaneously and the sample exhibited two melting endotherms: one at high temperature associated with folded-chain crystals and one at low temperature which appeared to correspond to fringed-micellar crystallization (see subsection on Crystallization, pp. 275–85 and ref. 64). The latter endotherm also corresponded to the melting of the gel at about 60°C, leaving a turbid suspension of folded-chain crystals up to 120°C. This indicated that fringed-micellar crystals provided the cross-links in the gel network. The crystallographic structure of these gel-forming crystals is not the usual 3_1-helical, Natta–Corradini structure but appears to be unique to the gel network, having a conformation close to full extension.[64] The possibility of isomeric irregularities has been considered.[64]

Wellinghof *et al.*[158] investigated the gelation of iPS in several solvents and found that the gelation temperature was close to the θ temperature and to the gelation temperature of aPS. They interpreted their results as indicating that a microphase separation was necessary for the crystallization associated with the gelation to occur. As phase separation proceeded, the local polymer concentration increased and small, fringed-micellar crystals formed. The gel formation and melting temperatures correlated with the polymer–solvent interaction energy (provided only non-specific polymer–solvent interactions were involved). The more unfavourable the interactions, the higher were the ucst and gelation temperatures. Guenet *et al.*[165] state that from thermodynamic considerations[153] any crystallization must occur *after* phase separation once the solution has been quenched into the miscibility gap. In addition to the slow appearance of turbidity in the gels (which has been associated with folded-chain crystallization) they also reported[165] a reversible turbidity on quenching iPS gels (in decalin) to between +5°C and −40°C. They ascribe this turbidity to the phase separation of chains not involved in network formation. It is perhaps worth noting a biological parallel to this reversible turbidity, namely the so-called cold cataract[166] which forms reversibly if the eye is cooled. This turbidity is again attributed to the phase separation and aggregate formation of proteins not involved in the gel structure. Indeed, it has been suggested that cataract formation in general may involve a phase separation of proteins released into solution from the gel structure by the action of proteases.[167]

Thus, a possible mechanism for gelation has emerged in which a continuous polymer-rich network forms through phase separation, possibly by spinodal decomposition, followed by a crystallization within this region, which stabilizes the structure. In some cases no crystallization or other stabilizing interaction has so far been detected.[159]

A parallel development has occurred in the polysaccharide field, and, in particular, for amylose gelation.[168] Amylose is the essentially linear component of starch, the other major component, amylopectin, being branched. Both are based on (1→4) α-D-glucose chains. Solutions of amylose cooled from 65°C to 32°C became turbid and gelled. Measurements of turbidity, shear modulus, and volume change during gelation occurred on the same timescale (see Fig. 23a), and the rates at which each of these parameters increased with time during gelation increased with concentration and decreasing temperature. Crystallization, as measured by X-ray diffraction, was found to occur at a much slower rate and apparently independently of the initial solution concentration (see Fig. 23b). Miles *et al.*[168] interpreted these observations as an initial phase separation into a polymer-rich network responsible for gelation followed by crystallization within this network. The polymer concentration within the polymer-rich phase being relatively insensitive to the initial solution concentration could explain the constant rate of crystallization after phase separation (see Fig. 23b). Hayashi and coworkers have also proposed phase separation as the mechanism for amylose[169] gelation and also for agarose[170] and gelatin.[171] From their observations of optical rotation and measurements of the phase diagrams, they have suggested that a coil-to-helix transition precedes the phase separation in amylose gelation, and that the phase separation results in coascervate particles, which subsequently associate to form the gel network. This can be thought of as a combination of the phase-separation and colloidal-gelation mechanisms. The coil-to-helix transition in such polymers results in stiff or rod-like regions along the polymer molecule, and an isotropic–liquid crystal phase separation could be raised as a possible mechanism.

So far, the role of phase separation in gelation has been limited to polymers which are regarded as flexible in solution. Thermoreversible gels can be formed from solutions of many rod-like polymers.[172–178] Miller *et al.*[176] have reviewed the phase separation and networks formed by some rod-like polymers. The nature of the restoring force in such networks of rods must clearly be different from that of a rubber-like, flexible polymer network. However, that network formation occurs via a phase separation remains a possibility.[177] Flory[61] has developed a detailed theory for the liquid crystal–isotropic phase behaviour of rod-like particles in a diluent.

The polypeptide, poly (γ-benzyl L-glutamate) (PBLG) exists in

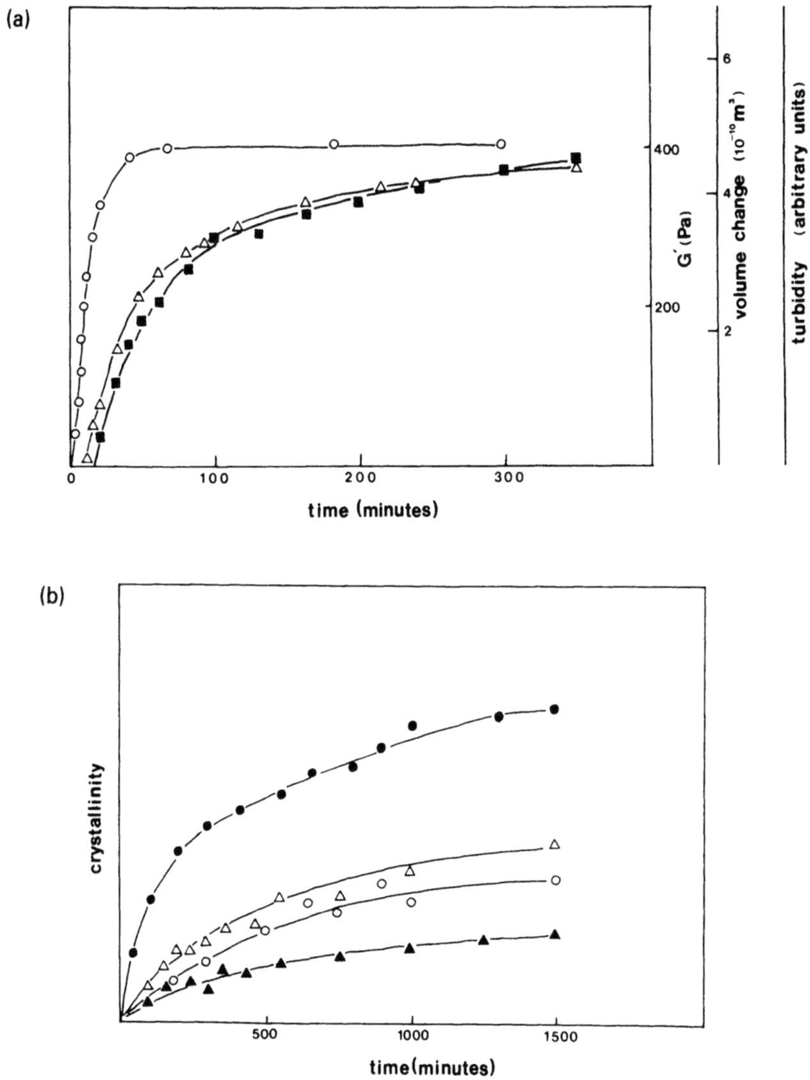

FIG. 23. (a) Plots of turbidity (○), volume change (△), and shear modulus (G') (■) against time during the gelation of a 2·4% amylose solution following to 32°C from 65°C. (b) Plots of crystallinity versus time after quenching for amylose solutions of concentration 2·4% (▲), 3·5% (○), 5% (△), and 7% (●). (After M. J. Miles *et al.*, *Carbohydr. Res.*, 1985, **135,** 257.)

solution in dimethylformamide in the rod-like α-helical conformation. Wee and Miller[179] have studied the gelation of PBLG and its phase behaviour. The liquid crystal–isotropic phase diagram was investigated using NMR, viscometry and polarizing microscopy. All features of the theoretical phase diagram were observed, including the broad biphasic region below which, on decreasing the temperature, it was found that at sufficiently high concentrations a gel network formed. Miller *et al.*[178] suggested that this network formation was controlled by the particular kinetics of the phase separation, and proposed spinodal decomposition. Light-scattering studies of the gelation kinetics strongly supported a non-nucleation mechanism.[176,180] A study[181] of the ternary system PBLG–DMF–H_2O showed that a small volume-fraction of water dramatically affected the phase diagram by raising in temperature the wide biphasic region. The non-equilibrium nature of gelation was again apparent from an ageing effect and also from the observation that whether gelation did indeed occur depended on how the biphasic region was entered. They noted that this gelation behaviour is not consistent with the previously suggested models.

An example of a ternary system in which the addition of a low molecular weight co-solute is thought to be responsible for polymer gelation is the system pectin–H_2O–sugar,[182] familiar from jam-making. A high-methoxy pectin is required (cf. low-methoxy pectin, subsection on Complementary Molecules, pp. 260–3). Several co-solutes, including a range of sugars and glycols, are effective in promoting gelation.[183] The current model[184,185] attributes the gelation to a demixing of the pectin resulting from the co-solute competing for the solvent. The effect of the low pH is probably to reduce the net charge on the pectin, allowing polymer–polymer association.

Competition for solvent forcing the demixing of one component has also been suggested[186] as the mechanism for the polymer–polymer–solvent synergistic gelling systems: κ-carrageenan–carob–water, agarose–carob–water, and xanthan–carob–water. Carob gum, a galactomannan, consists of a mannan backbone partially substituted with galactose residues, resulting in 'smooth' galactose-free regions and 'hairy' galactose-substituted regions along the molecule. A subsequent gelation mechanism proposed[187] that the 'smooth' regions somehow interacted, persumably through hydrogen bonding, with the helices of the κ-carrageenan, agarose or xanthan molecules. However, recent fibre-diffraction data[188] do not support this model for the carrageenan–carob–water system, but indicate that only the carrageenan is involved in the gel network. This would lend support to the former mechanism[186] of competition for solvent and demixing, followed by crystallization of, in this case, the carrageenan. A specific-interaction mechanism is perhaps more likely in the xanthan–carob case as the cellulosic-type backbone of xanthan has a periodicity similar to the two-fold mannan backbone.

In the context of mixed polymer gelation, it is worth mentioning the observation[64] that the addition of 1% isotactic polystyrene to a 15% solution of atactic polystyrene at room temperature results in the formation of a weak gel under conditions in which neither of the polymers would gel alone.

A distinction should be made between the phase separation in a ternary system in which the addition of one polymer component may result in a demixing of the other with the possibility of network formation, and the phase separation of the two polymers into two virtually pure-component solutions, each of which gels in its own right. As an example, the ternary phase diagrams for various combinations of proteins and polysaccharides have been explored,[189] and the behaviour of the phase-separated polymers manipulated to produce (anisotropic) gel composites.

Crystallization

Crystallization is the most commonly cited mechanism for junction-zone formation in polymer gel networks. A distinction must be made between crystallization from solution resulting directly in network formation and crystallization which occurs subsequently to gelation. This distinction is not always apparent in the reports of crystallization and gelation. For example, crystallization may occur in an already phase-separated network or it may involve polymer molecules not included in the network structure. Another possibility is that a specific interaction or phase separation results in the formation of a weak network but that this network is consolidated by subsequent crystallization. Here the relative importance of the various mechanisms becomes unclear. These form the grounds for the current debate on the relative importance of crystallization and phase separation in gelation. In the examples which follow crystallization has been favoured as the primary mechanism.

Fringed-Micellar Crystallite Junctions

These crystalline junction zones are three-dimensional and have a high functionality since each crystal may contain ~100 molecules. The conditions which generally favour micellar crystallization rather than the other morphologies were outlined in the subsection on Crystallization (pp. 251–5). Perhaps the most easily visualized example is that of a block copolymer consisting of sections of crystalline and non-crystalline polymers[190,191] in which the crystalline sections are insufficiently long to permit folding (see Fig. 24). The possibility of acrylonitrile crystalline junction zones in gels of the copolymer acrylonitrile–vinyl acetate was raised by Paul.[192] Similarly, for the copolymer ethylene–vinyl acetate in various solvents, Takahashi[193] observed the crystallization of the ethylene sections, and, based on a theory developed by Takahashi *et al.*,[194] calculated the stem length of ethylene in the crystallites to be 10 repeats irrespective of the mole fraction of ethylene in the copolymer. This copolymer approach has been extended to homopolymer polyethylene for which the crystalline regions are no longer interrupted by a block of non-crystalline polymer but by branches in the molecule[195] or by chlorinated regions.[196]

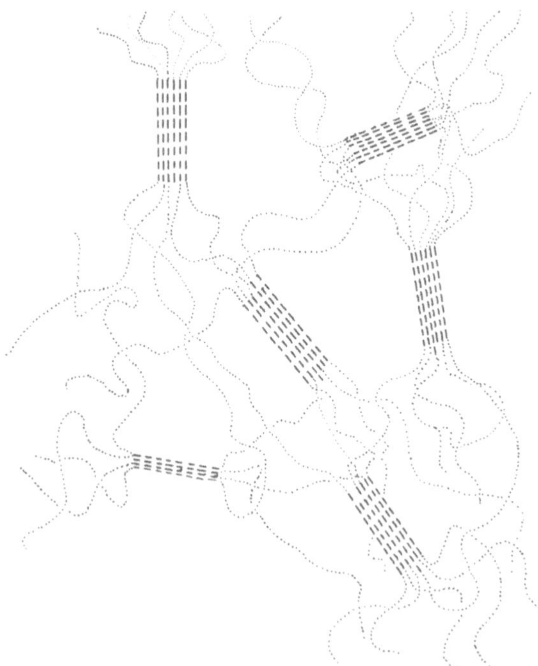

FIG. 24. Schematic representation of a network structure formed by copolymer molecules consisting of non-crystalline sections and sections which crystallize to form fringed-micellar crystallite junctions.

However, in the latter case, it was found that for the highly chlorinated, non-crystalline sample the gel setting and melting temperatures were the same, whereas the crystalline samples showed hysteresis. Again, the two processes of phase separation and crystallization were suggested, with the former apparently sufficient to form a gel.

A polymer which is chemically a homopolymer may be regarded as a copolymer in the sense that it contains sections of stereospecific isotactic or syndiotactic sequences, and these may crystallize to form junction zones in a gel. Because of the limited sequence length such crystallites will be fringed micelles. Takahashi and Kato[197] have taken this approach in studying the gel-melting behaviour of the 'tactic' copolymers: poly(vinyl chloride), polyacrylonitrile, and poly(vinyl alcohol). In these cases the copolymeric nature is an accidental consequence of the polymer synthesis producing different stereoisomers. In the case of PVC, it is regions of syndiotactic sequences that crystallize. However, the existence of crystallinity in commercial material was long disputed,[198] but X-ray evidence[199] has now clearly established the possibility of syndiotactic crystallization in such material. Further evidence for the role of crystallization in gelation comes from an investigation by Dorrestijn *et al.*[200] of randomly post-chlorinated PVC. They attributed the suppression of gelation to the effect of intramolecular chlorine distribution on crystallizability. Keller *et al.*[31,199,201–203] have studied extensively the gelation and crystallization of PVC. From meridional line-broadening measurements they estimated[199] that a minimum of 13 repeating units are involved in the crystal junction zones. This is in direct conflict with statistical analysis of NMR data on PVC which indicate a negligibly small amount of such syndiotactic-sequence lengths. Two possibilities exist: (1) the Bernouillian statistics used are inappropriate, because, for example, syndiotactic sequences are bunched; or (2) isotactic sequences can be incorporated into the syndiotactic lattice. Both are unattractive, and the problem remains unresolved.

The PVC gel network was studied as a function of polymer crystallinity and polymer–solvent interaction.[202] Using a range of solvents, a correlation between gel-forming ability, as judged by indentation-modulus values, and solubility parameter was observed. Twin maxima either side of the minimum corresponding to the δ-value for PVC confirmed that medium solvents gel PVC most readily. Good solvents do not permit sufficient interchain contact to produce sufficient junctions, whereas poor solvents presumably favour intrachain contact to such an extent that the tendency is to precipitate and again the number of junctions is low. In the stretched gels, the micellar crystals orient with their chain direction parallel to the stretch direction. The X-ray diffraction from these crystals can be characterized by diffuse equatorial maxima (reflecting their limited lateral size) and lattice parameter slightly larger (because of crystal imperfection and size) than those of the more ordered, folded-chain, platelet crystals. A second pattern oriented with the chain direction perpendicular to the stretch direction is often observed. This has the fingerprint of the folded-chain lamellar crystal, and has been interpreted as originating from lamellar crystals, not directly part of the gel network, orienting during stretching as morphological units.[31,201] This is supported by the fact that lath-shaped lamellar crystals orient with their long axis and corresponding crystal axis in the stretch direction (see Fig. 25), for example, **a** for PVC and **b** for polypropylene (PP). Isodimensional lamellar crystals such as isotactic polystyrene (iPS) do not show this double orientation.

The gelation of iPS has already been discussed in terms of a phase separation, but there is also strong evidence intimately linking crystallization and gelation. Indeed, the crystallographic structure of the micellar gel crystals is distinctly different from the traditional Natta–Corradini structure and appears to be unique to gels.[64,204] This new structure has a fibre-repeat close to a fully extended (all-*trans*) conformation but is actually a 12_1 helix, in contrast to the 3_1-helix of

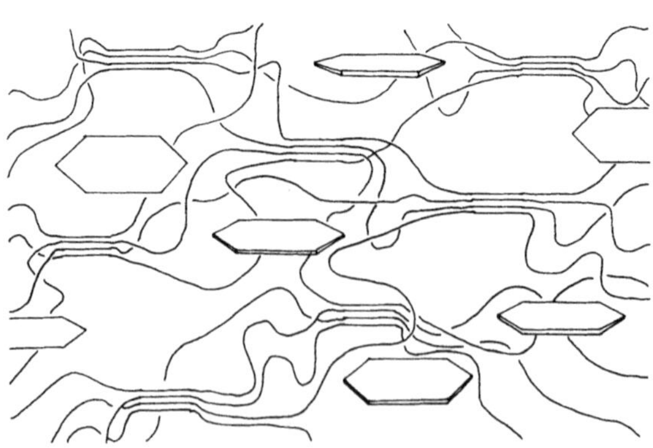

FIG. 25. Schematic representation of lath-shaped, folded-chain crystals not directly part of the gel network orienting during stretching with their long axis in the stretch direction. (After A. Keller, ref. 31.)

the traditional *trans–gauche* conformation. The exact structure remains uncertain, as does the reason why gelation crystallization alone should produce this structure. In some solvents, but not decalin, a transformation from the 12_1- to 3_1-helical structure has been reported. In the same study, Guenet *et al.*,[165] using a toluene extraction method, reported the network to consist of thread-like structures 10–20 nm in diameter and reminiscent of structures attributed to spinodal phase separation.

For flexible polymers there is competition between chainfolded and fringed-micellar crystals, the latter being favoured by large supercoolings.[64] For stiff polymers chainfolding is not possible and micellar crystals can form at relatively small supercoolings. As a result of relatively robust helical conformations, many polysaccharides are inherently stiff molecules and, in general, gelation occurs readily, particularly for the more crystalline polymers. The highly extended state of these polymers also results in gelation occurring down to very low concentrations. The bacterial polysaccharide, gellan gum,[205,206] forms gels at concentrations down to 0·04% by weight. Figure 26

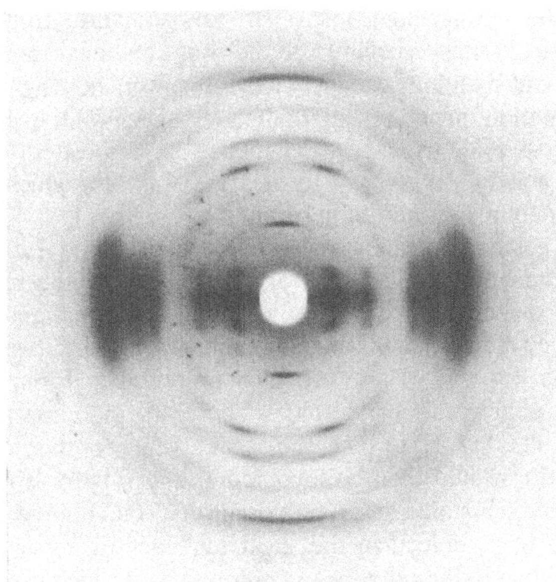

FIG. 26. X-ray fibre diffraction pattern from a stretched gel of the polysaccharide gellan. (Reproduced with permission of M. J. Miles and V. J. Morris.)

shows the X-ray fibre diffraction pattern from a stretched gellan gel formed at 1% polymer concentration. The diffuseness of the equatorial reflections remains even after annealing and suggests the limited lateral extent of the crystallites expected for fringed micelles. Native gellan contains one acetyl group per tetrasaccharide repeat unit. Deacetylated material produces stiffer and more crystalline gels. Thus, a relatively minor substituent can dramatically affect gel properties, and gellan appears to be by no means an isolated example. In the case of the polysaccharide from *Klebsiella* K54, the acetyl groups produce a change in molecular conformation. The native polymer, which is acetylated, is in a three-fold helical conformation and does not gel. Deacetylation causes a conformational transition to a four-fold or possibly eight-fold helix.[207] The polysaccharide, which is now crystallizable, readily forms gels under the appropriate ionic conditions. The deacetylated polysaccharide from K54 is identical to that produced by a different microorganism, an *Enterobacter* (NCIB 11870).[207,208] With only one acetyl group per eight sugars difference, it appears the microorganisms can control their rheological environment through changes in minor substituents.

Gelation accompanied by a conformational change also occurs for curdlan, the extracellular microbial polysaccharide from a mutant strain of *Alcaligenes faecalis*. Here the conformational transition is not brought about by minor chemical changes but by heating. The change in conformation, from an approximately seven-fold, triple-stranded helix to a six-fold, triple-stranded helix[209] is associated with crystallization in general, in contrast to the case of iPS for which the 12-fold helix is apparently unique to gelation.

Several gelation mechanisms have been proposed for the ι- and κ-carrageenan polysaccharides (for a fuller review see ref. 24). The prevailing suggestion[210] involves a major conformational transition from single, flexible molecules in solution to intertwined double helices (see Fig. 27). These are then proposed to form side-by-side aggregates under ionic conditions which allow the close approach of the molecules by screening the charges along their backbones. Crystallization would then be favoured although there is no report of such in the gel state. It is proposed that the double helices are interrupted by a 'defect' in the sugar sequence (a missing anhydride bridge resulting in the other chair conformation) which would limit the longitudinal dimension of the helices and any fringed-micellar crystallites that may form.

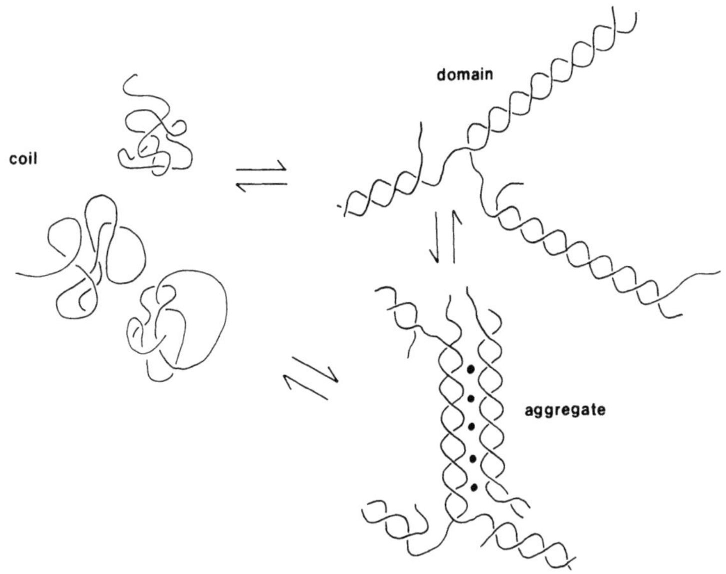

FIG. 27. The prevailing gelation mechanism proposed for ι- and κ-carrageenan involving the formation of intertwined double helices, the aggregation of which is mediated by cations (●).

Smidsrød and Grasdalen[211] have suggested an alternative model based on an intramolecular conformational change (rather than double-helix formation) on cooling. They have proposed that gelation results from the association of polymer molecules via selective binding of certain cations to the carrageenans, having obtained NMR data which indicated binding of certain Group I cations.[211–213] However, gelation has been reported with all Group I ions depending on cation concentration.[214] By considering the cations as solvent-structure breakers or makers,[215] it may be possible to explain the efficacy of the various ions in promoting gelation. A reduction in solvent quality could favour intra-helix hydrogen bonding resulting in a conformational transition. A poorer solvent would also favour polymer–polymer associations. The observed cation binding could have more to do with crystallization. Again, the possibility of demixing, possibly involving a conformational change, followed by crystallization arises.

Flow-Induced Gelation and Fibrous Crystallization

Gelation resulting from the shearing of a polymer solution has been reported both for intrinsically non-crystalline[112,216–221] and for crystalline polymers.[222–224] Eliassaf *et al.*[216] observed that stirred aqueous solutions of poly(methacrylic acid) set to a firm gel, which subsequently, after up to 1 h, reverted to the original liquid state. They suggested that the flow-field promoted intermolecular interactions and that network junctions were a result of hydrogen bonding, the relative stiffness of this molecule permitting crystallite-like accumulations of hydrogen bonds. These bonds continually break and reform, and under stationary conditions a higher degree of intramolecular bonding is again reached. Liquori *et al.*[112] suggested that for atactic poly(methyl methacrylate) (PMMA) or polystyrene crystallization is more directly involved and that there are regions of stereoregularity in the atactic polymers which form crystalline junctions under shearing (see the preceding subsection for quiescent gelation of atactic polymers).

Matsuo *et al.*[221] reported that constantly sheared solutions of PMMA in Arochlor showed a maximum in viscosity, streaming birefringence and cotangent of extinction angle with time. This sort of behaviour has been interpreted as the gradual build-up of a gel network from the shearing surfaces until it bridges them, whereupon it is broken up and the viscosity drops to a lower value again (see Fig. 28). The observation of gel particles in such systems has been reported.[218–220] It is worth noting experimental[225] and theoretical[226,227] developments related to surface gelation which must play an important role in this process.

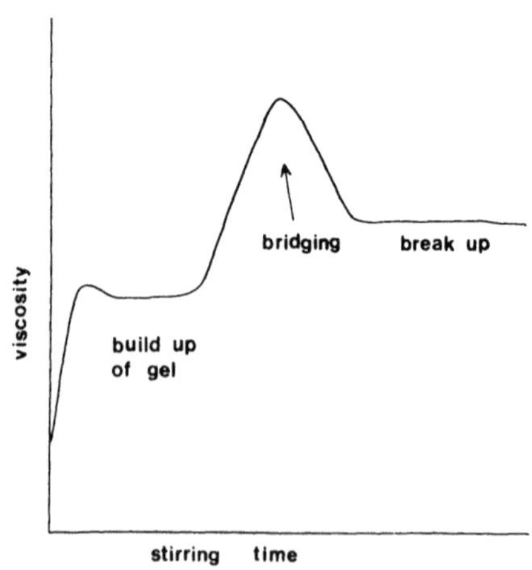

FIG. 28. Typical apparent viscosity versus time behaviour during stirring of a polymer solution (for example, in a Couette viscometer) resulting in a solution structure responsible for gel-network formation.

The development of a maximum in shear viscosity with time has also been observed for polyethylene solutions sheared in a Couette device at 135°C.[224] This temperature is above that at which crystallization usually occurs. If shearing was ceased and the solution cooled to 95°C, a clear gel formed. On cooling further the gel became turbid. An additional surprising effect was observed: under some conditions of shear and temperature, the solution could be held at the high temperature without shearing as a liquid for periods of hours, and, on cooling, a gel would still form. On cooling to below 90°C the gel became turbid. These phenomena were reversible on heating. With increasing temperature, firstly the gel became clear and secondly it melted. This is in contrast to gels formed under quiescent conditions. These gels firstly melted to a turbid suspension which on further heating became clear. In both types of gel the turbidity has been attributed to folded-chain crystallization. This suggests that if the junctions are crystalline, then the crystals involved in the junction of the quiescent gels are less stable and those in the flow-induced gels more stable than the folded-chain lamellar crystals. For flow-induced polyethylene gels, electron microscopy has shown the presence of fibrous crystals with a fine overgrowth acting as the junctions in the network.[222] In this case turbidity results from normal chain-folded overgrowth to form 'macro' shish-kebabs.

However, the structure of the solution after shearing at high temperature remains uncertain. The lifetime of the structure is too long for it to be simple entanglements but too short for the usual stable crystals.[32] Could this be related to the behaviour observed by Eliassaf *et al.*[216] for polymethacrylic acid? Keller[228] has suggested that since the time-course of properties such as viscosity during shearing is a feature common to crystalline and non-crystalline polymers, then the formation of a solution structure is a general phenomenon transcending crystallization and most likely involving an orientation-induced phase separation. This again highlights the intimate involvement of crystallization, phase separation, and specific interactions in the process of gelation. Wolf,[229] in considering the effect of flow in a polymer solution on phase separation, has treated the process as near-equilibrium and examined the thermodynamic consequences of adding the energy stored in the sheared state to the Gibbs energy of the rest state. The result was that the region of homogeneity was extended by the shear with the demixing curve exhibiting two maxima. This behaviour was confirmed for the system *trans*-

decalin/polystyrene.[229] However, at high strain rates other effects resulting from extension and orientation of polymer molecules may contribute to phase separation. Witten and Cohen[230] have calculated, using a mean-field approximation, how elongation produced by a flow-field can favour inter- rather than intra-molecular association. Perhaps surprisingly, the maximum in intermolecular association occurs for molecules having only a small number of associating groups along their length.

Keller[31] has listed the phenomena which characterize quiescent and flow-induced gelation under the headings of Class I and Class II, respectively. In particular, the production of high-modulus, high-strength fibres has been closely associated with Class II gelation.[231–234] The 'natural' draw ratio of bulk polyethylene is about six. At this elongation the polymer molecules are fully oriented but not fully stretched. This is a result of the high entanglement density in the bulk. Since there are fewer entanglements in gels, the maximum draw ratio, and thus the molecular extension and stiffness of the fibre on removal of the solvent, are much higher. Bulk material produced via the gel state also has a low entanglement density and can lead to high-stiffness material on stretching.

At present, there are no reports of high-stiffness fibres produced by stretching Class I gels. These are usually formed by intrinsically low-crystallinity polymers inappropriate for the extended-chain crystallization essential for high stiffness. Highly crystalline polymers such as polyethylene are expected to crystallize from solution in the fringed-micellar morphology only under extreme conditions. However, there is one brief report[235] of Class I gelation in polyethylene requiring a high concentration (8%) and low molecular weight. Polyethylene may also gel with network junctions based on folded-chain crystals (Class III gelation[233]), and this forms the final crystalline morphology of gelation.

Folded-Chain Crystallite Gelation

A gel network which involves folded-chain, lamellar crystals can be obtained by cooling a semi-dilute solution of ultra-high molecular weight polyethylene (UHMW-PE). The usual chain-folded crystallization occurs, but, because of the high molecular weight, complete disentanglement with the resulting crystal precipitation is prevented, and a gel network forms,[234,236] provided that each lamella is connected to at least two others. This connection may be simply a result of a polymer molecule crystallizing in two crystals, or may involve the freezing in of interlocking loose loops from the fold surface (see Fig. 10). Lower supercoolings result in fewer entanglements and thus higher drawability and stiffer fibres. Patents[237] for the production of such fibres are based on the optimization of a process which involves the spinning of a semi-dilute solution of UHMW-PE through an orifice into a quench bath. This produces gel filaments which can then be ultradrawn as the solvent is expelled.

Berghmans and Stoks[238] recently reported a study of the gelation mechanisms of solutions of poly(vinyl alcohol) (PVA) in ethylene glycol and, in particular, the roles of phase separation and crystallization. At high supercoolings, they suggest that gelation involves a phase separation into polymer-rich and polymer-deficient regions with subsequent crystallization in the former. This mechanism parallels that suggested for the gelation of amylose. However, gelation of PVA can also occur above the upper critical solution temperature, and this they attribute to the formation of junction zones consisting of folded-chain crystals. The interlocking-loop structure is not explicitly involved, but simply the existence of a given polymer molecule in more than one crystal is implied.

Topological Cross-Linking

Rings and Loops

A gel network cross-linked by interlocking ring molecules (Fig. 13) provides the simplest example of a truly topologically knotted network. This is the 'olympic' gel proposed by de Gennes, who also suggests a method for realizing such a structure.[239] In the subsection on Folded-Chain Crystallite Gelation (pp. 284–5), the suggestion was discussed that the gelation of UHMW-PE involved the interlocking of loose loops from the fold surfaces of neighbouring crystals. Although such a gelation mechanism would require folded-chain lamellar crystallization, it is interesting to note that the nature of the cross-link is actually topological, each loop-and-crystal pair forming a 'ring' interlocked with its neighbour.

Although not true topological knots, the cross-links in polymer entanglements, which behave as gels on sufficiently short timescales, are also topological in nature.

Intertwined Helices
In general, intertwined helices are not knotted. Nevertheless, if the intertwining is intermolecular, the nature of the cross-link is essentially topological, although in practice physical interactions, such as hydrogen bonding, are also involved. In the polysaccharide field, the cross-links in a carrageenan gel network were originally proposed to be intermolecular, intertwined, double helices,[184] but this has been superseded by a model based on the aggregation or crystallization of helices.[210]

There have been several mechanisms proposed for the gelation of gelatin.[25,240–242] In general, the mechanisms involve a partial reforming from solution of the collagen triple helix. One type of mechanism involves the aggregation[241] or crystallization[242] of these helices to form the junction zones. A recent mechanism, proposed by Hayashi and Oh,[171] involves a phase separation, following triple helix formation, to form coascervate particles which subsequently associate to form a gel network. This essentially brings us back to the original mechanism of Hardy.[123] If the strands of the triple helix are from different molecules, i.e. the intertwining is inter- rather than intra-molecular, then a topological cross-link could form. Using a modified rubber-elasticity theory and measuring the rate of gelation, Oakenfull[243] has determined that the primary gel network consists of junction zones each having three molecules passing through them, which is in surprisingly good agreement with the expectations of the intertwined-helices model.

CONCLUSION

No universal mechanism for gelation has yet emerged. In general, the gelation of a polymer solution is characterized by a limited demixing of the polymer and solvent in such a manner as to form a continuous network. The nature of the polymer cross-linking interaction seems to be dependent on the particular polymer–solvent system. Specific interactions, crystallization and phase separation continue to feature prominently. In many cases, the process of gelation may be complex, involving several mechanisms the relative kinetics of which determine the structure of the network that is formed. There are many examples for which crystallization is not the primary mechanism of network formation, but where it apparently consolidates a network initially formed as a result of a phase separation or specific interaction. It has also been demonstrated that the polymer molecule need not consist of blocks of interacting and non-interacting species for gelation to occur. Molecular entanglement can be sufficient to limit complete association and precipitation. However, it must surely remain surprising that polyethylene can gel, although the nature of the structure formed in solution during stirring at high temperatures is as yet unknown.

Any universal model would therefore need to operate at a more fundamental level and predict network formation in terms of the characterizable properties of interactions involved in a particular system. At the centre of such a model must be the non-equilibrium nature of gelation, involving characteristic lifetimes of the interactions, reptation, diffusion of the interacting species, and disentanglement times. Such a model would predict the conditions for gelation rather than precipitation. A theoretical development of this type would require considerable effort, but is justifiable by the wide range of polymer–solvent systems, both biological and synthetic, in which polymer gelation is of fundamental importance.

REFERENCES

1. Buchard, W., Schmidt, M. and Stockmayer, W. H., *Macromolecules,* 1980, **13,** 580 and 1265.
2. Gordon, M., *Macromolecules,* 1984, **17,** 514.
3. de Gennes, P. G., *Scaling Concepts in Polymer Physics,* 1979, Cornell University Press, Ithaca and London.
4. Stauffer, D., *Phys. Rep.*, 1979, **54,** 1.
5. Essam, J. W., *Rep. Prog. Phys.*, 1980, **43,** 833.
6. Coniglio, A., Stanley, H. E. and Klein, W., *Phys. Rev. Lett.*, 1979, **42,** 518.
7. Gonzalez, A. E. and Muto, S., *J. Chem. Phys.*, 1980, **73,** 4668.
8. Gonzalez, A. E., *Polymer,* 1983, **24,** 77.
9. See, for example, Hentschel, G., *Nature (London),* 1985, **315,** 719.
10. Hong, D. C. and Stanley, H. E., *J. Phys. A: Math. Gen.*, 1983, **16,** L525.
11. Stanley, H. E., in *Kinetics of Aggregation and Gelation,* F. Family and D. P. Landau, Eds, 1984, North-Holland, Amsterdam.
12. Stauffer, D., Coniglio, A. and Adam, M., *Adv. Polym. Sci.*, 1982, **44,** 103.
13. Lookman, T., Pandey, R. B., Jan, N., Stauffer, D., Moseley, L. L. and Stanley, H. E., *Phys. Rev. B,* 1984, 2805.
14. Gordon, M., in *Integration of Fundamental Polymer Science and Technology,* L. Kleintjens and P. J. Lemstra, Eds, 1986, Elsevier Applied Science, London.

15. BROCHARD, F. and DE GENNES, P. G., *J. Phys,* 1983, **44,** L785.
16. DE GENNES, P. G., *J. Phys. Chem.*, 1984, **88,** 6469.
17. DE GENNES, P. G., *C.R. Acad. Sci. Paris,* 1982, **294,** Sér. II, 827.
18. DE GENNES, P. G., *C.R. Acad. Sci. Paris,* 1983, **296,** Sér. II, 881.
19. NAGY, M., *Colloid Polym. Sci.*, 1985, **263,** 245.
20. See, for example, HARTENSTEIN, R., *Principles of Physiology,* 1972, Van Nostrand Reinhold, New York, p. 487.
21. TAYLOR, D. L. and CONDEELIS, J. S., *Int. Rev. Cytol.*, 1979, **56,** 57.
22. BLANSHARD, J. M. V. and MITCHELL, J. R., *Polysaccharides in Food,* 1979, Butterworth, Guildford.
23. CLARK, A. H. and LEE-TUFFNELL, C. D., in *Functional Properties of Food Macromolecules,* J. R. Mitchell and D. Ledward, Eds, 1985, Elsevier Applied Science Publishers, London, Chap. 5.
24. MORRIS, V. J., in *Functional Properties of Food Macromolecules,* J. R. Mitchell and D. Ledward, Eds, 1985, Elsevier Applied Science Publishers, London, Chap. 3.
25. STAINSBY, G., in *The Science and Technology of Gelatin,* A. G. Ward and A. Courts, Eds, 1977, Academic Press, London, Chap. 4.
26. See, in general, PHILLIPS, G. O., WEDLOCK, D. J. and WILLIAMS, P. A., Eds, *Gums and Stabilisers for the Food Industry,* Vols I and II, 1982 and 1984, Pergamon Press, Oxford; *Ibid., Idem.*, Vol. III, 1986, Elsevier Applied Science, London.
27. See, for example, BALAZS, E. A., In *The Structure of the Eye,* G. K. Smelser, Ed., 1961, Academic Press, New York, p. 294.
28. See, for example, MACK, E. J., SHARMA, K., KIM, S. W., *Top. Pharm. Sci., Proc. Int. Congr. Pharm. Sci. FIP, 43rd,* 1983, 265.
29. See, for example, REILAND, J., in *Methods in Enzymology XXII,* W. B. Jakoby, Ed., 1971, Academic Press, New York, p. 287.
30. See, for example, PAIN, R. H., in *The Mechanical Properties of Biological Materials,* J. F. V. Vincent and J. D. Currey, Eds, 1980, Society for Experimental Biology, London.
31. KELLER, A., in *Structure–Property Relationships of Polymeric Solids,* P. A. Hiltner, Ed., 1983, Plenum Press, New York.
32. NEWMAN, S., FRENZ, D. and THOMASEK, J., *Science,* 1985, **288,** 888.
33. See, for example, SAKKA, S., *Treatise Mat. Sci. Technol.*, 1982, **22,** 129.
34. See, for example, MACKENZIE, J. D., *Ultrastructure Processing Ceramics Glasses and Composites,* L. L. Hench and D. R. Ulrich, Eds, 1983, Wiley, New York, p. 15.
35. YOUNG, A. T., Proc. 32nd National Symposium and Topical Conference, Houston, Texas, 1985 (to be published in *J. Vac. Sci. Technol.*, 1986, **A4,** 1128.
36. ZWIJNENBURG, A. and PENNINGS, A. J., *Colloid Polym. Sci.*, 1976, **254,** 868.
37. BARHAM, P. J. and KELLER, A., *J. Mat. Sci.*, 1985, **20,** 2281.
38. SMITH, P., LEMSTRA, P. J., KALB, B. and PENNINGS, A. J., *Polym. Bull.*, 1979, **1,** 733.
39. FLORY, P. J., *J. Amer. Chem. Soc.*, 1941, **63,** 3083.
40. STOCKMAYER, W. H., *J. Chem. Phys.*, 1945, **11,** 45.
41. ODELL, J. A., KELLER, A. and MILES, M. J., *Polymer,* 1985, **26,** 1219.

42. See, for example, KAWAI, T., KELLER, A., CHARLESBY, A. and ORMEROD, M. G., *Phil. Mag.*, 1964, **10,** 779; Keller, A., in *Developments in Crystalline Polymers—1*, D. C. Bassett, Ed., 1981, Elsevier Applied Science, London, p. 37.
43. FLORY, P. J., *J. Chem. Phys.*, 1942, **10,** 51.
44. HUGGINS, M. L., *J. Phys. Chem.*, 1942, **46,** 151.
45. HILDEBRAND, J. H. and SCOTT, R. L., *The Solubility of Non-Electrolytes*, 1964, Dover, New York.
46. See, for example, OLABISI, O., ROBESON, L. M. and SHAW, M. T., *Polymer–Polymer Miscibility*, 1979, Academic Press, New York, p. 35.
47. See, for example, KONINGSVELD, R., KLEINTJENS, L. A. and SCHOFFELERS, H. M., *Pure Appl. Chem.*, 1974, **39,** 1.
48. CAHN, J. W., *J. Chem. Phys.*, 1965, **42,** 93.
49. VAN AARTSEN, J. J., *Eur. Polym. J.*, 1970, **6,** 919.
50. WALSH, D. J. and ROSTAMI, S., *Polymer*, 1985, **26,** 418.
51. DERHAM, K.-W., GOLDSBROUGH, J. and GORDON, M., *Pure Appl. Chem.*, 1974, **38,** 97.
52. MURAOKA, Y., INAGAKI, H. and SUSUKI, H., *Brit. Polym. J.*, 1983, **15,** 110.
53. FLORY, P. J., *J. Amer. Chem. Soc.*, 1965, **86,** 1833.
54. PATTERSON, D., *Macromolecules*, 1969, **2,** 672.
55. SUZUKI, K., TANIGUCHI, Y. and ENOMOTO, T., *Bull, Chem. Soc. Japan*, 1972, **45,** 336.
56. BUSK, G. C., *Food Technol.*, 1984, **38,** 59.
57. FRANKS, F., Ed., *Water, a Comprehensive Treatise*, 1972, Plenum Press, New York.
58. DE GENNES, P.-G., ref. 3, p. 125.
59. SOLC, K. and BATTJES, K., *Macromolecules*, 1985, **18,** 220.
60. LAURENT, T. C., PRESTON, B. N. and CARLSSON, B., *Eur. J. Biochem.*, 1974, **43,** 231.
61. FLORY, P. J., *Proc. R. Soc. London*, 1956, **A234,** 73.
62. JENCKEL, E. and HEUSCH, R., *Kolloid-Z.*, 1953, **130,** 89.
63. See, for example, WUNDERLICH, B., *Macromolecular Physics* Vol. 2, 1976, Academic Press, New York, p. 22.
64. GIROLAMO, M., KELLER, A., MIYASAKA, K. and OVERBERGH, N., *J. Polym. Sci., Polym. Phys. Ed.*, 1976, **14,** 39.
65. FRANK, F. C., in *Growth and Perfection of Crystals*, R. H. Doremus, B. W. Roberts and D. Turnbull, Eds, 1958, Wiley, New York, p. 529.
66. See, for example, ref. 63, p. 17.
67. See, for example, BASSETT, D. C., *Principles of Polymer Morphology*, 1981, Cambridge University Press, London, and New York, p. 146.
68. PENNINGS, A. J. and PIJPERS, M. F. J., *Macromolecules*, 1970, **3,** 261.
69. HOFFMAN, J. D., *Macromolecules*, 1985, **18,** 772.
70. REBINDER, P. A., *Sovr. Problemy Fiz. Khim.*, 1969, **3,** 334.
71. See, for example, FLORY, P. J., *J. Amer. Chem. Soc.*, 1941, **63,** 1097; HOCHBERG, A., TANAKA, T., and NICOLI, D., *Phys. Rev. Lett.*, 1979, **43,** 217; BEINERT, G., BELKEBIR-MRANI, A., HERZ, J., HILD, G. and REMPP, P., *Faraday Discuss. Chem. Soc.*, 1974, **57,** 27.
72. See, for example, ZIMM, B. H., PRICE, F. P. and BIANCHI, J. P., *J.*

Phys. Chem., 1958, **62,** 979; MALINSKY, J., KLABAN, J. and DUSEK, K., *J. Macromol. Sci.—Chem.*, 1971, **A5**(6), 1071; SHAH, A. C., PARSONS, I. W. and HAWARD, R. N., *Polymer,* 1980, **21,** 825.
73. See, for example, TANAKA, T., FILLMORE, D., SUN, S.-T., NISHIO, I., SWISLOW, G. and SHAH, A., *Phys. Rev. Lett.*, 1980, **45,** 1636.
74. See, for example, BLOMBÄCK, B. and MASAHISA, O., *Thrombosis Res.*, 1982, **25,** 51.
75. ANDERSEN, S. O., *Biochim. Biophys. Acta,* 1963, **69,** 249.
76. ANDERSEN, S. O., *Biochim. Biophys. Acta,* 1964, **93,** 213.
77. FRY, S. C., *Biochem. J.*, 1982, **203,** 493; *Idem, Ibid.*, 1982, **204,** 449.
78. O'NEILL, M. A. and SELVENDRAN, R. R., *Biochem. J.*, 1980, **187,** 53.
79. FRY, S. C., *Planta,* 1983, **157,** 111.
80. GEISSMANN, T. and NEUKOM, H., *Helv. Chim. Acta,* 1971, **54,** 1108.
81. MARKWALDER, H. U. and NEUKOM, H., *Phytochemistry,* 1976, **15,** 836.
82. HARTLEY, R. D. and JONES, E. C., *Photochem.*, 1976, **15,** 1157.
83. ROMBOUTS, F. M., THIBAULT, J.-F. and MERCIER, C., Eur. Pat. Appln. EP 124439, 1984.
84. ROMBOUTS, F. M., THIBAULT, J.-F. and MERCIER, C., French Patent Appln 8307208, 1983.
85. See, for example, MEADOW, P. M., in *Companion to Biochemistry,* A. T. Bull, J. R. Lagnado, J. O. Thomas and K. F. Tipton, Eds, 1974, Longman, London, p. 345.
86. SANDFORD, P. *Adv. Carbohydr. Chem. Biochem.* 1979, **36,** 265.
87. KOHN, R., FURDA, I., HAUG, A. and SMIDSRØD, O., *Acta Chem. Scand.* 1968, **22,** 3098.
88. KOHN, R., *Pure Appl. Chem.* 1975, **42,** 371.
89. MACKIE, W., *Biochem. J.*, 1971, **125,** 89.
90. MACKIE, W., PEREZ, S., RIZZO, R., TARAVAL, F. and VIGNON, M., *IJBM,* 1983, **5,** 329.
91. ATKINS, E. D. T., MACKIE, W. and SMOLKO, E. E., *Nature* (*London*) 1970, **225,** 626.
92. ATKINS, E. D. T., MACKIE, W., PARKER, K. D. and SMOLKO, E. E., *J. Polym. Sci., Part B,* 1971, **9,** 311.
93. ATKINS, E. D. T., NIEDUSZYNSKI, J. A., MACKIE, W., PARKER, K. D. and SMOLKO, E. E., *Biopolymers,* 1973, **12,** 1865.
94. ATKINS, E. D. T., NIEDUSZYNSKI, I. A., MACKIE, W., PARKER, K. D. and SMOLKO, E. E., *Biopolymers,* 1973, **12,** 1879.
95. GRANT, G. T., MORRIS, E. R., REES, D. A., SMITH, P. J. C. and THOM, D., *FEBS Lett.*, 1973, **32,** 195.
96. REES, D. A., *Biochem. J.*, 1972, **126,** 257.
97. MORRIS, E. R., REES, D. A., THOM, D. and BOYD, J., *Carbohydr. Res.*, 1978, **66,** 145.
98. KEY, P. Y. and SELLEN, D. B., *J. Polym. Sci., Polym. Phys. Ed.*, 1982, **20,** 659.
99. DARZINS, A. and CHAKRABARTY, A. M., *J. Bacteriol.*, 1984, **159,** 9.
100. SHERBROCK-COX, V., RUSSELL, N. J. and GACESA, P., *Carbohydr. Res.*, 1984, **135,** 147.
101. STEINNES, A., *Gordian,* 1975, **75,** 228.

102. Wuhrmann, K. and Pilnik, W., *Experientia,* 1945, **1,** 330.
103. Sterling, C., *Biochim. Biophys. Acta,* 1957, **26,** 186.
104. Walkinshaw, M. D. and Arnott, S., *J. Mol. Biol.,* 1981, **153,** 1055.
105. Walkinshaw, M. D. and Arnott, S., *J. Mol. Biol.,* 1981, **153,** 1075.
106. Rees, D. A., *Carbohydr. Polym.,* 1982, **2,** 254.
107. Thom, D., Dea, I. C. M., Morris, E. R. and Powell, D. A., *Prog. Food Nutrition Sci.,* 1982, **6,** 97.
108. Yoshida, T., Sakurai, S., Okuda, T. and Takagi, Y., *J. Amer. Chem. Soc.,* 1962, **84,** 3590.
109. Dumas, P., Spassky, N. and Sigwalt, P., *Makromol. Chem.,* 1972, **156,** 55.
110. Hatada, K., Shimizu, S., Terawaki, Y., Ohta, K. and Yuki, H., *Polym. J.,* 1977, **13,** 811.
111. Grenier, D. and Prud'homme, R. E., *J. Polym. Sci., Polym. Phys. Ed.,* 1984, **22,** 577.
112. Liquori, A. M., Anzuine, G., Coiro, V. M., D'Alagni, M., de Santis, P. and Savino, M., *Nature (London),* 1965, **206,** 358.
113.Ősscher, F., ten Brinke, G. and Challa, G., *Macromolecules,* 1982, **15,** 1442.
114. ten Brinke, G., Schomaker, E. and Challa, G., *Macromolecules,* 1985, **18,** 1925.
115. Schwarz, J. and Rehage, G., *Kolloid-Z.u.Z.Polymere,* 1973, **251,** 689.
116. Flory, P. J., *Principles of Polymer Chemistry,* 1953, Cornell University Press, Ithaca, New York.
117. Gordon, M., Irvine, P. and Kennedy, J. W., *J. Polym. Sci., Polym. Symp.,* 1977, 61.
118. Koningsveld, R. and Kleintjens, L. A., *Macromolecules,* 1985, **18,** 243.
119. de Gennes, P.-G., *J. Chem. Phys.* 1980, **72,** 4756.
120. Pincus, P. J., *Chem. Phys.,* 1981, **75,** 1996.
121. Hashimoto, T., Kumaki, J. and Kawai, H., *Macromolecules,* 1983, **16,** 641.
122. Russell, T. P., Hadziioannou, G. and Warburton, W. K., *Macromolecules,* 1985, **18,** 78.
123. Hardy, W. B., *Proc. R. Soc. London.,* 1899, **66,** 95.
124. Joly, M., *Trans. Faraday Soc.,* 1952, **48,** 279.
125. Joly, M., *Kolloid Z.,* 1956, **145,** 65.
126. Joly, M., *Kolloid Z.,* 1962, **182,** 133.
127. See, for example, Hofmann, U., *Kolloid-Z.,* 1952, **125,** 86.
128. van Olphen, H., *J. Colloid Sci.,* 1962, **17,** 660.
129. Callaghan, I. C. and Ottewill, R. H., *Faraday Discuss. Chem. Soc.,* 1974, **57,** 110.
130. Rebinder, P. A., *Sovr. Problemy Fiz. Khim.,* 1969, **3,** 334.
131. Abaturova, N. A., Vlodauets, I. N. and Rebinder, P. A., *Kolloid. Zhur.,* 1972, **34,** 315.
132. Rogovina, L. Z. and Slonimskii, G. L., *Russian Chem. Rev.* 1974, **43,** 503.

133. Astbury, W. T., Dickinson, S. and Bailey, K., *Biochem. J.*, 1935, **29,** 2351.
134. Senti, F. R., Copley, M. J. and Nutting, G. C., *J. Phys. Chem.*, 1945, **49,** 192.
135. Ferry, J. D., *Adv. Protein Chem.*, 1948, **4,** 1.
136. See, for example, Clark, A. H. and Lee-Tuffnell, C. D., in *Functional Properties of Food Macromolecules,* J. R. Mitchell and D. Ledward, Eds., 1986, Elsevier Applied Science Publishers, London, chap. 5.
137. Barbu, E. and Joly, M., *Faraday Discuss. Chem. Soc.*, 1953, **13,** 77.
138. Skoulios, A. and Finaz, G., *Compt. Rend.*, 1960, **251,** 739.
139. Skoulios, A., Tsouladze, G. and Franta, E., *J. Polym. Sci., Part c,* 1963, **4,** 507.
140. Friedel, G., *Ann. Phys. Paris,* 1922, **18,** 200.
141. Ren, S., *Kyoto Daigaku Nippon Kagakuseni Kenkyusho Koenshu,* 1961, **18,** 1.
142. Sirakashi, K., *Seni-i Gakkaishi,* 1964, **20,** 306.
143. Chen-Chow, P. and Frank, S. G., *Int. J. Pharm.*, 1981, **8,** 89.
144. Schmolka, I. R., *J. Amer. Oil Chem. Soc.*, 1960, **54,** 110.
145. Rassing, J., McKenna, W. P., Bandyopadhyay, S. and Eyring, E. M., *J. Mol. Liquids,* 1984, **27,** 165.
146. Lundberg, R., O'Brien, D. E., Makowski, H. S. and Klein, R. R., US Patent Appln. 106027, 1979.
147. Joanny, J. F., *Polymer,* 1980, **21,** 71.
148. Ref. 3, p. 151.
149. Heuser, E. and von Neuenstein, W., *Cellulosechemie,* 1922, **3,** 94.
150. Ringer, S., *J. Physiol.*, 1890, **11,** 464.
151. Jones, G., *J. Appl. Polym. Sci.*, 1962, **6,** 15.
152. Newman, S., Krigbaum, W. R. and Carpenter, D. K., *J. Chem. Phys.*, 1956, **60,** 648.
153. Cahn, J. W., *J. Amer. Ceram. Soc.*, 1969, **52,** 118.
154. Miles, M. J., Petermann, J. and Gleiter, H., *J. Macromol. Sci.—Phys.* 1976, **B12,** 523.
155. Pines, E. and Prins, W., *Macromolecules,* 1973, **6,** 888.
156. Wun, K. L., Feke, G. T. and Prins, W., *Faraday Discuss. Chem. Soc.*, 1974, **57,** 146.
157. Feke, G. T. and Prins, W., *Macromolecules,* 1974, **7,** 527.
158. Wellinghof, S., Shaw, J. and Baer, E., *Macromolecules,* 1979, **12,** 932.
159. Tan, H.-M., Moet, A., Hiltner, A. and Baer, E., *Macromolecules,* 1983, **16,** 28.
160. Boyer, R. F., Baer, E. and Hiltner, A., *Macromolecules,* 1985, **18,** 427.
161. Gan, J. Y. S., Francois, J. and Guenet, J.-M., *Macromolecules,* 1986, **19,** 173.
162. Kusuyama, H., Miyamoto, N., Chatani, Y. and Tadokoro, H., *Polym. Commun.*, 1983, **24,** 119.
163. Helms, J. B. and Challa, G., *J. Polym. Sci., Part 2*, 1972, **10,** 1447.

164. Lemstra, P. J. and Challa, G., *J. Polym. Sci., Polym. Phys. Ed.*, 1975, **13,** 1809.
165. Guenet, J.-M., Lotz, B. and Wittmann, J.-C., *Macromolecules,* 1985, **18,** 420.
166. Tadeusz, G.-K., Tardieu, A. and Delaye, M., *Biochim. Biophys. Acta,* 1984, **800,** 28.
167. Duncan, G. and Jacob, T., *Human Cataract Formation, Ciba Foundation Symposium,* Vol. 106, 1984, p. 132.
168. Miles, M. J., Morris, V. J. and Ring, S. G., *Carbohydr. Res.*, 1985, **135,** 257.
169. Hayashi, A., Kinoshita, K., Miyake, Y. and Cho, C., *Agric. Biol. Chem.*, 1983, **47,** 1699.
170. Hayashi, A., Kinoshita, K. and Yasueda, S., *Polym. J.*, 1980, **12,** 447.
171. Hayashi, A. and Oh, S.-C., *Agric. Biol. Chem.*, 1983, **47,** 1711.
172. Bawden, F. C. and Pirie, N. W., *Proc. R. Soc. London,* 1937, **B123,** 274.
173. Ingram, V. M., *Nature (London)*, 1956, **178,** 792.
174. Bur, A. J. and Fetters, L. J., *Chem. Rev.*, 1976, **76,** 727.
175. Bair, T. I., Morgan, P. W. and Killian, F. L., *Macromolecules,* 1977, **10,** 1396.
176. Miller, W. G., Chakrabarti, S. and Seibel, K. M., in *Microdomains in Polymer Solutions,* P. L. Dubin, Ed, Plenum, New York, 1985.
177. Tohyama, K. and Miller, W. G., *Nature (London)*, 1981, **289,** 813.
178. Miller, W. G., Kou, L., Tohyama, K. and Voltaggio, V., *J. Polym. Sci., Polym. Symp.*, 1978, **65,** 91.
179. Wee, E. L. and Miller, W. G., *J. Phys. Chem.*, 1971, **75,** 1446.
180. Russo, P., Magestro, P., Mustata, M., Saunders, M. J. and Miller, W. G., *Polym. Preprints,* 1986, **27,** 229.
181. Miller, W. G., Russo, P. S. and Chakrabarti, S., *J. Appl. Polym. Sci., Appl. Polym. Symp.*, 1985, **41,** 49.
182. Lockwood, B., *Food Process. Ind.*, 1972, **41**(Nov.), 47.
183. Kertesz, Z. I., *The Pectic Substances,* 1951, Interscience, New York.
184. Rees, D. A., *Adv. Carbohydr. Chem. Biochem.*, 1969, **24,** 267.
185. Rees, D. A. and Welsh, E. J., *Angew. Chem. Int. Ed. Engl.*, 1977, **16,** 214.
186. Kovacs, P., *Food Technol.*, 1973, **27,** 26.
187. Dea, I. C. M., McKinnon, A. A. and Rees, D. A., *J. Mol. Biol.*, 1972, **68,** 153.
188. Miles, M. J., Morris, V. J. and Carroll, V., *Macromolecules,* 1984, **17,** 2443.
189. Antonov, Y. A., Grinberg, V. Y., Zhurauskaya, N. A. and Tolstoguzov, V. B., *J. Texture Studies,* 1980, **11,** 199.
190. Takahashi, A., Nakamura, T. and Kagawa, I., *Polym. J.*, 1972, **3,** 379.
191. See, for example, Mandelbern, L., *Crystallization of Polymers,* 1964, McGraw-Hill, New York, pp. 113, 308.
192. Paul, D. R., *J. Appl. Polym. Sci.*, 1967, **11,** 439.

193. TAKAHASHI, A., *Polym. J.*, 1973, **4,** 379.
194. TAKAHASHI, A., NAKAMURA, T. and KAGAWA, I., *Polym. J.*, 1972, **3,** 207.
195. TAKAHASHI, A., SAKAI, M.,and KATO, T., *Polym. J.*, 1980, **12,** 335.
196. TAN, H. M., CHANG, B. H., BAER, E. and HILTNER, P. A., *Eur. Polym. J.*, 1983, **19,** 1021.
197. TAKAHASHI, A. and KATO, T., *Res. Rep. Fac. Eng. Mie Univ.*, 1976, **1,** 97.
198. STRAFF, R. S. and UHLMANN, D. R., *J. Polym. Sci., Phys. Ed.*, 1976, **14,** 35.
199. LEMSTRA, P. J., KELLER, A. and CUDBY, M., *J. Polym. Sci., Phys. Ed.*, 1978, **16,** 1507.
200. DORRESTIJN, A., LEMSTRA, P. J. and BERGHMANS, H., *Polym. Commun.*, 1983, **24,** 226.
201. GUERRERO, S. J., KELLER, A., SONI, P. L. and GEIL, P. H., *J. Polym. Sci., Polym. Phys. Ed.*, 1980, **18,** 1533.
202. GUERRERO, S. J. and KELLER, A., *J. Macromol. Sci.—Phys.*, 1981, **B20,** 167.
203. GUERRERO, S. J., MEADER, D. and KELLER, A., *J. Macromol. Sci.—Phys.*, 1981, **B20,** 185.
204. ATKINS, E. D. T., ISAAC, D. H., KELLER, A. and MIYASAKA, K., *J. Polym. Sci., Phys. Ed.*, 1977, **15,** 211.
205. MOORHOUSE, R., COLEGRAVE, G. T., SANDFORD, P. A., BAIRD, J. K. and KANG, K. S., in *Solution Properties of Polysaccharides,* D. A. Brant, Ed., *ACS Symp, Ser,* 1981, **150,** 111.
206. CARROLL, V., CHILVERS, G. R., FRANKLIN, D., MILES, M. J., MORRIS, V. J. and RING, S. G., *Carbohydr. Res.*, 1983, **114,** 181.
207. MILES, M. J. and MORRIS, V. J., unpublished data.
208. O'NEILL, M. A., MORRIS, V. J., SELVENDRAN, R. R., SUTHERLAND, I. W. and TAYLOR, T. T., *Carbohydr. Res.*, 1986, **148,** 63.
209. FULTON, W. S. and ATKINS, E. D. T., in *Fiber Diffraction Methods,* A. D. French and K. H. Gardner, Eds, American Chemical Society, Washington, D.C., *ACS Symp. Ser.*, 1980, **141,** 385.
210. MORRIS, E. R., REES, D. A. and ROBINSON, G., *J. Mol. Biol.*, 1980, **138,** 349.
211. SMIDSRØD, O. and GRASDALEN, H., *Carbohydr. Polym.*, 1982, **2,** 270.
212. SMIDSRØD, O., ANDERSON, I.-L., GRASDALEN, H., LARSEN, B. and PAINTER, T., *Carbohydr. Res.*, 1980, **80,** C11.
213. GRASDALEN, H. and SMIDSRØD, O., *Macromolecules*, 1981, **14,** 229.
214. BELTON, P. S., CHILVERS, G. R., MORRIS, V. J. and TANNER, S. F., *Int. J. Biol. Macromol.*, 1984, **6,** 303.
215. MORRIS, V. J. and CHILVERS, G. R., *Carbohydr. Polym.*, 1983, **3,** 129.
216. ELIASSAF, J., SILBERBERG, A. and KATCHALSKY, A., *Nature* (*London*), 1955, **176,** 1119.
217. LODGE, A. S., *Polymer,* 1961, **2,** 195.
218. PETERLIN, A. and TURNER, D. T., *Polym. Lett.*, 1965, **3,** 517.
219. PETERLIN, A., QUAN, C. and TURNER, D. T., *Polym. Lett.*, 1965, **3,** 521.

220. PETERLIN, A., PHILIPPOFF, W. and TURNER, D. T., *Polymer,* 1965, **6,** 35.
221. MASUO, T., PAVAN, A., PETERLIN, A. and TURNER, D. T., *J. Colloid Interface Sci.,* 1967, **24,** 241.
222. BARHAM, P. J., HILL, M. J. and KELLER, A., *Colloid Polymer Sci.,* 1980, **258,** 899.
223. KELLER, A. and BARHAM, P. J., *Plast. Rubb. Int.* 1981, **6,** 19.
224. NARH, K. A., BARHAM, P. J. and KELLER, A., *Macromolecules,* 1982, **15,** 464.
225. KIM, M. W., PEIFFER, D. G. and PINCUS, P. *J. Phys.-Lett.,* 1984, **45,** L-953.
226. BRAY, A. J. and MOORE, M. A., *J. Phys. A,* 1977, **10,** 1927.
227. CLERC, J. P., CARTON, J. P., ROUSSENQ, J. R. and STAUFFER, D., *J. Phys.-Lett.,* 1981, **42,** L-249.
228. KELLER, A., *IUPAC Macromolecules,* H. Benoit and P. Rempp, Eds, 1982, Pergamon Press, Oxford, pp. 171–190.
229. WOLF, B. A., *Macromolecules,* 1984, **17,** 615.
230. WITTEN, T. A. and COHEN, M. H., *Macromolecules,* 1985, **18,** 1915.
231. ZWIJNENBURG, A. and PENNINGS, A. J., *Colloid Polym. Sci.,* 1978, **259,** 868.
232. BARHAM, P. J., *Polymer,* 1982, **23,** 1112.
233. BARHAM, P. J. and KELLER, A., *J. Mat. Sci.,* 1985, **20,** 2281.
234. LEMSTRA, P. J. and KIRSCHBAUM, R., *Polymer,* 1985, **26,** 1372.
235. Ref. 233, ref. 50 therein.
236. SMITH, P., LEMSTRA, P. J. and BOOY, H. C., *J. Polym. Sci., Polym. Phys. Ed.,* 1981, **19,** 877.
237. DSM/STAMICARBON, US Patents 4344908 (1982), 4422993 (1983), 4430383 (1984), 4436689 (1984).
238. BERGHMANS, H. and STOKS, W., *Rolduc Polymer Conference: Integration of Fundamental Polymer Science and Technology,* Elsevier Applied Science, London, 1986.
239. Ref. 3, p. 132.
240. VEIS, A., *The Macromolecular Chemistry of Gelatin,* 1964, Academic Press, New York.
241. EAGLAND, D., PILLING, G. and WHEELER, R. G., *Faraday Discuss. Chem. Soc.,* 1974, **57,** 181.
242. GERNGROSS, O., HERRMANN, K. and LINDEMANN, R., *Kolloid Z.,* 1932, **60,** 276.
243. OAKENFULL, D., *Gums and Stabilisers for the Food Industry,* Vol. 3, G. O. Phillips, D. J. Wedlock and P. A. Williams, Eds, 1986, Elsevier Applied Science Publishers, London.

Chapter 6

ELECTRO-OPTIC APPLICATIONS OF LIQUID CRYSTALLINE POLYMERS

HARRY J. COLES

Department of Physics, Schuster Laboratory, The University, Manchester, UK

SUMMARY

In this chapter we review the main chemical structural features of side chain polymer liquid crystals and how these lead to nematic, cholesteric and smectic mesophases. Electro-optic phenomena exhibited by these phases are then considered in terms of potential device applications. For example consideration is given to the Freedericksz transition, the twisted nematic device, the guest–host effect, electro-hydrodynamic phenomena and laser writing in nematic and, where appropriate, smectic phases. As is discussed the major use of these side chain polymer liquid crystals is in the storage of field induced alignment. Since the side chain groups may be highly functional, i.e. optically non-linear, fluorescent, ferro-electric, etc. the stored alignment leads to the possibility of new types of functional polymer films. The factors that influence or control the alignment are considered in some detail and extensive data on typical polymer systems have been given. Finally it is shown that polymer–low molar mass liquid crystal mixtures lead to materials with improved visco-elastic properties that may be electro-optically switched on the millisecond timescale.

INTRODUCTION

Materials that exhibit liquid crystalline phases have been recognized for almost a century,[1] yet it is only over the last two decades that such materials have evolved from the laboratory curiosity stage to become the basis of a new science. Liquid crystalline phases have recently been called the fourth state of matter.[2] However, such a definition belies the richness of structural properties that characterizes the many different liquid crystalline phases and leads to the immense interest in their behaviour.

A liquid crystalline phase is generally defined in terms of the long-range intermolecular order exhibited by its constituent molecules. In a crystalline phase there are three degrees of long-range positional order. In a liquid crystalline phase at least one of these degrees is removed and in a liquid phase there in no long-range positional order. There may, however, exist short-range rotation or orientational order between the molecules if they exhibit a shape anisotropy, so that a true isotropic liquid phase is only exhibited when neither positional long-range nor orientational short-range order exists. Thus, if one or more of these degrees of positional or rotational order are manifest by the anisotropic molecules, then we have a useful definition of the liquid crystalline phase or mesophase between the isotropic liquid and crystalline phases. As shape anisotropy is a prerequisite in this definition we might expect macroscopic properties such as refractive index, electric conductivity, viscosity and elasticity also to be anisotropic. Since liquid crystalline phases are also fluid, external fields (magnetic, electrical, optical, mechanical and thermal) may be used to induce changes in these properties. It was the interdependence of the optical and electrical properties and the quest for new electro-optic devices that led to the recent resurgence of interest in liquid crystalline phases.[3]

Depending on the degree of order in the system, three general classes of liquid crystalline phases may be defined, i.e. nematic, cholesteric (or chiral nematic) and smectic. These are shown schematically in Fig. 1.

In the nematic phase the molecules show only short-range orientational order and there is no long-range positional order between the molecules. Thermal motion of the freely translating molecules ensures that the alignment direction is only an average property, defined by the so-called director $\hat{\mathbf{n}}$. This does, however, lead to locally anisotropic refractive indices, dielectric constants and magnetic susceptibilities which are referred to the director direction, e.g. the birefringence Δn is given by $n_{\parallel} - n_{\perp}$ where the subscripts indicate parallel and perpendicular to the director respectively. Although individual molecules may be polar, the nett antiparallel ordering in the system ensures that macroscopically the nematic phase is non-polar.

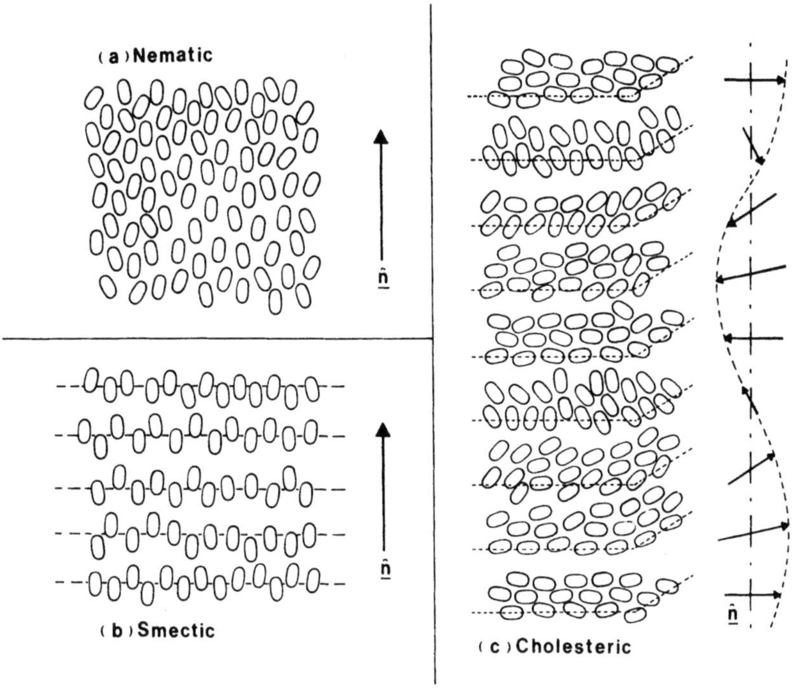

FIG. 1. Schematic representation of (a) nematic, (b) smectic and (c) cholesteric (or chiral nematic) liquid crystalline phases. In the nematic phase only orientational correlations are present with a mean alignment in the direction of the director **n̂**. In the smectic phase there are additional layer-like correlations between the molecules in planes perpendicular to the director. The planes, drawn as broken lines, are in reality due to density variations in the direction of the director. The interplane separation then corresponds to the period of these density waves. In the cholesteric phase the molecules lie in planes (defined by broken lines) twisted with respect to each other. Since the molecules in one plane exhibit nematic-like order with a mean alignment defined by the director **n̂**, the director traces out a right- or left-handed helix on translation through the cholesteric medium in a direction perpendicular to the planes. When the period of this helix is of the order of the wavelength of light, the cholesteric phase exhibits bright Bragg-like reflections. In these illustrations the space between the molecules (drawn as ellipsoids for simplicity) will be filled with the alkyl chains, etc., to give a fairly high packing density.

The cholesteric or chiral nematic phase may be considered as a perturbed planar nematic phase in which the director in each plane is slightly rotated with respect to that of an adjacent plane. Asymmetry in the chemical structure causes this rotation which results in the director tracing out a left- or right-handed helix. When such helix pitch lengths are of the order of the wavelength of light, the planes give rise to an intense Bragg-like coloured reflection. Thermally induced changes in the pitch length lead to a change in this coloured reflection.

The smectic phase exhibits both orientational and some positional order. The positional order ensures that the centres of the molecules are arranged on average in planes. These planes are also aligned on average for non-tilted phases, perpendicular to the orientation director of the molecules in each layer. The molecular packing in each plane may be disordered (i.e. the smectic A phase) or ordered (smectic B, H, etc.). It is now recognized that there are many different layer-like smectic phases, depending on the order in and between the planes, in which the director is either perpendicular to the planes or tilted at some angle. Further, although non-disturbed smectic phases are macroscopically non-polar, alignment techniques may be used to produce ferroelectric phases. These smectic phases have recently been extensively reviewed.[4]

Over the last two decades research has concentrated on the production of low molar mass liquid crystal materials with electro-optical properties suitable for use at ambient temperatures. Since one highly desirable property was electro-optic switching on a reasonably fast timescale (i.e. faster than the eye response time of ~100 ms) and because this switching time depends on cooperative molecular reorientation, attention was focused on the synthesis of relatively small molecules of low mean viscosity of the type

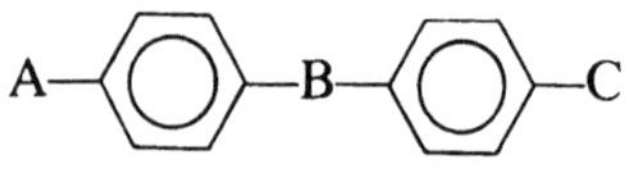

For such typical structures, A would represent a strongly dipolar group such as cyano or nitrile to give the desired electrical structure, B might be an azo or ester linkage to give a good rigid conjugation between the two aromatic rings and therefore a high birefringence, whilst C would typically be an alkyl or alkyloxy chain to promote high geometrical anisotropy and phase stability. Use of different substituents[5] has given rise to many thousands of different liquid crystalline systems with mesophases exhibited over a wide temperature range.[6] Despite this wide range of materials, it is only over the last decade that electro-optic devices have become firmly established with the discovery of the cyanobiphenyl family of compounds.[7] These materials, which have the linking group B removed, have given rise to a series of stable enantiotropic room-temperature liquid crystal mixtures[8] that form the basis of current electro-optic displays. Modification to the macroscopic properties, principally the birefringence, viscosity and clearing temperatures, have been induced by replacing one or more of the aromatic rings by cyclohexane or bicyclo-octane ring systems[5] and in the case of the electrical properties by replacing the polar head group by a non-cyano group or by the use of lateral substituents on the ring structure. Thus, the field of low molar mass liquid crystal research is rich with structural variants that form stable room-temperature mesophases of very different electrical and optical properties.

The possibility of creating new polymer materials by combining the unique properties of low molar mass liquid crystals with those of high molar mass compounds, has led to the new science of polymer liquid crystals. When polymer liquid crystals were originally recognized in the classic work of Robinson[9] in the early 1950s, this work was concerned with synthetic polypeptides in solution, i.e. lyotropic systems, It is only recently that thermotropic polymer liquid crystals, where the mesomorphism is induced as a function of temperature alone, have been synthesized and led to the production of films, coatings and fibres with highly unusual and specific functions.[10–12] Essentially, the low molar mass compounds can be incorporated into the polymer in two ways. Firstly, the low molar mass mesogens are joined head-to-tail, with suitable spacer groups to form a main-chain polymer, i.e. the mesogenic groups are all contained in and create the polymer backbone. These materials form basic structural units of systems forming high- or ultra-high-modulus fibres.[10] Secondly, the mesogenic groups are linked, via a suitable reactive substituent, to an existing polymer backbone to form a side-chain polymer or comb-like system. In such systems the mesogenic side-group maintains a degree of orientational freedom, dependent on the coupling strength to the backbone, and its alignment may be altered by the application of external electric or other fields.[11,12] This realignment, which takes place on a timescale orders of magnitude faster for side-chain polymers in comparison with the main-chain systems, allows the optical properties of the material to be readily altered by use of external electric fields. Since mesogenic groups may be synthesized that are highly birefringent, dichroic, fluorescent or exhibit phases with high optical scattering powers, their use in side-chain polymers leads to an immense range of new electrically switchable polymers.

In this chapter we shall only be concerned with electro-optic and thermo-optic switching effects in thermotropic side-chain polymer liquid crystals. We will consider briefly the synthesis and structure of such compounds and show how the nematic, cholesteric and smectic phases arise. Since the optical properties of each of these phases are different, and may be altered depending on alignment within the phase, each gives rise to different electro-optic effects. If these are coupled to the use of dye additives or substituents, then it will be realized that a wide range of electro-optic devices based on dichroism or fluorescence as well as birefringence or scattering power may be fabricated. These will be considered and discussed in terms of their performance and potential applications. Finally, possible uses of polymer–low molar mass liquid crystal solutions will be considered in terms of electro-optic device applications.

SIDE-CHAIN POLYMER LIQUID CRYSTALS: STRUCTURAL CONSIDERATIONS

The most general preparative procedures used in the synthesis of side-chain polymer liquid crystals are (a) homopolymerization or copolymerization of the mesogenic (or non-mesogenic) groups themselves[12] or (b) attachment of the mesogenic units via a reactive site to an existing polymer backbone.[11] As shown schematically in Fig. 2, for either route, the nett effect is a semi-flexible backbone with pendant side-groups, A, B, C, etc.

The most common backbones (1) so far considered are polyacrylate, -methacrylate and -siloxane systems. Since the polysiloxane has the greatest backbone flexibility, it has the lowest glass transition temperature, T_g. This has consequences for device performance. The flexible alkylene spacer (2) serves to decouple partially[13] the motion of the side-groups and the backbone and has allowed room-temperature mesophases to be readily produced.[14] The side-group moieties (3) confer the particular dielectric and optical properties on the polymer liquid crystal. Changes in the nature of the side-group, the length of the alkylene spacer and the odd–even effect either in the spacer or in the mesogenic head-group for an alkyl or alkyloxy termination lead to different polymeric mesophases.[14] Further, in the copolymer systems, the side-groups can be used to provide composite effects analogous to eutectic mixtures in low molar mass liquid crystals.[8] Non-mesogenic

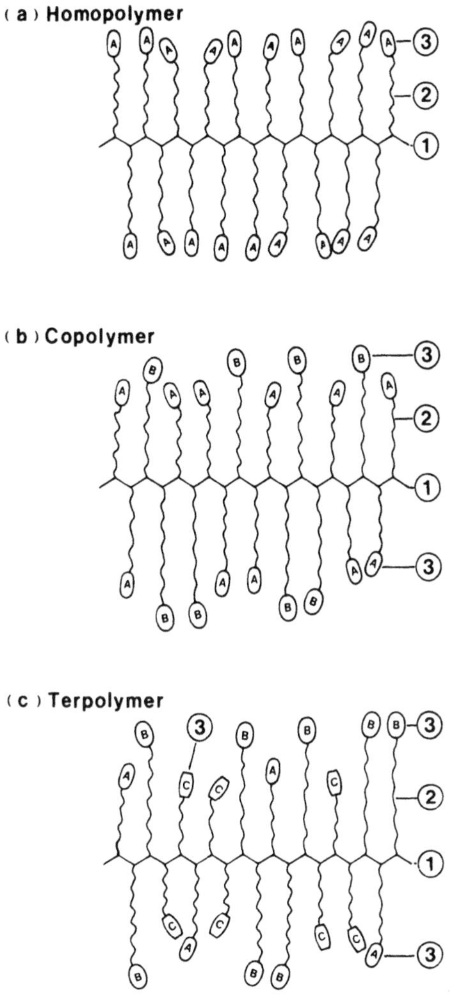

FIG. 2. Schematic illustration of side-chain polymer crystals where (1) represents the polymer backbone, (2) represents the alkylene spacer and (3) represents the different side-group moieties. In a real polymer the backbone would not generally be in the rigid configuration shown. For example, with an SiO backbone the alkylene chains are attached to each Si atom (for a fully substituted system) and rotational isomerism due to free rotation about the linking oxygen would ensure a random backbone configuration. Typical spacer groups consist of between 4 and 12 methylene units and the side-group moieties contain several bulky phenylene units. Therefore, in three-dimensional space the molecules would look more like a 'bottle brush' with random bristles attached to a flexible backbone stem.

dye side-groups may be combined with mesogenic groups to form dyed polymeric mesophases.[15] Thus the plethora of structures and mesophases available in the low molar mass systems is enriched by the polymerization process to give dichroic, fluorescent and even optically non-linear mesophases.[16]

In our own work, carried out under an extensive collaboration with Professor George Gray, F.R.S., and Drs David Lacey and Peter Gemmell of Hull University (UK), we have studied a number of nematic, cholesteric and smectic polymer mesophases with a view to exploiting their electro-optical properties.[16,17] Some of the typical structures studied are given in Fig. 3. These are only a few of the structures now available and the reader is referred to ref. 14 for a more complete range of polysiloxane systems as well as to the recent reviews[11,12] which also cover the acrylate and methacrylate backbone systems.

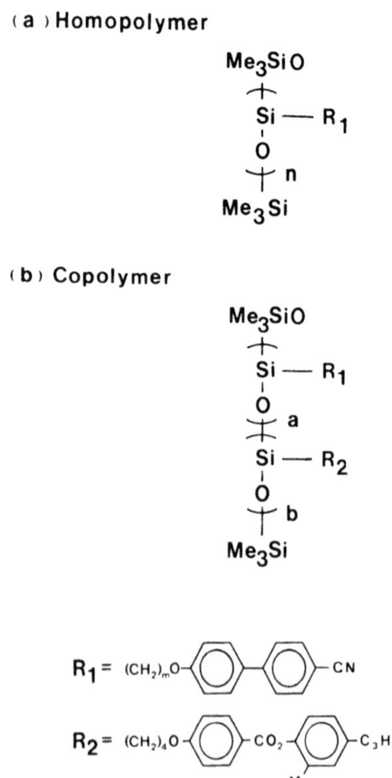

FIG. 3. Typical polysiloxane chemical structures. For the homopolymer $n = 50$ and for the copolymer $a + b = 50$.

TABLE 1
HOMOPOLYMER TRANSITION DATA FOR THE 50 SiO-UNIT SIDE-CHAIN POLYSILOXANE SYSTEM OF FIG. 2(a) WITH

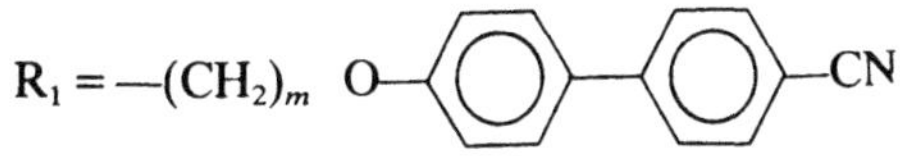

Spacer *m*	Phase type	Transition temperatures[14] (°C) T_g	T_c
3	Smectic A	40	152
4	Smectic A	28	132·5
5	Smectic A	14·5	169·5
6	Smectic A	13·5	165·5

Since the cyanobiphenyl derivatives[7] had proved to be excellent materials for electro-optic displays,[18] equivalent homopolymer polysiloxane liquid crystals, based on these low molar mass mesogens, were synthesized. Their properties are summarized in Table 1. Since the homopolymers only exhibited smectic phases, copolymers were prepared[14] using destabilizing 2-methyl-substituted ester side-groups. As shown in Table 2, at sufficiently high ester-group concentrations, a

TABLE 2
COPOLYMER TRANSITION DATA FOR THE 50 SiO-UNIT SIDE-CHAIN POLYSILOXANE SYSTEM OF FIG. 3(b)

Spacer *m*	Substituent ratio *a*:*b*	Phase type	Transition Temperatures[14] (°C) T_g	T_c
5	1:1	Smectic A	7	122
5	8:17	Nematic	4	53
5	7:3	Smectic A	7	151
6	1:1	Smectic A	4	86

nematic phase was observed. If the substituent (R_1) used for the homopolymer was replaced by a group combining the essential features of the copolymer (i.e. a cyano head-group, an ester linkage and a lateral methyl group) as follows:

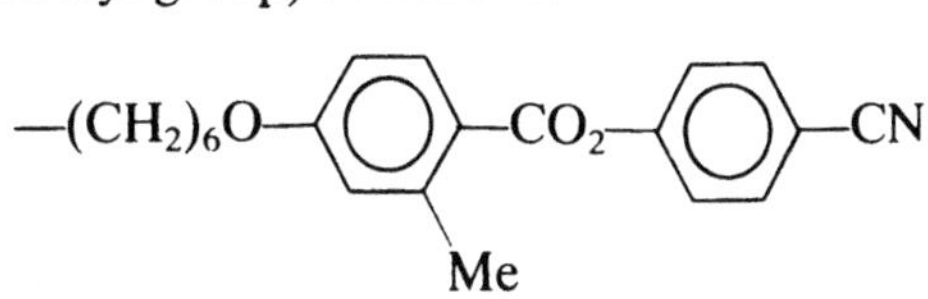

then a nematic phase, with $T_g = 9°C$ and $T_{NI} = 46°C$ was observed.[14] Combination of this side-group (as R_1) with a chiral alkene (R_2) of the following type

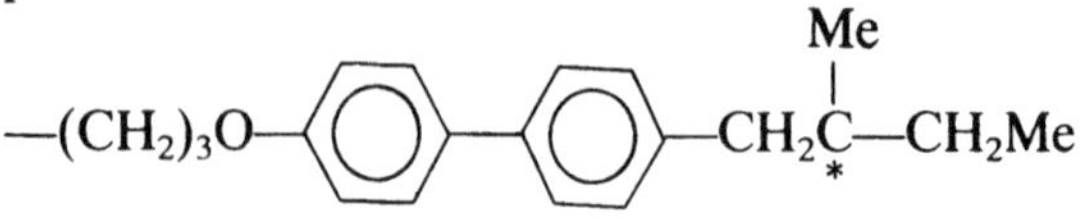

on the polysiloxane backbone gave (for a ratio of $b:a$ of 3:17) a room-temperature cholesteric polymer which reflected light in the ultraviolet and had $T_g = 5°C$ and $T_{ChI} = 31°C$. In the notation used herein T_{KN}, T_{KS} or T_{KCh} refer to the crystalline (K) to nematic (N), smectic (A) or cholesteric (Ch) phase transitions, respectively. The nematic, cholesteric or smectic to isotropic phase transitions are similarly denoted T_{NI}, T_{ChI} or T_{SI}, respectively. These mesophase to isotropic phase transition temperatures define the start of the phase transition at which the isotropic phase first starts to appear as the sample is heated whilst T_c denotes the true clearing temperature above which no residual mesophase is observed microscopically.

The data contained in Tables 1 and 2 illustrate a number of points to be considered when choosing low molar mass mesogen/polymer conbinations. Firstly, the polymerization process leads to a mesophase of higher order than that exhibited by the low molar mass compound alone, e.g. for the side-group on its own with $m = 6$, the mesogen has $T_{KN} = 35°C$ and $T_{NI} = 52°C$ whereas the homopolymer has $T_g = 13{\cdot}5°C$ and $T_c = 165{\cdot}5°C$ (for an S_{A-I} phase transition). Secondly, as shown, the polymer clearing temperatures are much higher than those of the low molar mass precursors. Thirdly, increasing the length of the flexible spacer lowers T_g and there is also an odd–even effect in T_g and T_c. Whilst the homopolymers have a strong tendency to form smectic A (S_A) phases, this effect can be reduced either by making copolymers (Table 2) or by using mesogenic molecules with lateral substituents (which also decreases T_g). In the latter case, room-temperature nematic homopolymers of a high dielectric anisotropy result, and addition of small amounts of a chiral substituent forces the nematic phase to distort into a classical cholesteric phase. Typical optical textures observed with such nematic, cholesteric and smectic polymer mesophases are shown in Fig. 4. It is worth noting that for polymer liquid crystals these textures are often only observed after annealing the polymer films a few degrees below T_c (for either a nematic or smectic A to isotropic phase transition) for hours or days, whereas low

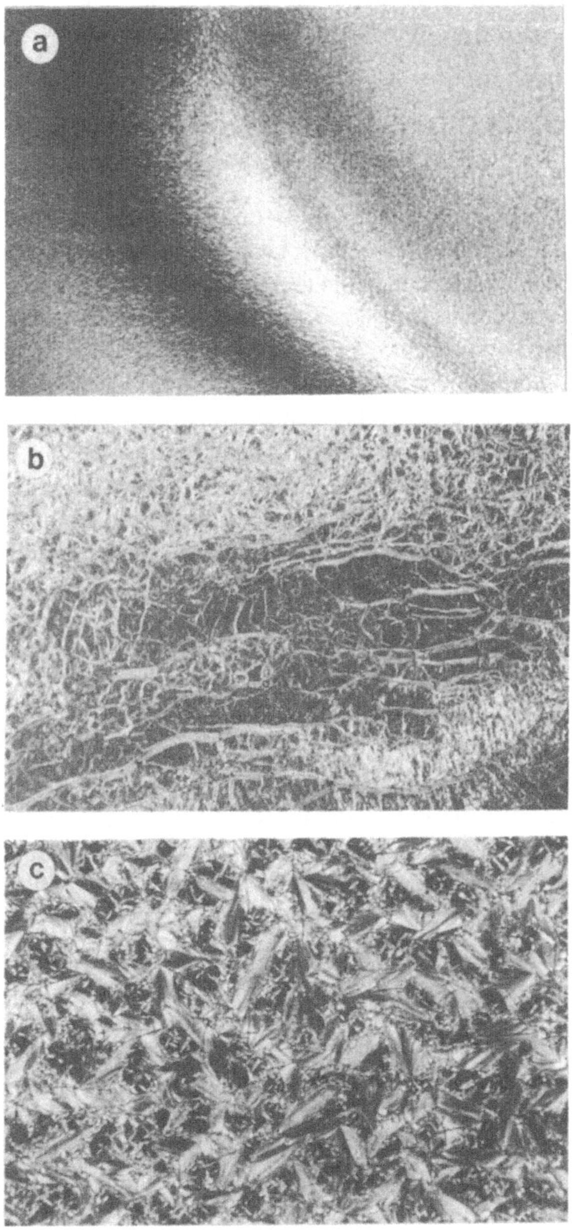

FIG. 4. Typical polysiloxane side-chain liquid crystal optical textures for (a) nematic, (b) cholesteric and (c) smectic A phases.

molar mass textures tend to appear instantaneously. Lack of patience can lead to a serious misclassification of the polymer texture!

ELECTRO-OPTIC PHENOMENA IN SIDE-CHAIN POLYMER LIQUID CRYSTALS

The preconditions for the use of polymer liquid crystals in display applications are that they exhibit bulk optical properties dependent on the molecular orientation in the mesophase and that this orientation may be altered on application of an external field. In this chapter we shall be concerned with electric or optical fields only. The particular optical property, i.e. (a) the birefringence, (b) the dichroism or (c) the scattering power, defines the display construction in terms of the use of polarized (a and b) or non-polarized (b and c) light, whereas the ability to switch from one orientation to another depends on the anisotropic electric permittivity and the orientational elastic constants. The dynamics of the induced orientation will depend, additionally, on the viscosity constants of the material.

Nematic Phases

The initial research[12,19,20] on electro-optic phenomena in side-chain polymer liquid crystals concentrated on systems that exhibited nematic phases so that a ready comparison could be made with low molar mass mesogens. Such measurements have established that electro-optic devices are feasible and have allowed elastic constants to be deduced from applications of the continuum theory. This theory,[21] originally derived for low molar mass nematic liquid crystals, defines a relationship for the free energy density F in terms of the elastic constants (k_{ii}) and the director $\hat{\mathbf{n}}$ such that:

$$F = \tfrac{1}{2} k_{11}(\operatorname{div} \hat{\mathbf{n}})^2 + \tfrac{1}{2} k_{22}(\hat{\mathbf{n}} \cdot \operatorname{curl} \hat{\mathbf{n}})^2 + \tfrac{1}{2} k_{33}(\hat{\mathbf{n}} \cdot \operatorname{curl} \hat{\mathbf{n}})^2 \qquad (1)$$

where k_{11}, k_{22} and k_{33} are the orientational elastic constants that describe the splay, twist and bend deformations respectively (Fig. 5). In a nematic mesophase, due to the anisotropic electric and magnetic properties, minimization of eqn (1) implies that the preferred orientation is with the director everywhere parallel. If an electric field ($\mathbf{E}$) is now applied across the nematic sample, an additional term, F_{E}, contributes to the free energy and is given by

$$F_{\mathrm{E}} = -\tfrac{1}{2} \varepsilon_0 \, \Delta\varepsilon (\hat{\mathbf{n}} \cdot \mathbf{E})^2 \qquad (2)$$

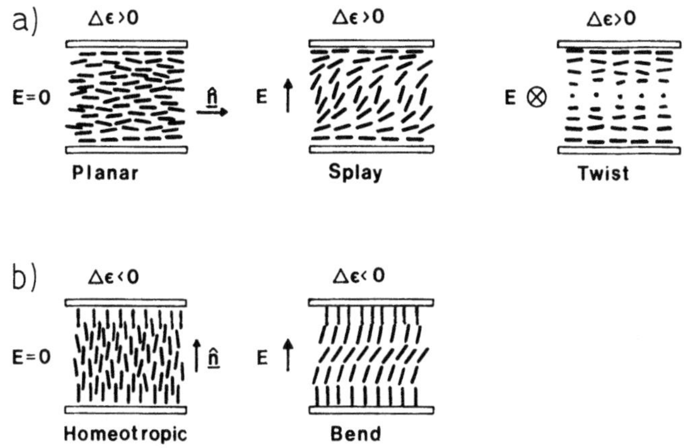

FIG. 5. The three principal elastic deformations induced by electric fields in nematic liquid crystals. In (a) the molecules are aligned in a planar texture with the director $\hat{\mathbf{n}}$ parallel to the substrate for $\mathbf{E}=0$. In (b) the initial alignment is homeotropic with the director perpendicular to the substrate. The arrow ↑ or circle ⊗ indicates that E is applied either perpendicular or parallel to the substrate, respectively.

where $\Delta\varepsilon$ is the dielectric anisotropy given by $\varepsilon_{\parallel}-\varepsilon_{\perp}$ and the subscripts refer to directions parallel and perpendicular to the director respectively. If $\Delta\varepsilon$ is positive, the director tends to align in the direction of $\mathbf{E}$, whereas if $\Delta\varepsilon$ is negative the predominant alignment is perpendicular to the field. The elastic constants of eqn (1) define a resistance to the field-induced deformation and hence there is competition between the dielectric and elastic forces such that on removal of the field the orientation of the director reverts back to its original direction. Since in the quiescent state the optical properties need to be well defined if polarized light is to be used, the director is normally aligned unidirectionally at the surface boundaries of the display device (Fig. 5). This surface alignment is caused by a combination of surface and topographical considerations at the surface–liquid crystal interface[22] but is the nett effect of minimization of the free energy (eqn (1)).

As shown in Fig. 5, there are two principal modes of surface alignment: (a) planar (or homogeneous) and (b) homeotropic. Although subtle changes may be induced in threshold voltages by pretilting[23] the surface molecules, the type of surface alignment forms the basis for orientation dependent optical effects in nematic polymers.

Surface Alignment

In low molar mass liquid crystals the methods of producing good surface alignment are well documented.[24] However, considerable difficulty has been experienced with nematic polymers (see ref. 12, p. 230), although Finkelmann *et al.*[25] report that good planar alignment was achieved uisng oriented polyimide surface coatings. Since this is the most important surface alignment for positive-$\Delta\varepsilon$ materials, the initial problems appear to have been overcome, although even using polyimide surface layers required the nematic polymers to be annealed for several hours at a temperature just below the clearing point. Further, the efficiency of the aligning surface will depend on the length of the alkylene spacer (Fig. 2), i.e. the decoupling of the mesogen and backbone, and the type of polymer backbone itself. In the work of Finkelmann *et al.*,[25] a polysiloxane backbone was used, whereas Shibaev's group[12] employed relatively stiff acrylate and methacrylate backbones.

Freedericksz Transition

The simplest display application of a nematic polymer liquid crystal, using polarized light, involves the splay Freedericksz transition shown in Fig. 5(a) for a material with positive $\Delta\varepsilon$ and planar surface alignment. The device is constructed using two polarizers crossed with each other, placed above and below the cell, with their polarization direction at 45° to the director of the planar texture. Above a threshold voltage given by

$$V_c = \pi\left(\frac{k_{11}}{\Delta\varepsilon\,.\,\varepsilon_0}\right)^{1/2} \tag{3}$$

at which the electric aligning forces just balance the elastic resistance, the induced orientation leads to a change in the birefringence as the director moves to a field-induced homeotropic texture. In the latter case the system becomes uniaxial with the director perpendicular to the electrode coated glass surfaces, i.e. parallel to **E**. Thus the transmission will be zero in the field-on state and, depending on the wavelength of light, viewing angle, cell thickness, Δn of the material and the temperature, will pass through several maxima and minima before attaining some value in the off state. A similar effect can, in principle, be observed using materials with negative $\Delta\varepsilon$ and the bend deformation of Fig. 5(b), in which case the off state would be non-transmissive.

The Freedericksz transition for positive-$\Delta\varepsilon$ nematic polymers was studied in some depth by Shibaev and coworkers,[12] who established voltage-, temperature- and frequency (f)-dependent response times as well as a frequency-dependent threshold voltage. However, since these authors could not achieve field-free surface alignment, the transition was from a nematic random scattering texture to a field-aligned homeotropic texture. Using crossed polarizers, they observed only a decrease in light transmission on application of the field. Since the initial surface alignment was undefined, it was not possible to determine elastic constants from these data, although the measurements established qualitatively typical threshold voltages (~5–80 V_{rms}) and response times (~1–240 s). In this work, since a.c. fields are applied across the samples, voltages are defined in terms of their root mean square value, i.e. V_{rms}.

The importance of surface orientation was also highlighted in an extensive study[26] of polyacrylic esters having either different alkylene spacer lengths or polar head-groups. In this work comparative data for V_c and response times have been given and show that the longer spacer chains lead to low threshold voltages (~5 V_{rms}) of the same order as for low molar mass compounds, whilst short alkylene chains lead to high threshold voltages (~60 V_{rms}). This implies, using eqn (3), an increase in k_{11} of two orders of magnitude on decreasing the spacer length from $m = 6$ to $m = 2$. However, as commented in ref. 12, this comparison does not take into account the degree of polymerization of the polymer backbone which also influences the threshold voltage. Clearly, measurements need to be carried out on well-defined fractions so that quantitative comparisons may be made, although it is evident that short spacer chains imply high elastic constants. Response times for the polymers were of the order of seconds or more at temperatures around 150°C. Such times are 40–50 times slower than structurally equivalent low molar mass compounds and show that the mobility of the mesogenic side-groups is strongly altered by fixation to the polymer backbone. In a parallel study on polysiloxane backbone copolymers (with $m = 6$), Pranoto and Haase[27] determined a value for $k_{11} \sim 3 \times 10^{-12}$ N at a temperature ~70°C above T_g and viscosities of the order of 10^3–10^4 cP. As will be seen below, the magnitude of the elastic constant is of the same order as that recorded for low molar mass compounds, whilst the viscosities of the polymers are two to three orders of magnitude larger.

The overall conclusion from this work is that the threshold voltages and elastic constants are slightly increased on polymerization of the side-groups unless the spacer chains are short, in which case the elastic constants increase by one to two orders of magnitude. The viscosity for both short and long spacers is increased by one or two orders of magnitude and therefore the Freedericksz transition response times (τ) are equivalently longer ($\tau \propto \gamma$).

Twisted Nematic Device

The majority of commercial low molar mass liquid crystal displays are based on the twisted nematic electro-optic effect.[28] Not surprisingly, therefore, one of the first attempted applications of nematic polymer materials was in a similar device.[25] In such a cell, Fig. 6, two

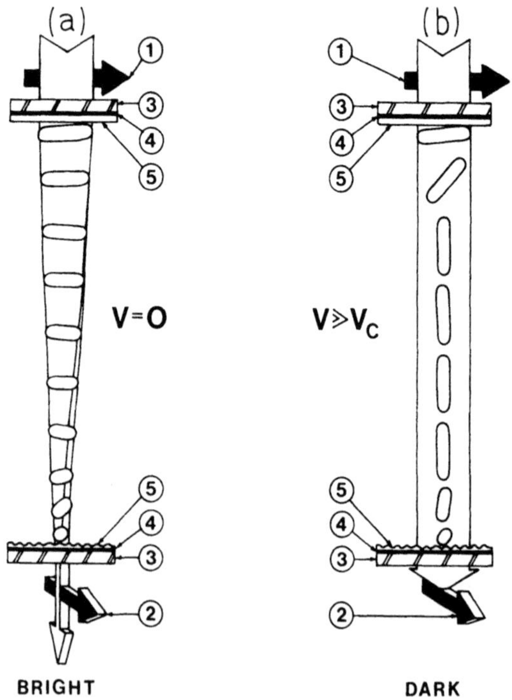

FIG. 6. Schematic operation of a twisted nematic cell for (a) no voltage and (b) a voltage several times the threshold (V_c) applied across the nematic layer. The cell components are as follows: (1) and (2), polarizer and analyser respectively with their polarization directions orthogonal as indicated by the dark arrows; (3) glass substrates; (4), conducting transparent electrodes; and (5), alignment layers with the rubbing directions orthogonal between the upper and lower substrates.

electrodes with alignment layers (typically rubbed polyimide) are arranged so that the aligning directions are orthogonal. This has the effect of twisting the director by 90° on translation from one boundary surface to the other through the nematic material. Provided that the Mauguin[29] inequality is obeyed (i.e. $d\,\Delta n > 2\lambda$, where d is the inter-electrode distance, λ is the wavelength of light and Δn is the liquid crystal refractive index anisotropy), then the plane of polarization of the incident light will be rotated through 90° by the optical waveguiding effect of the twisted nematic material. With the polarizer arrangement shown in Fig. 6, this leads to a transmission of light through the cell. If a voltage is now applied across the cell, then as shown by Schadt and Helfrich[28] and provided a critical threshold voltage (V_{th}) is exceeded, the director reorientates in the direction of the electric field. In this field-aligned (homeotropic) state the refractive index anisotropy experienced by the incident light is effectively zero and the polarization guiding effect is lost. Thus, because of the crossed polarizer configuration, the transmission of the cell is switched to zero. This field-induced orientation and loss of guiding therefore produces an electro-optic switch or display device.

The threshold voltage for deformation of the twisted nematic cell is given[28] by

$$V_{th} = \pi\left(\frac{k_{11} + (k_{33} - 2k_{22})/4}{\varepsilon_0\,\Delta\varepsilon}\right)^{1/2} \tag{4}$$

where k_{ii} $(i = 1, 2, 3)$ are the splay, twist and bend constants as previously defined and $\Delta\varepsilon$ is the dielectric anisotropy at the frequency of the applied field. [In order to eliminate ionic conductivity effects, the field is normally applied as an a.c. pulse, in which case V_{th} is an rms value as before.] This threshold voltage thus defines the voltage at which the nematic layers undergo a twist Freedericksz transition (see subsection on Freedericksz Transition, pp. 310–12) and is dependent on the elastic and dielectric constants of the nematic polymer material. These constants, through eqn (4), only relate to static device properties. If, however, the polymer materials are to be useful for such a device, then the dynamic response of a twisted nematic cell has to be considered.

The times for a twisted nematic electro-optic cell to switch on or rise (τ_r) or to switch off or decay (τ_d) have been defined by Jakeman and Raynes[30] as follows:

$$\tau_r = \frac{\eta d^2}{\varepsilon_0\,\Delta\varepsilon V^2 - k\pi^2} \tag{5}$$

and

$$\tau_d = \frac{\eta d^2}{k\pi^2} \tag{6}$$

where $k = k_{11} + (k_{33} - 2k_{22})/4$, η is a bulk viscosity, V is the aligning voltage, and $\Delta\varepsilon$ and d are as previously defined. Since the shapes of the rise and decay curves are fairly complex,[31] these equations are approximations and τ_r and τ_d are defined in terms of the time taken for a 90% change in the transmission of the cell to occur after application or cessation, respectively, of V. As can be seen from eqns (5) and (6), the dynamic response times depend directly on the bulk viscosity and inversely on factors such as the square of the applied field, the dielectric anisotropy and the elastic constants. Therefore, practically it is important to determine the nematic polymer bulk viscosity in addition to the elastic and dielectric constants.

In an extensive work on polysiloxane polymer liquid crystals, Finkelmann *et al.*[25] examined their performance in a twisted nematic electro-optical cell. In parallel with the polymer measurements, analogous studies were carried out on low molar mass nematic compounds of equivalent dielectric anisotropy ($\Delta\varepsilon$). Both the low molar mass nematic materials and the polymer side-group moieties were mixtures of methoxy- (or alkyloxy-) and chlorine-terminated benzoate esters. Cells were prepared using rubbed polyimides and in the polymer case the samples were annealed for several hours to obtain defect-free aligned textures. Measurement of the threshold voltages, for both the low molar mass compounds and the polymer, showed:

(1) With increasing $\Delta\varepsilon$ the threshold voltage decreased in accordance with eqn (4).
(2) For materials of comparable $\Delta\varepsilon$, the threshold voltages, and therefore the elastic constants of the low molar mass compounds and the polymers, were comparable for long spacer groups. It was observed that the polymers had a slightly lower threshold than the low molar mass materials. However, since the low molar mass mixture contained an OC_6H_{13} termination in comparison with the OCH_3 termination on the polymer, it is not clear that the lower threshold is significant. Such a long chain might, in the monomer case, be expected to increase V_{th}.
(3) For polymers with shorter spacer lengths than $m = 6$ (where m

is the number of methylene units), the threshold voltage increased dramatically. For example, under equivalent conditions $m = 6$ gave $V_{th} = 4\ V_{rms}$, $m = 5$ gave $V_{th} = 7{\cdot}3\ V_{rms}$, $m = 4$ gave $V_{th} > 50\ V_{rms}$ and for $m = 3$ no field effects could be induced.

From the threshold measurements one concludes that, for long spacer lengths where the side-chain mesogenic moiety motion is essentially decoupled from the backbone motion, the low molar mass compounds and the polymers have similar elastic constants. If, however, the flexible spacer length in the polymer is short, then the elastic constants (as related through k in eqn (4)) increase markedly. This result is also implicit in the work of Ringsdorf and Zentel on side-chain polyacrylic esters.[26]

Few experimental data are available on the response times of the polymeric systems. For the polymer system most like the low molar mass compound with roughly similar voltage thresholds ((2) above), values of $\tau_r = 70 \rightarrow 80$ s and $\tau_d = 190 \rightarrow 390$ s were obtained. These times, which correspond to reduced temperatures T^* of $0{\cdot}998 \rightarrow 0{\cdot}952$, were obtained for samples with $d = 15\ \mu$m and an applied voltage of ten times the threshold, i.e. $V = 45 \rightarrow 80\ V_{rms}$ for the two extremes of T^* (where T^* is measuring temperature/clearing temperature). Although no dynamic data were presented for the low molar mass compounds, data typical for cyanobiphenyl liquid crystals[32] would suggest $\tau_r \simeq 10$ ms and $\tau_d \simeq 100$ ms for $T^* = 0{\cdot}98$ and $V = 10\ V_{rms}$ for $d = 15\ \mu$m. If, following eqn (5), V were increased to $45\ V_{rms}$, τ_r would decrease by a factor of 20. In the studies of Freedericksz transitions, Ringsdorf and Zentel[26] determined τ_r and τ_d values of the order of $200\ \text{ms} \rightarrow 2$ s and $200\ \text{ms} \rightarrow 10$ s respectively for $d = 20\ \mu$m, $V = 8 \rightarrow 80\ V_{rms}$ and $T = 80°C \rightarrow 190°C$, whilst Pranoto and Haase[27] recorded values of $\tau_r \simeq 1$ s and $\tau_d = 25$ s for $T^* = 0{\cdot}98$, $V = 30\ V_{rms}$ and $d = 13\ \mu$m for similar polysiloxanes to those studied by Finkelmann *et al.*[25] From these three sets of work it is clear that the polymers exhibit response times several orders of magnitude greater than their equivalent low molar mass counterparts. Following eqns (5) and (6) and the approximate equivalence of k for both systems, these increased response times can only be due to the increased viscosity of the polymers. This point will be returned to in the subsection on Polymer-Low Molar Mass Nematic Mixtures (pp. 338–8).

In conclusion to this section, it is evident that nematic side-chain polymer liquid crystals are not useful for conventional twisted nematic liquid crystal display devices due to the slow response times resulting from the high polymer viscosity. However, since the threshold voltages are of the same order of magnitude in polymer and low molar mass systems, if the polymer has a suitably long methylene spacer, then electro-optic effects may be obtained with acceptable applied voltages on a slow timescale. Since many of these polymers exhibit a glass transition between the nematic and crystalline phases, such electro-optic effects may be stored. This would appear to be the main area of application of pure nematic sidechain polymers. One disadvantage of the systems studied in twisted nematic cells to date is the high operating temperatures required to write the electro-optic information. In all of the work reported above the alignment was carried out at temperatures $\geq$100°C for glass transitions (T_g) at or near to ambient temperatures. Since for a practical storage device, using nematic polymers, T_g will be greater than room temperature, this suggests a fundamental limitation to the usefulness of these materials, i.e. high operating temperatures for response times on a reasonable timescale. It is worth pointing out, however, that since nematic polymers do have such high viscosities, the viscosity itself may be used to store optical information effectively for several months in the nematic phase above T_g.[33,34] Provided the moderately high temperatures do not pose manufacturing problems, then nematic polymers in twisted nematic cells can be used as polarization plane rotators. The degree of rotation will be defined by the degree of twist imposed by the angle between the rubbing directions of the surface alignment at the cell electrodes. Since one electrode could be made to be rotatable at the construction stage, such a device would appear to have a potential use as a variable polarization rotator in optical experiments.

Guest–Host Effect

As the previous sections have shown, nematic polymer liquid crystals may be oriented by surface forces and in electric fields. It has been shown recently that such field-induced changes in orientation may also be used to orient pleochroic dyes[35,36] through the guest–host effect.[37] In such an effect either guest dyes dissolved in a nematic polymer[35,36] host or side-chain dye moieties in a nematic copolymer[38] system (where A is a nematic moiety and B is a dye in Fig. 2b) undergo a cooperative realignment as the nematic director responds to the applied field. Since the pleochroic dye has its absorption transition dipole defined relative to its chemical structure, rotation of the dye produces a change in absorption of the guest–host cell when illuminated by polarized light (Fig. 7). In the off state, the guest dye and the nematic host are aligned in a planar texture by the surface forces. If the dye absorption transition moment has the same direction as the input polarization plane, then the device appears coloured (Fig. 7a). Application of a field to a material of positive $\Delta\varepsilon$ induces a homeotropic alignment (Fig. 7b). In this state the absorption moment is then aligned predominantly in the direction of propagation of the light, i.e. perpendicular to the polarization plane, and little absorption takes place. Thus the display appears only slightly coloured. On removal of the field, the orientation of the dye and host relaxes back, because of the strong surface forces, to a coloured state (Fig. 7c). In this explanation it has been assumed that the absorption moment is parallel to the dye molecular axis and the nematic director. Negative dyes with a transverse absorption dipole moment could be similarly used. In this case the 'off state' would be clear and the on state would be coloured due to the rotational degeneracy of the transition dipoles in the homeotropic state. Generally, the positive dye with a positive

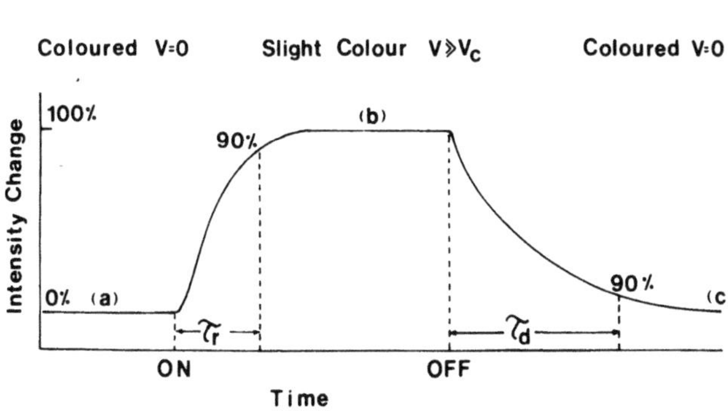

FIG. 7. Schematic representation of the guest–host Heilmeier cell ($\Delta\varepsilon$ positive). The open arrow indicates the polarization plane of the incident light whilst the open and closed ellipses represent the host mesogens and guest dyes respectively. The initial texture is planar.

$\Delta\varepsilon$ host gives the highest optical contrast for this planar-to-homeotropic deformation.

Recently Ringsdorf and Schmidt[38] and Ringsdorf *et al.*[39] have studied both guest–host mechanisms in detail. The polymer liquid crystal structures studied by these authors are shown in Fig. 8. The results for the azo dye system serve to illustrate the usefulness of the guest–host effect. For dyed copolymer systems, Fig. 8(a), inclusion of up to 40% by weight of the dye had no significant effect on the phase transitions of the polymer (i.e. $T_g = 33°C$, $T_{NI} = 132°C$). In a Heilmeier cell, Fig. 7, with one polarizer, a cell spacing of 12 μm, at a T^* of 0·96 (i.e. $T = 115°C$) values of $\tau_r = 1{\cdot}2$ s and $\tau_d = 44$ s were recorded for a 30 V_{rms} applied voltage. For $V = 5\ V_{rms}$, τ_d remained virtually unchanged whilst τ_r increased to 12·2 s. As in the case of the twisted nematic devices, τ_r and τ_d are defined for 90% intensity changes (see Fig. 7). Clearly the response times are similar to those recorded for the twisted nematic cells and are again determined principally by the polymer viscosity. Similar problems of high operating temperatures are encountered with these guest–host cells to those met with the twisted nematic devices. It the dye was dissolved in the nematic polymer host, similar results would be obtained. In this case, however, there may be a limitation on the solubility of the dye which would result in a reduced optical contrast. No data have been presented for

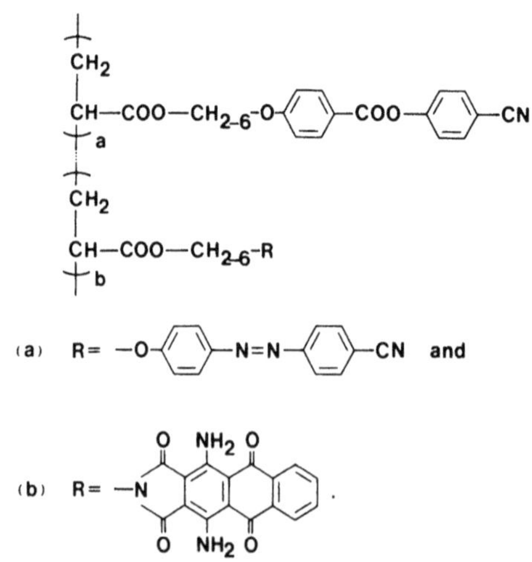

FIG. 8. Molecular structure of typical dyed copolymers with (a) an azo dye moiety and (b) an anthraquinone side-group.

the elastic constants in these systems, but from the applied voltages used they must be similar to the pure nematic polymers. One interesting application of these materials results from storage below T_g or in the highly viscous nematic phase near T_g. Since the dye transition moment is aligned and fixed in space, such guest–host systems when aligned by surface forces or electric fields may be used as dichroic filters. For positive materials, the surface-aligned planar texture gives the largest absorption whilst the electrically aligned homeotropic texture will be virtually clear. By applying different voltages for shorter times than those needed for full orientation, these guest–host systems should be capable of producing variable dichroism through the field-controlled alignment. Cooling the sample rapidly should then allow this effect to be stored. These materials act essentially like a molecular 'venetian blind' provided the planar textures are well defined. Unfortunately, azo dyes under constant illumination do gradually decompose. For this reason, Ringsdorf *et al.*[39] recently extended their work to the anthraquinone dye copolymers shown in Fig. 8(b). Unlike the azo dyes, which form nematic structures in their own right, the anthraquinone dyes have a tendency to disturb the nematic phase. This results in an increase in T_g and a reduction in T_{NI} for the copolymer until $T_g = T_{NI}$ for 39% by weight dye concentration for the equivalent cyano copolyester. However, considering that the maximum solubility of an anthraquinone dye in a low molar mass nematic liquid crystal is only a few per cent by weight, the polymer presents considerable advantages for high optical contrast even though the copolymer backbone tends to reduce the degree of order slightly in comparison with equivalent low molar mass nematic materials. The behaviour of copolymer dyes in such low molar mass materials will be considered in the subsection on Guest Dye Polymer–Host Low Molar Mass Nematic Systems (pp. 334–5).

Cholesteric Phases

Since thermodynamically cholesteric phases are similar to nematic systems and may be considered as tightly twisted or chiral nematics, the electro-optic response times might be expected to be of the same order for both systems. Little work has been carried out on the electro-optics of side-chain chiral polymer systems. In our own work[33] we have used a tightly twisting cholesteric polymer to give a high scattering texture in the cholesteric phase. The sample was then heated into the biphasic region and an electric field applied transversely across the sample. This induced a homeotropic optically clear texture which, on cooling back into the cholesteric phase, remained stored for several months. The pitch of the polymer was such as to give Bragg reflections in the ultraviolet spectral range. Since the applied voltages and the operating conditions were similar to those used for smectic polymers and the latter systems gave much better optical contrast, viewing angle and long-term storage, the cholesteric electro-optic effects will not be considered further herein.

The main use of cholesteric polymers has been in the production of bright circular-polarization reflecting filters.[40] As with low molar mass cholesterics, the polymers give rise to Bragg-like interference reflection colours (Fig. 9). The different colours are produced by heating the materials to a specific temperature to give a specific cholesteric helix pitch and therefore reflection colour. Rapid quenching to below T_g then allows this pitch or colour to be stored since the high polymer viscosity prevents the helix deforming before it cools to below T_g. Whilst this storage phenomenon is not electro-optic but thermo-optic, it does illustrate potential for these materials for information (colour) data storage.

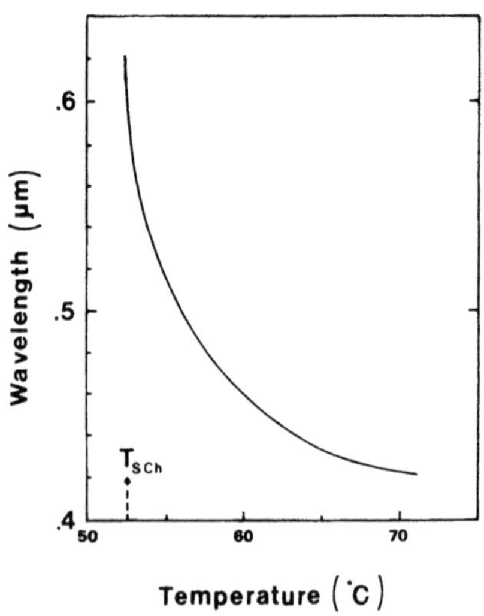

FIG. 9. Schematic dependence of the wavelength of selectively reflected light on temperature for a typical side chain polymer liquid crystal. T_{SCh} denotes the cholesteric to smectic A phase transition and above this temperature the reflected colour changes markedly on heating. Rapid cooling of the material to below T_g stores the reflectance colour obtained during the heating cycle.

Electro-hydrodynamic Effects

The electro-optic effects described above all refer to director reorientation induced through the dielectric properties ($\Delta\varepsilon$) of the polymer. If, however, a change in the optical properties is induced by a current flow resulting from the anisotropic electrical conductivity ($\Delta\sigma$) of the material, then the so-called electro-hydrodynamic effects are observed.[41,42]

In a preliminary study of electro-hydrodynamic effects, low-frequency electric fields (~200–800 Hz) were applied to a cyano-containing polyacrylate sample held at a temperature just below its clearing point.[43] For a sample that initially had a homeotropic texture, application of the field induced a turbulent texture which scattered light strongly. It was observed that this effect was frequency- and temperature-dependent. At higher frequencies ($\geqslant$1 kHz), the effect became very small and transformed to dielectric realignment giving rise to a homeotropic texture. At high temperatures (just below T_c) the electro-hydrodynamic effect was more marked and easier to induce due to the increased mobility of the ionic impurities. If the sample was held at one temperature, it was found that simply switching the applied a.c. field frequency from a range of 200–400 Hz to 1–4 kHz gave a transformation from a scattering to a clear texture. Depending on the frequency and temperature used, these changes took several seconds for the scattering texture, and between 5 and 20 s for the homeotropic texture, to be induced with 200 V_{rms} applied voltages. Although the optical transmission of these two induced textures only differed by a factor of 5, this frequency effect could have applications for variable optical density devices.

Laser Writing

As described in the subsection on Twisted Nematic Device (pp. 312–16), electric fields may be applied to nematic side-chain polymer liquid crystals to give rise to a homeotropic optically clear texture. If a region of this clear texture is heated into the isotropic phase and then allowed to cool, in the absence of an applied field, then a polydomain scattering texture is formed since the orientation is destroyed in the isotropic phase (Fig. 10). This effect has also been demonstrated by two groups using focused laser beams as the heat source[44,45] to give scattering on clear homeotropic textures. The scattering texture in the nematic phase may then be stored for long times if the sample is cooled to below the glass transition temperature. As shown by Shibaev

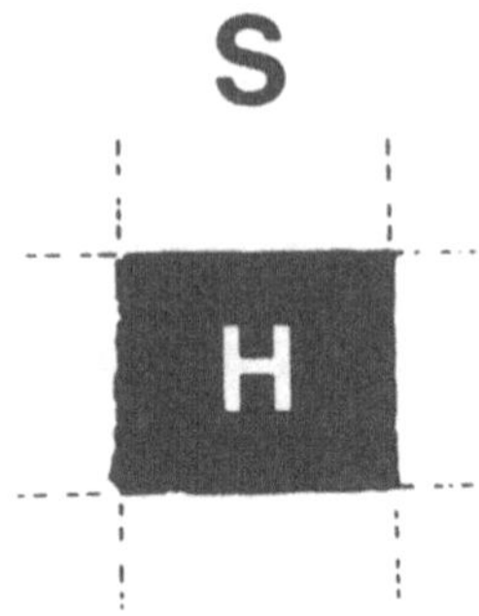

FIG. 10. Aligned homeotropic texture (H) in contrast to the polydomain scattering texture (S). The broken lines mark the electrode edges (3 mm × 3·5 mm actual size) and the sample is shown as viewed through crossed polarizers.

et al.,[44] this scattering texture may be erased in the nematic phase by applying an a.c. field to reinduce the homeotropic alignment. In the nematic phase near T_c, this 'erasure process' takes several seconds with an applied field of 100 V_{rms} at 500 Hz. In the work of ref. 44, a spatial resolution of the order of 0·1 mm was obtained with samples held at 105·5°C (i.e. ~0·5°C below T_c). Typical laser power densities of 130 W cm^{-2} applied for several seconds produced a temperature jump of 0·6°C which transformed the material 0·1°C into its isotropic phase and therefore a non-aligned state. On cooling, this region gave a polydomain scattering texture in these 40 μm-thick cyano-terminated polyacrylate films.

In our own work on polysiloxane nematic films, we used dye additives to increase the laser power absorption. Laser writing on the micron scale was achieved with energy densities ~1 nJ μm^{-2} and pulses of ~2 μs duration in samples held at ambient temperatures (Fig. 11). As will be discussed in detail in the subsection on Laser Writing (pp. 331–3), such effects may also be observed in smectic polymer phases. However, it is worth emphasizing that with dye additives temperature jumps of up to 100°C may readily be achieved and thus laser writing may be carried out at room temperature, rather than near T_c. Because of the high nematic viscosity, the optical information may be stored for several months. Erasure has also been achieved[33,45] by selective heating and cooling in the presence of an a.c. field.

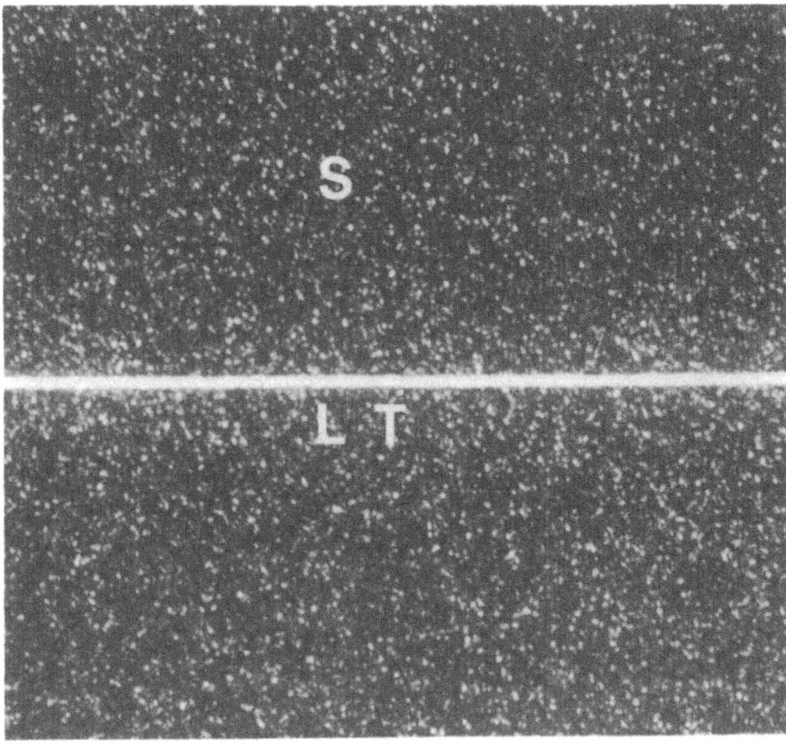

FIG. 11. A laser-written track (LT) against a scattering nematic texture (S). The sample is shown under high magnification through crossed polarizers. The scale is given by the laser track diameter which is 2·5 μm.

Smectic Phases

It is evident from the above discussion that, apart from fast laser writing, the electro-optic response time of a typical nematic side-chain polymer is several orders of magnitude slower than for an equivalent low molar mass system. Since such polymer response times are often of the order of seconds or more, this suggests that their major application is liable to be in the long-time storage of optical information at temperatures below T_g. Since T_g then has to be above ambient temperatures, this implies that the nematic devices will have to be used at very high temperatures (*ca* 100–200°C) to induce effects on a reasonable timescale. Clearly this is a considerable disadvantage from a device construction viewpoint. We realized recently[33,46] that such problems could be overcome if we used the higher-order polymer smectic phase for the data storage, since T_g could then be well below ambient temperatures. This has allowed various devices to be constructed based on the high scattering power inherent in non-aligned smectic phases combined with the high optical clarity of the aligned smectic homeotropic texture. In this section we shall only be concerned with the lowest-order smectic A phase since devices based on the higher-order phases have yet to be demonstrated.

Thermo-optic Properties

The high light-scattering power of smectic polymers may readily be illustrated by carrying out thermo-optic analysis (TOA) on thin (~20 μm) samples. In this technique the light transmission of the sample is studied through crossed polarizers as a function of temperature (Fig. 12). At low temperatures the non-aligned scattering texture is immobile and the transmission is low due to the high turbidity. At some critical temperature, T_w, the texture becomes highly birefringent and mobile and the light transmission increases. This threshold temperature for mobility has important implications for optical data storage. As the temperature is increased further at T_{SI}, 'black' isotropic regions appear in the mesophase (Fig. 13). These regions correspond to polymers transformed into the isotropic phase which therefore appear 'black' through crossed polarizers. As the temperature is increased further, the size of these isotropic regions increases and the light transmission therefore decreases, until at T_c all birefringent texture is lost. This corresponds to a minimum in transmitted light intensity. This fluid–biphasic behaviour between T_w and T_c

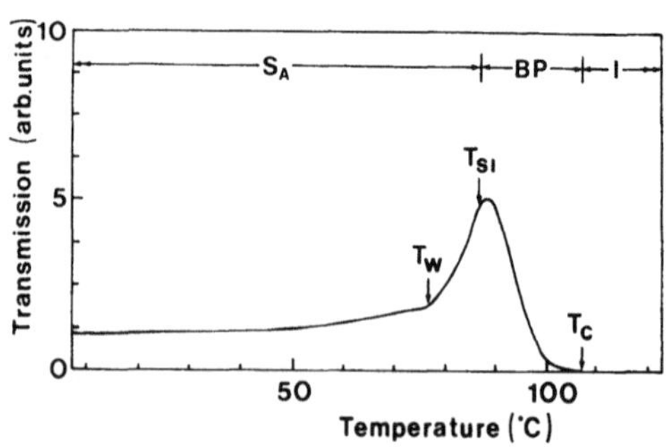

FIG. 12. Thermo-optic curve for a typical polysiloxane smectic polymer where S_A and I denote the smectic and isotropic phases respectively and BP indicates the extent of the stable S_{A-I} biphasic region.

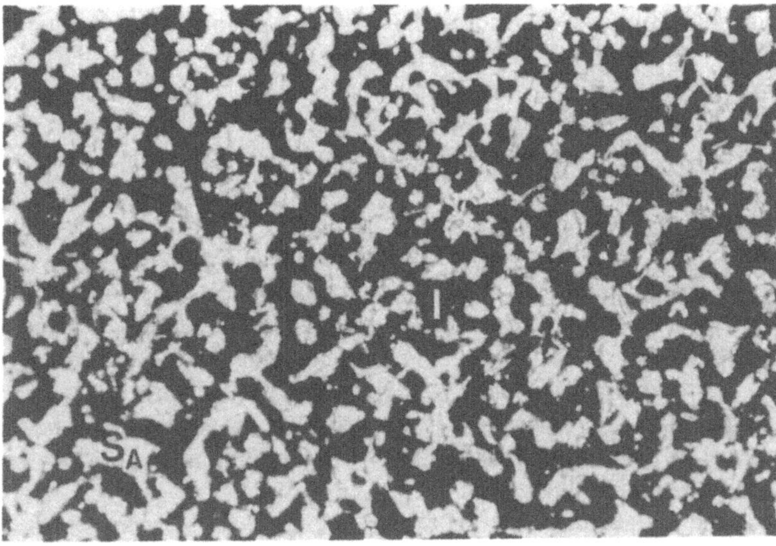

FIG. 13. Biphasic optical texture showing the coexistence of the smectic A (S_A) and isotropic (I) phases.

appears to be typical of smectic polymer liquid crystals and it is believed that the width of this transition is related to the distribution in molar masses inherent in such polymer systems. Low molar mass polymers would be expected to have phase transition temperatures lower than high molar mass material of the same chemical structure.[47] The polydispersity factor, $Z(=M_w/M_n)$, was ~2 for the sample shown in Fig. 12. Although polymers subjected to different cooling rates would show different absolute degrees of scattering at low temperatures (below T_w), the TOA curves seem to be typical for these smectic polymers. The polymer illustrated in Fig. 12 takes several hours of annealing before typical focal conic textures, Fig. 4, are observed. If the methylene spacer linkage between the mesogenic moiety and the siloxane backbone is increased, the decoupling between the backbone and the moiety is increased and focal-conic textures on the 5–10 μm size scale form more rapidly.[34] Since, however, for a high scattering power the texture should be on the size scale of a micron or less, the high degree of decoupling would not appear to be desirable unless rapid cooling rates are used. This has important implications for laser writing (see below). It is worth stressing that the 'fine grained' scattering texture in polysiloxane side-chain polymer liquid crystals is obtained without the use of surface-aligning agents. The polymer backbone appears to restrict the rapid formation of the large-area focal-conic textures, in contrast to the case with low molar mass smectic materials. Thus, unlike low molar mass liquid crystals and the nematic polymer devices described above, there appears to be no need for surface alignment layers in smectic polymer side-chain systems.

Electric Field Effects

The optical microscopy or TOA of the smectic side-chain polymers shows the important effect that viscosity has on the behaviour of these systems, i.e. the existence of T_w and the fluid region, the rate of growth of texture on cooling and the immobility of the texture below T_w. Since smectic materials will be more viscous than the nematic phases, this suggests again that the materials should be useful for the storage of electro-optic information if this information can be written on an acceptable timescale.

In the initial work[33] carried out on smectic side-chain polysiloxanes, it was established that application of electric fields induced either (a) a scattering texture,[49] or (b) a clear homeotropic alignment.[50] For a fixed rms voltage and cell thickness the scattering texture was observed at low frequencies whilst the clear texture was observed at frequencies of $\geqslant$500 Hz. As shown in Fig. 14, the response time (defined as the time for the light intensity to drop by 50% for a cell viewed through crossed polarizers) was also frequency-dependent.[51] In these experiments the field was applied across the cell in the viewing direction. As is evident from Fig. 14, the response times are fairly slow (~5–40 s), although clearly there is an optimum frequency for each polymer to minimize this response time. These measurements were carried out between T_w and T_{SI}. Further work was then carried out to determine how the response time varied with (a) temperature, (b) voltage and (c) cell thickness. The results in Fig. 15 show that

(1) there is a definite temperature threshold (T_w) below which electro-optic effects are not induced on an acceptable timescale,
(2) there is a voltage threshold for the induced homeotropic (or scattering) texture below which no changes are observed,
(3) the response time decreases rapidly with increasing voltage,
(4) the response time decreases with decreasing sample thickness, and
(5) the threshold voltage decreases with decreasing sample thickness.

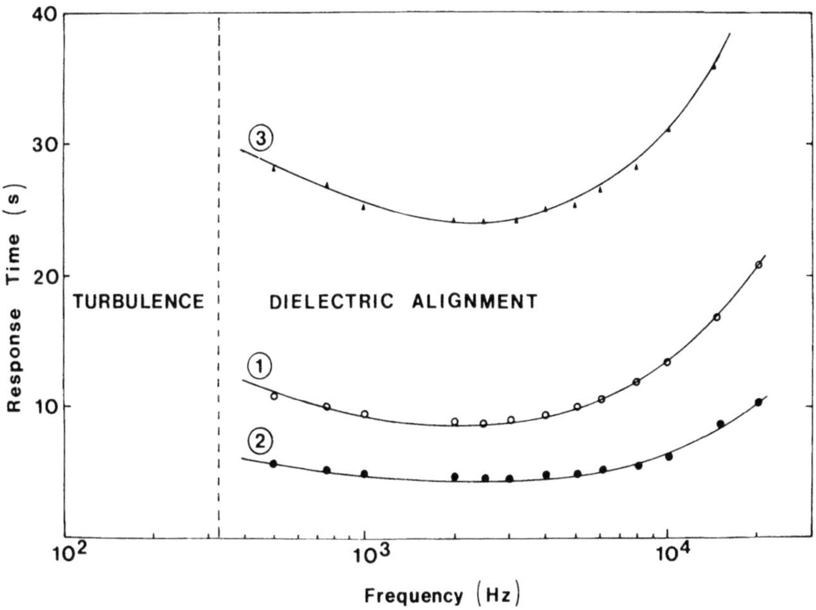

FIG. 14. Response time versus frequency of the applied field ($V = 140\ V_{rms}$ and $d = 21\ \mu m$) for (1), a homopolymer; (2), a copolymer; and (3), a terpolymer. Each of these polymers had a 50-unit polysiloxane backbone with various combinations of cyanobiphenyl and benzoate ester side-groups and the data refer to temperatures between T_w and T_{SI}.

If optimum values of V, T, f and d are chosen, then it has been shown[52] that electro-optic phenomena may be induced on the millisecond-to-second timescale in these smectic polysiloxane polymers. It is important to note, however, that these electro-optic responses were achieved in the fluid–biphasic region above T_w. Whilst it is possible to reduce T_w to below room temperature by modifying the mesomorphic polymer structure, to produce a room-temperature device, the form of the response time–temperature relationship and the relatively limited temperature range (~5–10°C) over which fast responses occur do not suggest a practical device. Such an application would require temperature stabilizing or compensation in its construction which would not be advantageous in comparison with current low molar mass devices.

It has been noted that, if the electrically induced texture is cooled below T_w, this texture remains stored in the smectic phase.[33,46] Since the homeotropic texture is optically clear, this then provides a very high optical contrast. As shown recently,[52] the optical contrast in such a polymer is at least an order of magnitude higher than that observed with equivalent low molar mass components (Fig. 16). These data refer to the homeotropic texture. It should be noted, however, that by using a combination of pulse width, voltage and frequency, a well-defined grey scale can be established for these polymer storage effects. The stored texture may be highly scattering, intermediate or clear.[48] Whilst the polymer storage is highly durable and textures have been stored for several years, it may be erased and rewritten simply by heating the sample back to the fluid–biphasic region. Since no aligning layers are required, the polymer devices may be very simply constructed, requiring only electrode-etched substrates. This represents a considerable advantage for large-scale production of devices in comparison with low molar mass compounds and nematic polymers.

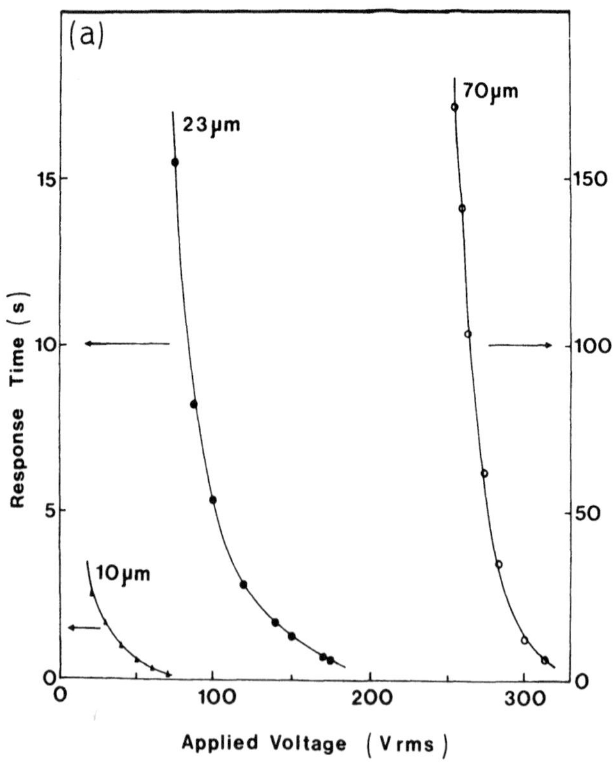

FIG. 15. (a) Response time as a function of the applied voltage for a smectic polysiloxane copolymer. The cell thicknesses were as indicated, the a.c. frequency was 2·5 kHz and measurements were carried out at $T_{SI} - 1°C$.

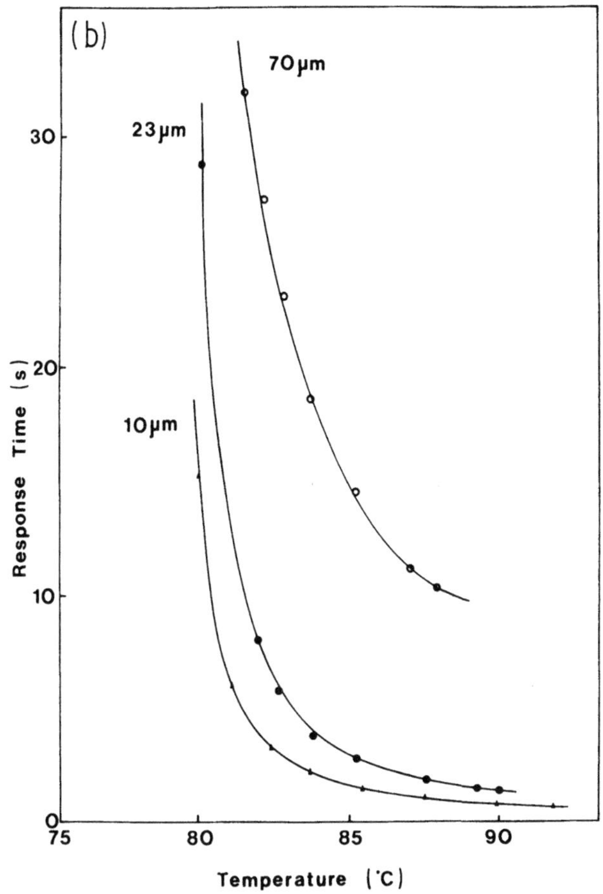

FIG. 15—*contd.* (b) Response time as a function of temperature for the same copolymer as in (a). The applied rms voltages for the 70, 23 and 10 μm cells were 300, 100 and 35 V_{rms} (at 2·5 kHz).

Guest–Host Storage Effects

As in the case of nematic polymers, guest dyes may be incorporated into smectic polymer hosts to give a coloured display.[52,53] In the latter case, the random alignment of the texture ensures a random distribution of the absorption transition moments of the dye molecules. This leads to very even and strongly coloured devices. The guest dye molecules may either be incorporated in the polymer structure itself as a side-group or be dissolved in the polymer matrix. Application of electric fields, as

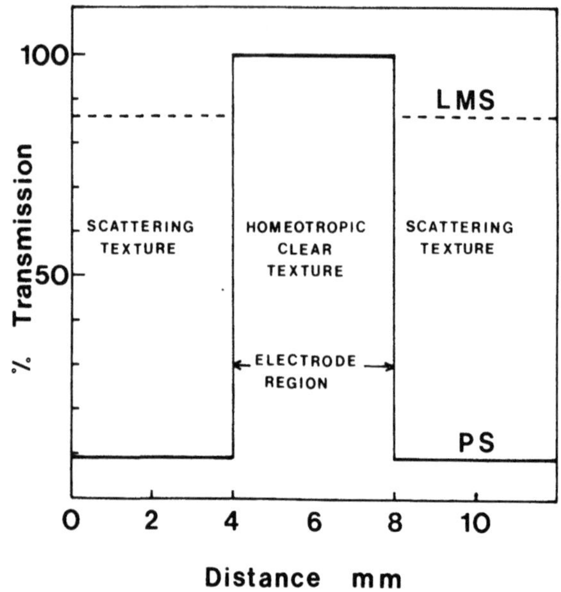

FIG. 16. Comparison of the transmission characteristics of a smectic copolysiloxane (PS) and a low molar mass (LMS) smectic cyanobiphenyl. For the polymer the sample thickness was 10 μm and for the low molar mass material it was 23 μm. Both systems were measured under identical experimental conditions.

above, induces a polymer realignment which in turn aligns the dye molecules.

In the homeotropic state the absorption moment for a positive dye will be along the viewing direction and the sample appears clear. Cooling below T_w then leads to switchable transparent or coloured storage devices (Fig. 17). In these devices variable colour or contrast may be obtained by a suitable combination of pulse width, voltage and frequency as before. Since the dye is either incorporated in the polymer structure as a mesogenic moiety or dissolved in low concentrations (~1–3% by weight), the viscosity, and therefore response times, of the guest host systems are essentially the same as for the non-dyed smectic polymers.[52] Thus the inclusion of dyes into the smectic polymers adds colour to the storage device. It has also been found that the brightness of the guest–host display may be improved by using fluorescent nitrostilbene guest dyes.[33] Since these dyes also have high non-linear optical coefficients, this suggests a new method of alignment for studying non-linear optical processes. As for the

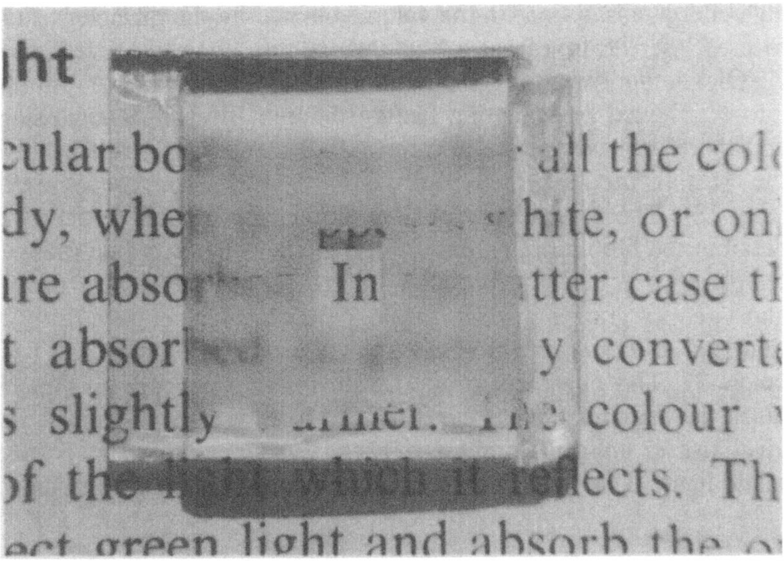

FIG. 17. Guest–host dyed smectic polysiloxane liquid crystal polymers. The centre clear region corresponds to the electrode area over which a field has been applied to induce the non-absorbing homeotropic texture. The polymer sample has a uniform thickness of ~21 μm throughout the test cell and the background lettering shows the high optical contrast. The traces of lettering showing through the scattering texture correspond to laser-induced alignment on the ~20 μm scale which gives a clear texture.

non-dyed smectic polymers, the construction of the electro-optic cells is very simple, involving only electrode-etched substrates. However, a further advantage with the dyed cells is the very wide viewing angle coupled with the very high optical contrast. It is envisaged that such materials will find wide applicability in cosmetic applications (e.g. decorative windows and electrically switchable venetian blinds) as well as in optical data storage and laser writing.

Laser Writing

It has been shown that moderate laser powers may be used to induce textural changes in both nematic and smectic polymer systems. However, in order to achieve laser writing on an acceptable timescale with low pulse energies, the polymer should be doped with a dye, as in the guest–host effect, to increase the absorption at the wavelength of the writing laser. It has been demonstrated that laser writing may be carried out at room temperature in dyed smectic polymers. The high-energy absorption, in the volume defined by the diameter of the focused laser beam, causes a local temperature jump to above T_w and T_c. On cooling (with the laser beam removed) this volume transforms into an aligned or scattering texture different from its surroundings and therefore gives optical contrast. If the beam is translated during the 'writing process', then highly defined laser tracks are observed (Fig. 18). The mechanisms of laser writing have been discussed in detail in the literature.[33,45] Essentially, the laser beam profile causes a temperature gradient in the laser track. This results in local density and viscosity gradients which together with the temperature distribution lead to a realignment of the fine-grained texture on cooling back below T_w. This alignment of the textures is shown clearly in Fig. 18. Since the alignment at the centre of the beam is 'pseudo-planar', this gives rise to anisotropic absorption and, under polarized monochromatic light, these regions behave as coloured polarizers. Because of

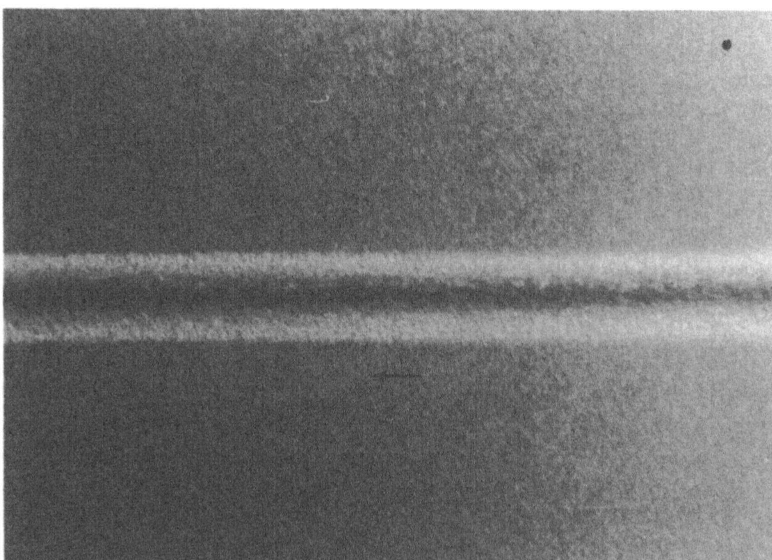

FIG. 18. Optical micrograph of a laser written track in a smectic polysiloxane. The arrow indicates the direction of tracking of the laser perpendicular to the sample and the track diameter of 10 μm indicates the scale. The sample is shown as viewed through crossed polarizers and with white light illumination. The dark and light bands indicate the highly birefringent nature of the laser aligned texture.

the high degree of planar alignment, the laser tracks also appear highly birefringent. This is shown under white light illumination in Fig. 18. Since the laser-written tracks are due to thermally induced processes within the beam waist, and only occur for volumes heated to above T_w, very high spatial resolution and beam diameters are achieved. Continuous tracks of 0·5 μm diameter have been achieved at writing speeds of 1·5 m s^{-1} and energy densities of $\leqslant$1 nJ μm^{-2}. Thus only lasers of a few milliwatts continuous working power are required. It has been noted that laser writing may be achieved in aligned or scattering textures, with or without electric fields applied across the smectic polymer samples. In this way a variety of clear-on-scattering, scattering-on-scattering and scattering-on-clear effects have been observed. Thus both positive- and negative-contrast displays may be produced. All of these laser-written effects were erasable on heating the samples to above T_w and cooling in the absence of a field or by applying a low-frequency a.c. field for a long time. This gave a coloured scattering background texture. If a clear texture was required then high-frequency a.c. fields ($f \geqslant 1$ kHz) were applied.

As with the electrically addressed undyed polymer liquid crystal displays, the laser-written devices are durable, show high optical contrast, exhibit a temperature threshold and may be erased and rewritten. Additionally, they have fast response times, in that only short microsecond laser pulses are required, they require low energy densities and have high spatial resolution. Since both positive and negative contrast have been achieved with varying degrees of 'grey' scale, the systems are multistable rather than bistable. Such devices should therefore find applications not only in high-resolution laser writing, but also in optical data storage, memories and novel optical components. Obviously, an optically clear laser track of diameter ~1 μm or more acts as a very good diffraction slit. The temperature threshold effect allows the same laser to be used for the writing (high powers) and reading (low power) processes. Finally, it is worth re-emphasizing that these effects are stored in the smectic phase since better operating parameters are obtained if T_g is suppressed to low temperatures. Such optically induced effects could, however, also be stored below T_g if necessary.

Polymer–Low Molar Mass Liquid Crystal Solutions

From the research published to date it would appear that the main electro-optic uses of side-chain polymer liquid crystals will be related either to storage phenomena or to slow update displays. The high polymer viscosity appears to preclude their use in fast electro-optic displays even though almost all of the electro-optic switching phenomena observed for low molar mass compounds have been observed in equivalent polymer liquid crystals. Although in the early polymer devices high temperatures were required (primarily to utilize below T_g storage while maintaining an acceptable viscosity) the polymers do exhibit interesting properties. As shown by Ringsdorf and Schmidt,[38] high dye concentrations could be dissolved to give good guest–host nematic devices. In our own work on smectic phases[33,48–53] high optical contrast was obtained and, in several studies on nematic polymers,[25–27] reasonable threshold voltages and elastic constants were noted. Since many of the materials used in current low molar mass electro-optic displays are mixtures of many different mesogens designed to optimize the important device parameters, such as V_{th}, k_{ii}, $\Delta\varepsilon$, Δn, η_1, contrast, viewing angle, multiplexibility, etc., the question naturally arises as to whether polymers might be similarly used in low molar mass compounds. Such work has only been carried out over the last year or two and so data are relatively scarce. In this section we will consider two such electro-optic applications of polymer liquid crystals.

Guest Dye Polymer–Host Low Molar Mass Nematic Systems

As shown by a number of groups,[54–57] liquid crystal side-chain polymers can be dissolved in low molar mass liquid crystals. Since such mixtures may then be more readily aligned by surface and electric fields, they have potential for displays. Ringsdorf *et al.*[39] have shown that nematic phenyl benzoate–anthraquinone side-group copolymers with an acrylate backbone are miscible in binary cyanophenyl benzoate ester nematic mixtures. Planar textures were prepared and order parameters of ~0·58 were obtained for the anthraquinone side-group of the polymer. Although in such systems the order parameter of a dissolved anthraquinone monomer in the nematic mixture was slightly higher (0·61) than for the dyed polymer case, higher quantities of dye could be dissolved and the temperature dependence of the order parameter was reduced in the dyed polymer–low molar mass mixtures. Both of these phenomena lead to improved device performance. No dynamic data were presented for these particular mixtures, although similar work on azo dye containing mixtures[38] shows that response times of a few seconds could be obtained with moderately low voltages (10–30 V_{rms}) and 12 μm-thick cells. Since high dye concentrations can be used, it would be interesting to examine such mixtures in commercially available 7 μm-thick cells, as this should lead to response times of the order of several hundred milliseconds. Further improvements could be made if polysiloxane polymers with much lower glass transitions were studied. Thus, these guest–host polymer–low molar mass mixtures have considerable potential for electro-optic applications.

Polymer–Low Molar Mass Nematic Mixtures

Several groups have studied the behaviour of polymer–low molar mass nematic mixtures recently.[54–57] The majority of this work has been

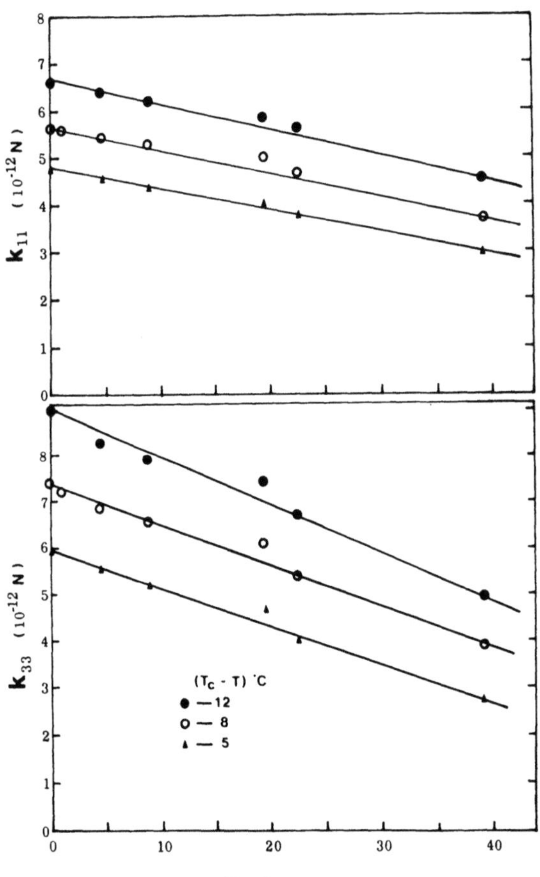

FIG. 19. Elastic constants k_{11} and k_{33} (i.e. splay and bend) as a function of concentration for a polysiloxane smectic copolymer of the type shown in Fig. 3(b) dissolved in a low molar mass cyanobiphenyl liquid crystal host.

directed logically towards phase and miscibility studies. Little work has been carried out to determine how the parameters that control the behaviour of nematic media in devices (i.e. elastic constants, viscosity, etc.: see eqns (4)–(6)) are altered with increasing polymer concentration.

In our own work[56–59] on solutions of polysiloxane cyanobiphenyl benzoate ester side-chain copolymers in cyanobiphenyls, we have studied how k_{ii}, $\Delta\varepsilon$, Δn, V_{th} and γ_1 alter as a function of polymer concentration up to approximately 40% by weight. Typical data for k_{ii} and γ_1 are shown in Figs 19 and 20. The important observations are that k_{ii} decreased by different amounts with increasing polymer concentration whilst γ_1 increased strongly.[57] As discussed in detail elsewhere,[59] the changes in the static properties, i.e. k_{33}, k_{11}, $\Delta\varepsilon/\varepsilon_\perp$ and V_{th}, lead to improved multiplexing capability in twisted nematic devices with a decreased temperature dependence. The 38% by weight

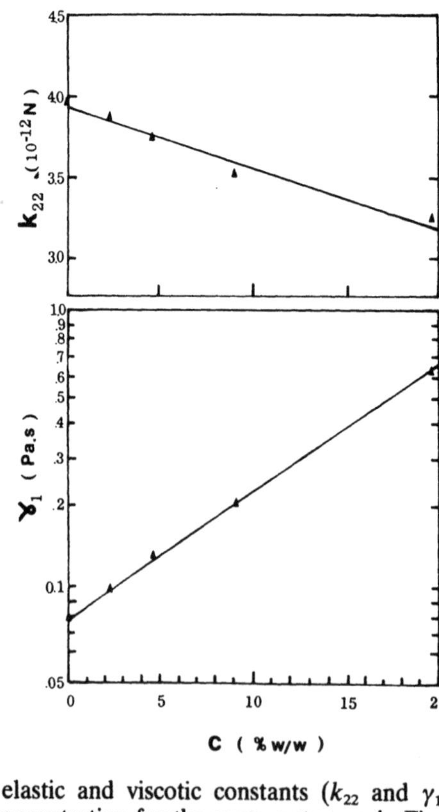

FIG. 20. Twist elastic and viscotic constants (k_{22} and γ_1) as a function of concentration for the same system as in Fig. 19.

solution leads to an improvement in the number of lines that can be multiplexed from 46 to 59. However, the increase in twist viscosity limits the usefulness of extending this effect since a 20% by weight solution increases the twist viscosity γ_1 by a factor of 8 in comparison with the pure low molar mass compound.[57] The effect of polymer concentration on the response time (90% change) of a twisted nematic cell is shown in Fig. 21. The rise and decay times for the 38% by weight mixture are ~1·5 s and 15 s respectively for $T = 29°C$, a 12 μm-thick cell and 5 V_{rms}. If a higher voltage of 20 V_{rms} was applied, then τ_r should decrease to ~200 ms under the same conditions; τ_d would not alter appreciably. However, at 22·5% by weight, τ_d and τ_r are ~200 ms and 2 s for the 12 μm-cell at $T = 29°C$ and $V = 5\ V_{rms}$. An increase in the applied voltage to 20 V_{rms} would decrease τ_r to ~25 ms under the same conditions. Thus for solutions up to ~20% by weight the increase in viscosity does not appear to be too critical for device performance.

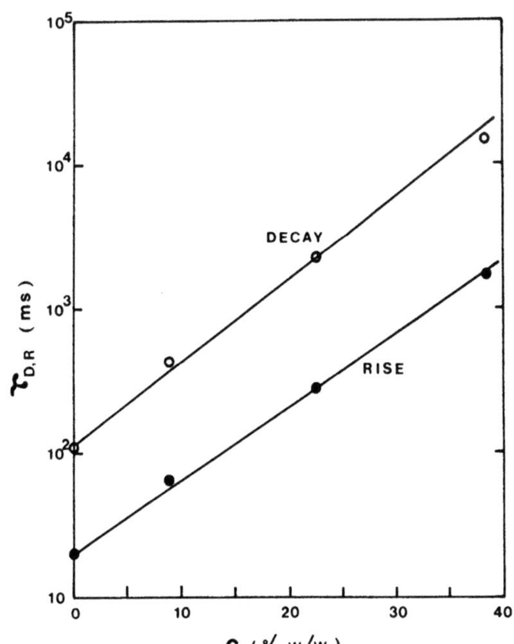

FIG. 21. Electro-optic response times for a twisted nematic device as a function of polymer concentration. The polymer was a polysiloxane backbone system (as in Figs 19 and 20) dissolved in a nematic cyanobiphenyl compound. The applied voltage was 5 V_{rms}, the cell thickness was 12 μm and data are given for $T = 29°C$.

In the work described above, polysiloxane polymers of 50 SiO units were used. Further improvements in performance might be expected using polymers with different backbone lengths, mesogenic moieties, spacer lengths and degrees of substitution. This suggests an interesting area for future research.

CONCLUSIONS

In this chapter we have tried to outline the chemical structure that leads to the formation of the various mesophases exhibited by side-chain polymer liquid crystals. Electro-optic phenomena in these systems have then been considered in terms of potential device applications. Since such studies have only been carried out over the last five years or so it is evident that there will be many more applications discovered over the next decade as a better understanding of the composite polymer–mesomorphic properties is attained.

For pure polymer liquid crystals, it has been shown that their most obvious use will be in storing field-induced alignment either of mesogenic moieties or other anisotropic material. Since in this way it is possible to align fluorescent or non-linear, non-mesogenic dyes, one could readily imagine other functional molecules being aligned and studied. Thus one should consider the optical properties of these materials not only for devices but perhaps as a medium for readily controlling the alignment of other materials (e.g. piezoelectric systems, ferroelectrics, etc.). Nonetheless, the optical properties of the polymer have been shown to have potential for high-contrast electric-field and laser-induced durable storage displays. As further work is carried out on the chemical synthesis and physicochemical properties, new materials might be expected to be produced. Currently, it is not clear how the backbone flexibility, molar mass, polydispersity, degree of side-chain substitution, length of the side-chain spacer, type of mesogenic moiety and its dipolar structure, etc., will influence the macroscopic behaviour of polymer–low molar mass liquid crystal solutions. What is clear is that with such a wealth of variables to consider, the science of polymer liquid crystals is certain to produce new materials and applications over the coming years.

REFERENCES

1. REINTZER, F., *Monatsh Chem.*, 1888, **9**, 421.
2. SAEVA, F. D. Ed., *Liquid Crystals: The Fourth State of Matter,* 1979, Marcel Dekker, New York.

3. HILSUM, C. and RAYNES, E. P. Eds, *Phil. Trans. R. Soc. London,* 1983, **A309,** 69–239.
4. GRAY, G. W. and GOODBY, J. W., *Smectic Liquid Crystals,* 1984, Leonard Hill, Glasgow.
5. GRAY, G. W., in *Polymer Liquid Crystals,* A. Ciferri, W. R. Krigbaum and R. B. Meyer, Eds, 1982, Academic Press, London, pp. 1–33.
6. KELKER, H. and HATZ, R., *Handbook of Liquid Crystal,* 1980, Verlag Chemie, Weinheim.
7. GRAY, G. W., HARRISON, K. J. and NASH, J. A., *Electron. Lett.,* 1973, **9,** 130.
8. STURGEON, B., see Ref. 3, pp. 213–38.
9. ROBINSON, C., *Trans. Faraday Soc.,* 1956, **52,** 571; *Discuss. Faraday Soc.,* 1957, **25,** 29.
10. DOBB, M. G. and MCINTYRE, J. E., *Adv. Polym. Sci.,* 1984, **60/61,** 61.
11. FINKELMANN, H. and REHAGE, G., *Adv. Polym. Sci.,* 1984, **60/61,** 99.
12. SHIBAEV, V. P. and PLATE, N. A., *Adv. Polym. Sci.,* 1984, **60/61,** 173.
13. FINKELMANN, H., RINGSDORF, H. and WENDORFF, J. H., *Makromol. Chem.,* 1978, **179,** 273.
14. GEMMELL, P. G., GRAY, G. W. and LACEY, D., *Mol. Cryst. Liq. Cryst.,* 1985, **122,** 205.
15. RINGSDORF, H. and SCHMIDT, H.-W., *Makromol. Chem.,* 1984, **185,** 1327.
16. COLES, H. J. and SIMON, R., British Patent GB 2146787B, 1983.
17. COLES, H. J. and SIMON, R., in *Recent Advances in Liquid Crystalline Polymers,* L. L. Chapoy, Ed., 1984, Elsevier Applied Science Publishers, London, pp. 324–34.
18. GRAY, G. W., HARRISON, K. J., NASH, J. A., CONSTANT, J., HULME, D. S., KIRTON, J. and RAYNES, E. P., in *Liquid Crystals and Ordered Fluids,* Vol. 2, J. F. Johnson and R. S. Porter, Eds., 1974, Plenum, New York, pp. 617–43.
19. RINGSDORF, H. and ZENTEL, R., *Makromol. Chem.,* 1979, **180,** 803.
20. FINKELMANN, H., KIECHLE, V. and REHAGE, G., *Mol. Cryst. Liq. Cryst.,* 1983, **94,** 343.
21. FRANK, F. C., *Disc. Faraday Soc.,* 1958, **25,** 19.
22. BERREMAN, D. W., *Phys. Rev. Lett.,* 1972, **28,** 1683.
23. DOORN, C. Z., GERRITSMA, C. J. and DE KLERK, J. J. M. J., in *The Physics and Chemistry of Liquid Crystal Devices,* G. J. Sprokee, Ed., 1980, Plenum, New York, pp. 95–104.
24. COGNARD, J., *Mol. Cryst. Liq. Cryst.,* 1982, *Suppl.* 1, 1.
25. FINKELMANN, H., KIECHLE, U. and REHAGE, G., *Mol. Cryst. Liq. Cryst.,* 1983, **92,** 49.
26. RINGSDORF, H. and ZENTEL, R., *Makromol. Chem.,* 1982, **183,** 1245.
27. PRANOTO, H. and HAASE, W., *Mol. Cryst. Liq. Cryst.,* 1983, **98,** 299.
28. SCHADT, M. and HELFRICH, W., *Appl. Phys. Lett.,* 1971, **18,** 127.
29. MAUGUIN, C., *Bull. Soc. Fr. Miner.,* 1911, **34,** 71.
30. JAKEMAN, E. and RAYNES, E. P., *Phys. Lett.,* 1972, **39A,** 69.
31. BLINOV, L. M., *Electro-Optical and Magneto-Optical Properties of Liquid Crystals,* 1983, Wiley, New York, pp. 128–30.

32. Hopwood, A. I., Ph.D. Thesis, Manchester University, 1985.
33. Coles, H. J. and Simon, R., British Patent GB 2146787B, 1983.
34. Simon, R. and Coles, H. J., *Polymer,* 1986, **27,** 811.
35. Talroze, R. V., Shibaev, V. P., Sinitzyn, V. V., Plate, N. A. and Lomonosov, M. V., *Polym. Prepr. Amer. Chem. Soc., Div. Polym. Chem.*, 1983, **24,** 309.
36. Finkelmann, H., Benthack, H. and Rehage, G., *J. Chim. Phys. Chim. Biol.*, 1983, **80,** 163.
37. Heilmeier, G. H., Castellano, J. A. and Zanoni, L. Z., *Mol. Cryst. Liq. Cryst.*, 1969, **8,** 293.
38. Ringsdorf, H. and Schmidt, H.-W., *Makromol. Chem.*, 1984, **185,** 1327.
39. Ringsdorf, H., Schmidt, H.-W., Baur, G. and Kiefer, R. in *Recent Advances in Liquid Crystalline Polymers,* L. L. Chapoy, Ed., 1985, Elsevier, London, pp. 253–60.
40. Finkelmann, H. and Rehage, G., *Makromol. Chem. Rapid Commun.*, 1980, **2,** 317.
41. Kirgbaum, W. R., Lader, H. J. and Ciferri, A., *Macromolecules,* 1980, **15,** 554.
42. Kirgbaum, W. R., and Lader, H. J., *Mol. Cryst. Liq. Cryst.*, 1980, **62,** 87.
43. Plate, N. A., Talroze, R. V. and Shibaev, V. P., *Pure Appl. Chem.*, 1984, **56,** 403 and refs therein.
44. Shibaev, V. P., Kostromin, S. G., Plate, V. A., Ivanov, S. A., Vetrov, V. Yu. and Yakovlev, A. I., *Polym. Commun.*, 1983, **24,** 364.
45. Coles, H. J. and Simon, R., *Polymer,* 1985, **26,** 1801.
46. Coles, H. J. and Simon, R., *Mol. Cryst. Liq. Cryst. Letts.*, 1985, **1,** 75.
47. Finkelmann, H. in Ref. 3, p. 107.
48. Simon, R. and Coles, H. J., *Liquid Crystals,* 1986, **1,** 281.
49. Simon, R. and Coles, H. J., *Mol. Cryst. Liq. Cryst. Lett.*, 1984, **102,** 43.
50. Coles, H. J. and Simon, R., *Mol. Cryst. Liq. Cryst. Lett.*, 1984, **102,** 75.
51. Coles, H. J. and Simon, R., in *Polymeric Liquid Crystals,* A. Blumstein, Ed., 1985, Plenum, New York, pp. 351–68.
52. Coles, H. J., *Faraday Discuss. Chem. Soc.*, 1985, **79,** 201–14.
53. Coles, H. J. and Simon, R., *Mol. Cryst. Liq. Cryst. Lett.*, 1986, **3,** 37.
54. Ringsdorf, H., Schmidt, H.-W. and Scheller, A., *Makromol. Chem. Rapid Commun.*, 1982, **3,** 745.
55. Finkelmann, H., Kock, H.-J. and Rehage, G., *Mol. Cryst. Liq. Cryst.*, 1982, **89,** 23.
56. Casagrande, C., Fabre, P., Veyssie, M., Weill, C. and Finkelmann, H., *Mol. Cryst. Liq. Cryst.*, 1984, **113,** 193.
57. Coles, H. J. and Sefton, M. S., *Mol. Cryst. Liq. Cryst. Lett.*, 1985, **1,** 19; Coles, H. J. and Hopwood, A. I., *ibid.*, 1985, **1,** 165.
58. Sefton, M. S. and Coles, H. J., *Polymer,* 1985, **26,** 1319.
59. Hopwood, A. I. and Coles, H. J., *Polymer,* 1985, **26,** 1312.

INDEX

GPSR Compliance

The European Union's (EU) General Product Safety Regulation (GPSR) is a set of rules that requires consumer products to be safe and our obligations to ensure this.

If you have any concerns about our products, you can contact us on ProductSafety@springernature.com

In case Publisher is established outside the EU, the EU authorized representative is:

Springer Nature Customer Service Center GmbH
Europaplatz 3
69115 Heidelberg, Germany

Batch number: 10607198

Printed by Printforce, the Netherlands